Globalization

The Transformation of Social Worlds

Globalization

The Transformation of Social Worlds

THIRD EDITION

D. STANLEY EITZEN
Colorado State University

MAXINE BACA ZINN
Michigan State University

WADSWORTH
CENGAGE Learning™

Australia • Brazil • Japan • Korea • Mexico • Singapore • Spain • United Kingdom • United States

WADSWORTH
CENGAGE Learning

Globalization: The Transformation of Social Worlds, Third Edition
D. Stanley Eitzen and Maxine Baca Zinn

Publisher/Executive Editor: Linda Schreiber-Ganster

Acquisitions Editor: Erin Mitchell

Assistant Editor: John Chell

Editorial Assistant/Associate: Mallory Ortberg

Managing Editor, Media: Bessie Weiss

Media Editor: Melanie Cregger

Marketing Manager: Andrew Keay

Marketing Assistant/Associate: Dimitri Hagnere

Marketing Communications Manager: Laura Localio

Content Project Management: PreMediaGlobal

Art Director: Caryl Gorska

Manufacturing Buyer: Rebecca Cross

Manufacturing Manager: Marcia Locke

Rights Acquisitions Specialist: Dean Dauphinais

Rights Acquisitions Director: Bob Kauser

Cover Designer: Riezebos Holzbaur/Tim Heraldo

Cover Image: Getty Images

Compositor: PreMediaGlobal

For product information and technology assistance, contact us at **Cengage Learning Customer & Sales Support, 1-800-354-9706**

For permission to use material from this text or product, submit all requests online at **www.cengage.com/permissions**. Further permissions questions can be emailed to **permissionrequest@cengage.com**.

Library of Congress Control Number: 2011921012

Student Edition:

ISBN-13: 978-1-111-30158-3

ISBN-10: 1-111-30158-1

Wadsworth
20 Davis Drive
Belmont, CA 94002-3098
USA

Cengage Learning is a leading provider of customized learning solutions with office locations around the globe, including Singapore, the United Kingdom, Australia, Mexico, Brazil, and Japan. Locate your local office at **international.cengage.com/region**.

Cengage Learning products are represented in Canada by Nelson Education, Ltd.

For your course and learning solutions, visit **www.cengage.com.**

Purchase any of our products at your local college store or at our preferred online store **www.cengagebrain.com.**

Instructors: Please visit **login.cengage.com** and log in to access instructor-specific resources.

Contents

Preface

Sociology as it is practiced, taught, and written about in the United States is, with few exceptions, centered on the United States. Sociological textbooks, for example, give occasional attention to comparative research but rarely examine the global forces that have dramatic and rapid effects everywhere. Of course, the United States is a central player in the world arena with the implementation of its foreign policy, the reach and power of U.S.–based corporations, the spread of its popular culture, and its overuse of natural resources.

Equally important, there are social forces outside its boundaries that have powerful effects on the United States. The terrorism manifested on September 11, 2001, is an obvious example. So, too, are financial disasters, wars, and extreme weather events that impact U.S. financial markets, agricultural prices, jobs, and the safety of its citizens. This book will consider the reciprocal effects of the U.S. and other parts of the world. Moreover, the selections will examine how the world is ever more interconnected, thereby affecting people everywhere around the globe, not just in the United States. Some examples:

- Global warming brings climate change, extreme weather conditions worldwide, and the migration of people, animals, and diseases as tropical forests are cut down in developing nations while developed nations consume excessive amounts of fossil fuels.

- Outbreaks of new diseases and drug-resistant diseases threaten everyone.

- Low-wage economies attract capital, moving jobs from place to place in a "race to the bottom," leaving disarray and unemployment where the jobs have vanished and dislocations and worker exploitation where those jobs are relocated.

- Global market forces are shaping stratification and inequality within societies.

- The ever-larger inequality gap both between societies and within them increases the likelihood of social unrest, parochialism, and terrorism worldwide.

- Extreme poverty affects immigration patterns (legal and illegal).
- Transnational criminal networks traffic drugs, sex workers, child labor, and hazardous wastes.

As we consider these and other topics, several themes recur. First, societies and groups within them experience globalization differently. Therefore, we have selected essays that focus on the consequences of globalization among developed countries and developing countries. These essays examine how the effects of globalization vary within nations as a result of social class, race, and gender. Second, because the consequences of globalization differ, selections will highlight the debates on whether the forms globalization take are beneficial or not. Third, many of the selections will have a critical edge. Implicitly, the authors of these articles ask: who benefits and who does not from the changes brought by globalization? Fourth, we want to emphasize that individuals, groups, and societies are not passive actors in the globalization drama. Powerful nations, corporations, and media impose their will from the top down, but there are also many groups actively organizing and shaping globalization from the bottom up. A final theme, as elaborated in the introductory essay, highlights the effects of globalization on how sociologists conceptualize "the social." In times past, sociological concepts and processes were framed within the boundaries of the nation-state. That has changed and is changing as sociologists and other social scientists reorient their concepts and theories to reflect the realities of an increasingly interconnected world.

This third edition includes twenty-six new selections. The new articles, for the most part, reflect an emphasis on grassroots resistance and social movements aimed at changing the outcomes of globalization toward more humane goals. They also, like the articles retained from the previous edition, combine facts, interpretation, and personal accounts. We have dropped two chapters and replaced them with a chapter on transnational migration and a concluding chapter entitled "Rethinking Globalization."

We have tried to make this collection reader-friendly. The selection of each article was guided by such questions as: Is it interesting? Is it informative? Is it thought-provoking? Does it communicate without the use of unnecessary jargon and overly sophisticated methodologies? We hope that you, the readers of this book, approach the subject of globalization motivated to understand the complexities of our changing social world. The essays included in this collection are intended to add to your understanding and to whet your appetite for further exploration into the intricacies and mysteries of transnational social life.

ACKNOWLEDGMENTS

We would like to thank the following reviewers for the third edition for their helpful comments:

Maria O'Malley, Briar Cliff University, Sioux City, IA

Brian Klocke, SUNY Plattsburgh, Plattsburgh, NY

Patricia Ahmed, University of Kentucky, Lexington, KY

Bennett M. Judkins, Emory and Henry College, Emory, VA

David Piacenti, University of Richmond, Richmond, VA

John Mitrano, Central Connecticut State University, West Hartford, CT

Kathleen Guidroz, Mount St. Mary's University, Gettysburg, PA

Rita Nnodim, Massachusetts College of Liberal Arts, North Adams, MA

Rebecca Ford, Florida State College at Jacksonville, Middleburg, FL

Shana L. Porteen, Western Illinois University, Macomb, IL

Nayda Pares-Kane, Monroe Community College, Rochester, IL

Jennifer Holz, University of Akron – Wayne College, Orrville, OH

Rosalind Kopfstein, Western Connecticut State University, Redding, CT

Carol Poll, FIT/SUNY, Sea Cliff, NY

Thomas Waller, Tallahassee Community College, Tallahassee, FL

Thanks also to Paula Miller of Michigan State University. We are also grateful to Kelly Eitzen Smith, Alan Zinn, Elizabeth Higginbotham, and Margaret L. Andersen for their assistance on this project.

D. Stanley Eitzen
Maxine Baca Zinn

Chapter 1

Globalisation

An Introduction

D. STANLEY EITZEN
MAXINE BACA ZINN

GLOBALIZATION DEFINED

Globalization refers to the greater interconnectedness among the world's people—to "the increasing scale, extent, variety, speed, and magnitude of international cross-border, social, economic, military, political, and cultural inter-relations (Wiarda, 2007:3). Put another way, globalization is a process whereby goods, information, people, money, communication, fashion (and other forms of culture) move across national boundaries.

There are several implications of this view of globalization. First, it signals "that we are all part of a steadily shrinking and interdependent world. Modern communications, transportation, and the Internet have all served to tie more and more countries and peoples together in newer and more complex ways" (Wiarda, 2007:3). Second, globalization is not a thing or a product, but rather a process. It involves such activities as immigration, transnational travel, using e-mail and the Internet, marketing products in one nation that are made else-where, moving jobs to low-wage economies, transnational investments, satellite broadcasts, the pricing of oil, coffee, wheat, and other commodities, and finding a McDonald's and drinking a Coke or Pepsi in virtually every major city in the world. Third, it follows that globalization is not simply a matter of economics, but also has far reaching political, social, and cultural implications. Fourth, glob-alization refers to changes that are increasingly remolding the lives of people worldwide. Globalization is not just "something out there," but is intimately connected to the everyday activities of institutions, families, and individuals within societies (Hytrek, and Zentgraf, 2008). And fifth, not everyone experiences globalization in the same way. It expands opportunities and enhances prosperity for some while leading others into poverty and hopelessness. Periods of rapid social change, we know, "threaten the familiar, destabilize old boundaries, and upset established traditions. Like the mighty Hindu god Shiva, globalization is not only a great destroyer, but also a powerful creator of new ideas, values, identities, practices, and movements" (Stegner, 2002:ix).

GLOBALIZATION THEN

Globalization is not a new phenomenon. For thousands of years people have traveled, traded, and migrated across political boundaries, exchanging food, artifacts, and knowledge. Consider the world around 1000 A. D. (the following is from Sen, 2002; and *U.S. News & World Report,* 1999). The Vikings plundered and traded, establishing settlements in northern France, Britain, Iceland, Greenland, and Russia. At the crossroads between East and West, the Byzantine Empire traded with foreigners from both sides of the globe. India was linked by maritime routes to Africa, the Middle East, and Southeast Asia. China (during the Song Empire) used sea routes to trade cotton goods, spices, and horses. The Islamic world was the first civilization to trade with the other major empires in Europe, Asia, and Africa. These cross-boundary interactions involved not only trade but the transfer of inventions, knowledge, and other cultural forms. For example, the ninth century Arab mathematician Mohammad Ibn Musa-al-Khwarismi, who gave the Western world algorithms and algebra, is "one of many non-Western contributors whose works influenced the European Renaissance and, later, the Enlightenment and the Industrial Revolution" (Sen, 2002:A3).

There have been other periods in which globalization processes accelerated. In sixteenth-century Europe, for example, trade and exploration expanded to all parts of the globe with Europeans settling in different regions. The late 1800s and early 1900s were characterized by great waves of immigration and high levels of trade and finance across national borders. The period following World War II was the precursor to contemporary globalization. The disintegration of the British, French, Dutch, Belgian, and Spanish colonial empires during this time resulted in the establishment of no fewer than eighty-eight new nations. Later, the Soviet Union collapsed, creating eighteen separate countries. These "new" countries were able to sell their raw materials and products on the world market and to purchase goods. They could now also establish local industries to compete with those in other countries. In addition, new technological innovations arising from the war laid the foundation for the transportation and communication advances of the current age. The third change in the post–World War II era was the emergence of transnational political and financial institutions. In 1945, the World Bank and the International Monetary Fund were created to help rebuild Europe and Japan. These two financial institutions continue to play an important role in the developing world. The United Nations, with its organizational units such as the World Health Organization and UNESCO, seeks to reduce tensions among nation-states and to find transnational solutions to political and social problems.

The World Court exists to adjudicate international disputes and to try war criminals.

THE CHARACTERISTICS OF GLOBALIZATION NOW

Transnational connections have existed for centuries, but since the 1960s the pace of these interconnections has increased exponentially. Speed of movement (in terms of both physical travel and travel via communications technology), as well as the volume of goods, messages, and symbols transported, have increased dramatically. For example, the *daily* turnover in foreign exchange markets has risen from $800 billion in the mid-1990s to almost $4 trillion in 2009 (James, 2009:22). So, too, has space seemed to shrink as travel and communication time has decreased. The indicators of this increasingly rapid change occur along a number of dimensions (the following is dependent in part on Brecher, Costello, and Smith, 2000:2–4; Scholte, 2000: 20–25; and Beynon and Dunkerley, 2000:5–7).

Production

Globalization has transformed the nature of economic activity. From the 1970s onward, transnational corporations have built factories and bought manufactured products from low-wage countries on a vastly expanded scale. A "global assembly line" has emerged in which products from athletic shoes to electronics are made by low-wage workers in low-wage countries and sold in developed countries. U.S., Japanese, and European transnational corporations have invested many billions in China and elsewhere to build state-of-the-art factories. For example, Dell, which makes personal computers, has components manufactured by hundreds of suppliers and subsuppliers in Mexico, Taiwan, Malaysia, Korea, and China. Nokia, the Finnish company known primarily for its cellular phones, uses components and assembled products produced in 10 different countries. The result has been the decline of manufacturing in developed countries and the migration of production jobs to low-wage economies. The U.S., where the average manufacturing wage is $16 an hour, lost one-fifth of its manufacturing jobs from 2000 to 2008 (Greenhouse, 2009). These jobs migrated to places like China, where the average manufacturing wage is 61 cents an hour. The job losses occur in a transnational domino effect as jobs that once moved to Mexico because of their low wages, for instance, have moved to even lower-wage economies. This migration of jobs has been called "the race to the bottom."

Markets

In the past, corporations limited their sales to domestic or perhaps regional markets. Now goods and services are marketed to the entire world. Nokia, for example, sells its products in 130 countries. Sometimes a transnational corporation will locate a factory in a country where it markets products. This is the strategy of the Japanese automobile manufacturers Honda and Toyota, which have major plants in the United States, where their products are popular.

Technology: The Tools of Globalization

New technologies—robotics, fiber optics, container ships, computers, communications satellites, and the Internet—have transformed information storage and retrieval, communication, production, and transportation. Microelectronic-based systems of information, for example, allow for the storage, manipulation, and retrieval of data in huge quantities (the amount of unique information generated worldwide each year is measured in exabytes, one exabyte being 1 followed by 18 zeroes). Information can be sent in microseconds via communications satellites throughout the world. As *Business Week* put it, "Anyone with a computer is a citizen of the world..." (1999:71). In short, "[t]echnological advancement in transportation and communications has not merely made the world smaller, *for many purposes it has made geography irrelevant* (emphasis added)" (Peoples and Bailey, 2003:36).

Corporate Restructuring

Major corporations have always operated internationally. Beginning in the 1980s, they reorganized internally to take advantage of the global economy. They merged with other corporations and developed strategic alliances with fellow multinationals. They arranged for companies in other societies to do various tasks ("corporate outsourcing"). Consider Boeing, which in the manufacture of its next-generation Dreamliner developed partnerships with corporations in Japan (Mitsubishi, Kawasaki, and Fuji), Sweden (Saab), Italy (Alenia Aaeronautica), France (Latecoere and Messier-Dowty), Germany (Diehl Luftfahrt Electronik) and the United Kingdom (Smiths Aerospace). Thus, Boeing is responsible for about one-third of the production of the Dreamliner while combining the remaining pieces supplied by its global partners (*Business Week,* 2009). The result is a decentralization of production for Boeing, but a consolidation of resources and power among these partnering corporations across national boundaries.

The 2000 largest global corporations in 2009 represented 62 countries, employed 76 million people, and had total sales of $30 trillion, assets of $124 trillion, and profits of $1.4 trillion (*Forbes,* 2010:96).

Neoimperialism

Following World War II, the imperialist powers gave political independence back to their colonies. Globalization, however, has kept these countries dependent economically on Western Europe, Japan, and the United States. "Globalization has taken from poor countries control of their own economic policies and concentrated their assets in the hands of first world investors. While it has enriched some third world elites, it has subordinated them to foreign corporations, international institutions, and dominant states" (Brecher, Costello, and Smith, 2000:3–4).

Changing Structure of Work

With globalization, worker security everywhere has declined. "All over the world, employers have downsized, outsourced, and made permanent jobs into contingent ones. Employers have attacked job security requirements, work rules, worker representation, healthcare, pensions, and other social benefits, and anything else that defined workers as human beings and employers as partners in a social relationship, rather than simply as buyers and sellers of labor power" (Brecher, Costello, and Smith, 2000:3; Hacker, 2006). In this pro-employer environment, labor unions have lost their power. Employers, when faced with employee demands for higher wages or better benefits, can simply threaten to move the operation to a setting where wages and benefits are lower.

Global Institutions

Organizations such as the World Trade Organization (WTO), the World Bank, and the International Monetary Fund (IMF) are involved in fostering transnational trade and providing economic development in underdeveloped countries. These kinds of organizations are powerful forces accelerating the globalization process. Whether their contributions have positive results is open to debate, as noted throughout this book.

Neoliberal Ideology and Policies

Contemporary globalization is fueled by a prevailing ideology known as neoliberalism, or the Washington Consensus. This ideology dates back to John Locke and Adam Smith, who argued that market forces will bring prosperity, liberty, and democracy if unhindered by government intervention. In terms of policy, neoliberals promote privatization, deregulation, and the dismantling of the welfare state. Most significant, neoliberalism promotes free trade—that is, the idea that state borders should be open to trade without tariffs and other restrictions. This ideology of unfettered capitalism is behind agreements such as NAFTA (the North American Free Trade Act), and it informs the policies of the World Trade Organization and the International Monetary Fund. Proponents of this ideology believe that free trade "expands economic freedom and spurs competition, [thus raising] the productivity and living standards of people in countries that open themselves to the global marketplace" (Cato Institute, cited in Ervin and Smith, 2008:2).

Governance

The sovereignty of the nation-state has for the most part been diminished by globalization (an exception is the U.S., which resists efforts by international organizations to control it). There are suprastate organizations that regulate transnational trade and international law. As a consequence of accepting neoliberal ideology, national governments do not hinder corporate decisions regarding outsourcing and the movement of capital, even though these decisions go against the welfare of their citizens. In effect, economic, political, and cultural change is now beyond the control of any national government (Benyon and Dunkerley, 2000:6).

Permeable Borders: The Transnational Movement of People, Environmental Dangers, Pandemics, and Crime

Insularity is no longer possible, as environmental pollution through the air, water, or food supply anywhere affects people elsewhere. Diseases are difficult to contain, as evidenced by the AIDS and swine flu pandemics. Criminal networks easily function and flourish when borders are permeable. They engage in the distribution of illegal drugs, prostitution, traffic in slaves, and sweatshops. Terrorism also becomes transnational when borders are porous.

Worldwide, more than 200 million people are living outside their country of birth or citizenship. Wars, droughts, floods, and changing climates are pushing people out of their homelands. So, too, is the hope of jobs luring them elsewhere.

Transnational migrant labor is typically from poor countries to rich ones. Over half of the world's legal and illegal immigrants are women. This "feminization of migration" reflects a worldwide gender revolution in which millions of women migrate across the globe to serve as nannies, maids, and sex workers. (Ehrenreich and Hochschild, 2002; Hondagneu-Sotelo, 2003).

One consequence of this flow of people across borders is the reverse flow of money, as many immigrants send money back to relatives in their native land. In 2007, this flow of money from the developed world to the developing world was an estimated $300 billion annually (DeParle, 2007).

The transnational migration of people is more than immigrant labor. It also includes involuntary migration (e.g., sex trafficking) and the transitory crossing of borders as tourists or to seek medical treatment.

Global Culture

National culture, traditionally, has been tied to place and time. It is the knowledge, symbols, and stories that people share within a national consciousness, giving identity to a nation and its people. Global culture, on the other hand, is de-ethnicized and de-territorialized, existing outside the usual reference to geographical territory. It is created and sustained by the media, corporate advertising, and the entertainment industry. The result is a single world culture "centered on consumerism, mass media, Americana, and the English language. Depending on one's perspective, this homogenization entails either progressive cosmopolitanism or oppressive imperialism" (Scholte, 2000:23). The westernized consumer lifestyle is symbolized by similar products (Nike shoes, fashion, pop music, Disney products, movies, Coca-Cola, McDonald's, ESPN, and CNN) that are found everywhere (Benyon and Dunkerly, 2000: 13–21).

The global culture is not as uniform, homogenizing, and universal as it would seem, however. There are often clashes between local and global cultures. Cultural diversity abounds. Religious fundamentalists in many parts of the world, most notably the Middle East, passionately resist modernity in general and the intrusion of the West in particular. Many people embrace their national identity and culture. Moreover, global communications and markets are often adapted to fit diverse local contexts. "Through so-called 'glocalization,' global news reports, global products, global social movements and the like take different forms and make different impacts depending on local particularities" (Scholte, 2000:23).

GLOBALIZATION: RECONFIGURING THE SOCIAL

The discipline of sociology emerged in the eighteenth and nineteenth centuries as the scholarly study of society. Understandably, since the world was divided into nations during this period, the focus of sociologists was on society as the nation-state, with geographical boundaries and social institutions unique to that society. Discussions of place included the local (community), urban and rural, and society. Social problems were examined and solutions offered for problems at the local, societal and regional levels.

Globalization—the transformation of world society in terms of flows of people, goods, capital, and ideas across national boundaries, linkages, institutions, culture, and consciousness (Lechner and Boli, 2000:2)—accelerated in the last decades of the twentieth century, resulting in sociologists and other social scientists beginning to think globally. Just as they explore the sources of social problems in the ordinary, everyday, normal workings of social institutions (e.g., institutional racism and sexism) and class relations, sociologists now also explore the ways that globalization, most particularly the world capitalist system, contributes to problems of people within and across national boundaries. For example, the various manifestations of inequality across and within societies and the degradation of the environment are consequences of the actions by transnational corporations. The pace of globalization has quickened at the beginning of the twenty-first century, and many sociologists are in the process of rethinking their eighteenth- and nineteenth-century roots to confront and understand the globalized and globalizing world. This reconceptualizing of the social is a significant shift in worldview. The selections in this reader will implicitly provide an overview of the new sociology as it seeks to understand the new social worlds resulting from globalization. This requires that we look at globalization at many different levels—from the largely invisible but powerful processes operating at the macro level, to global practices at institutional levels and their impact on the daily experiences of women, men, and children of different classes, races/ethnicities, and international contexts.

STRUCTURE OF THE BOOK

Chapter 1 provides an extended definition of globalization. Chapter 2 examines the debates surrounding globalization, such as: Does globalization bring diverse people together or divide them further? Does globalization increase or decrease global inequality? Should trade across borders be free? Are profit-seeking global corporations the source of good or ill? (Put another way, is world capitalism the answer to local and global problems?). Chapter 3 focuses on the various forms of transnational migration—transitory (tourist), migrant labor, and forced migration such as sex trafficking. Chapter 4 looks at the economic side of globalization, engaging such topics as neoliberal ideology, transnational commerce, and the flow of work and jobs across borders. Chapter 5 examines global power and politics, focusing on the declining sovereignty and significance of nation-states, institutions of transnational governance, the new world order. Chapter 6 centers on cultural globalization as it is manifested in the media, consumerism, and tourism. Chapter 7 considers the restructuring of social institutions and social arrangements resulting from globalization, focusing on gender, families, and relationships. Chapter 8 looks at the globalization of terror—the threats, the tools of global terrorism, and why the U.S. is so vulnerable to terrorists acts. Chapter 9 examines other social problems that have resulted from globalization—environmental degradation, global pandemics, poverty/hunger, and transnational criminal networks. Chapter 9 considers efforts by groups and movements organized to change the negative consequences of globalization. The concluding chapter reexamines globalization and its effects in social life.

REFERENCES

Beynon, John, and David Dunkerley (eds.). 2000. *Globalization: The Reader* (New York: Routledge).

Brecher, Jeremy, Tim Costello, and Brendan Smith. 2000. *Globalization from Below: The Power of Solidarity* (Cambridge, MA: South End Press).

Business Week. 1999. "The Internet Age," (October 4):71.

Business Week. 2009. "Big Changes for Boeing's Dreamliner." Online: http://articles .moneycentral.msn.com/Investing/Extra/big-changes-for-boeing.

Clifford, Mark L. 2002. "How Low Can Prices Go?" *Business Week* (December 2): 60–61.

DeParle, Jason. 2007. "Migrant Money Flow: A $300 Billion Current." *New York Times* (November 18). Online: http://www.nytimes.com/2007/11/18weekinreview/ 18deparle.html?

Ehrenreich, Barbara, and Arlie Russell Hochschild. 2002. "Introduction." In *Global Women*, Barbara Ehrenreich and Arlie Russell Hochschild (eds.). (New York: Metropolitan Books).

Ervin, Justin, and Zachary A. Smith. 2008. *Globalization: A Reference Book.* Santa Barbara, CA: ABC-CLIO.

Forbes. 2010. "The Global 2000," (May 10):92–105.

Greenhouse, Steven. 2009. *The Big Squeeze: Tough Times for the American Worker.* New York: Anchor Books.

Hacker, Jacob S. 2006. *The Great Risk Shift* (New York: Oxford University Press).

Hondagneu-Sotelo, Pierrette. 2003. "Gender and Immigration: A Retrospective and Introduction." In *Gender and U.S. Immigration: Contemporary Trends*, Pierrette Hondagneu-Sotelo (ed.). (Berkeley: University of California Press), 3–19.

Hytrek, Gary, and Kristine M. Zentgraf. 2008. *America Transformed: Globalization, Inequality, and Power* (New York: Oxford University Press).

James, Harold. 2009. "The Late, Great Globalization." *Current History* 108 (January), 20–25.

Lechner, Frank J., and John Boli (eds.). 2000. *The Globalization Reader* (Oxford, UK: Blackwell Publishers).

Martin, Philip, and Jonas Widgren. 2002. "International Migration: Facing the Challenge," *Population Bulletin* 57 (March):entire issue.

Peoples, James, and Garrick Bailey. 2003. *Humanity: An Introduction to Cultural Anthropology* (Belmont, CA: Wadsworth). The section on globalization was accessed at http://www.wadsworth.com/anthropology_d/resources/terrorism/booklet.html.

Sen, Amartya. 2002. "How to Judge Globalism," *The American Prospect* (Winter):A2–A6.

Scholte, Jan Aart. 2000. *Globalization: A Critical Introduction* (New York: Palgrave).

Steger, Manfred B. 2002. *Globalism: The New Market Ideology* (Lanham, MD: Rowman & Littlefield).

U.S. News & World Report. 1999. "The Year 1000: What Life Was Like in the Last Millennium," special double issue (August 16 and 23):38–94.

Chapter 2

Debating Globalization
Introduction

While globalization is not a new phenomenon, it is now much greater in scope and accelerating more rapidly than at any previous time in history. The world's people are interconnected as never before. Scholars disagree on the effects and implications of this hugely important process. The major debates surrounding globalization include the following: Have transnational corporations superseded nation-states? Does globalization enhance or undermine democracy? Is the culture of the West, spread through movies, television, advertising, consumerism, and the English language, making the world more homogeneous? With globalization, is the local no longer relevant? Each of these questions is subsumed under this fundamental question: Is globalization a good or a bad thing? More precisely, who benefits and who suffers from globalization? Answering this question involves asking many others: Will the world's inhabitants be more secure or insecure because of globalization? Is the environment more secure because of globalization or more likely to deteriorate further? Will the world's poor be uplifted by globalization or will their condition worsen further? Will the workers of the world benefit from technological advances and economic development or will wages be decreased as jobs go to the lowest-wage nations in a race to the bottom?

The first essay in this chapter, by noted British sociologist Anthony Giddens, illustrates the broad sweep of globalization and its consequences. For Giddens, globalization is a transformational force, not only "out there," separate from individual lives, but also "in here," influencing us on a personal level.

Next, *New York Times* columnist and two-time Pulitzer Prize recipient Thomas L. Friedman introduces his book *The World Is Flat,* arguing that technology is leveling the playing field of global competition. According to Friedman, this "flattening" effect is positive, as people everywhere gain the ability to connect, cooperate, and compete in a new era of prosperity and innovation.

In the next essay, however, Pankaj Ghemawat challenges Friedman's thesis that new technologies level the playing field of global competitiveness. Ghemawat argues that even in a globalized world, geographic boundaries define our movements and constrain cross-border integration.

Finally, Jeremy Brecher, Tim Costello, and Brendan Smith take a firm stand on the negative consequences and contradictions of "globalization from above." The authors argue that globalization has negative consequences by aggravating old problems and creating new ones.

1

Globalisation

ANTHONY GIDDENS

Giddens provides some of the debates about globalization with regard to economics, culture, and the nation state. He suggests that for good and bad, globalization has shifted our life circumstances, becoming "the way we live."

A friend of mine studies village life in central Africa. A few years ago, she paid her first visit to a remote area where she was to carry out her fieldwork. The day she arrived, she was invited to a local home for an evening's entertainment. She expected to find out about the traditional pastimes of this isolated community. Instead, the occasion turned out to be a viewing of *Basic Instinct* on video. The film at that point hadn't even reached the cinemas in London.

Such vignettes reveal something about our world. And what they reveal isn't trivial. It isn't just a matter of people adding modern paraphernalia—videos, television sets, personal computers and so forth—to their existing ways of life. We live in a world of transformations, affecting almost every aspect of what we do. For better or worse, we are being propelled into a global order that no one fully understands, but which is making its effects felt upon all of us.

Globalisation may not be a particularly attractive or elegant word. But absolutely no one who wants to understand our prospects at century's end can ignore it. I travel a lot to speak abroad. I haven't been to a single country recently where globalisation isn't being intensively discussed. In France, the word is *mondialisation*. In Spain and Latin America, it is *globalization*. The Germans say *Globalisierung*.

The global spread of the term is evidence of the very developments to which it refers. Every business guru talks about it. No political speech is complete without reference to it. Yet even in the late 1980s the term was hardly used, either in the academic literature or in everyday language. It has come from nowhere to be almost everywhere.

Given its sudden popularity, we shouldn't be surprised that the meaning of the notion isn't always clear, or that an intellectual reaction has set in against it. Globalisation has something to do with the thesis that we now all live in one world—but in what ways exactly, and is the idea really valid? Different thinkers have taken almost completely opposite views about globalisation in debates that have sprung up over the past few years. Some dispute the whole thing. I'll call them the sceptics.

According to the sceptics, all the talk about globalisation is only that—just talk. Whatever its benefits, its trials and tribulations, the global economy isn't especially different from that which existed at previous periods. The world carries on much the same as it has done for many years.

Most countries, the sceptics argue, gain only a small amount of their income from external trade. Moreover, a good deal of economic exchange is between regions, rather than being truly world-wide. The countries of the European Union, for example, mostly trade among themselves. The same is true of the other main trading blocs, such as those of Asia-Pacific or North America.

Others take a very different position. I'll label them the radicals. The radicals argue that not only is globalisation very real, but that its consequences can be felt everywhere. The global market-place, they say, is much more developed than even in the 1960s and 1970s and is indifferent to national borders. Nations have lost most of the sovereignty they once had, and politicians have lost most of their capability to influence events. It isn't surprising that no one respects political leaders any more, or has much interest in what they have to say. The era of the nation-state is over. Nations, as the Japanese business writer Kemchi Ohmae puts it, have become mere "fictions." Authors such as Ohmae see the economic difficulties of the 1998 Asian crisis as demonstrating the reality of glob-alisation, albeit seen from its disruptive side.

The sceptics tend to be on the political left, especially the old left. For if all of this is essentially a myth, governments can still control economic life and the welfare state remain intact. The notion of globalisation, according to the sceptics, is an ideology put about by free-marketeers who wish to dismantle welfare sys-tems and cut back on state expenditures. What has happened is at most a rever-sion to how the world was a century ago. In the late nineteenth century there was already an open global economy, with a great deal of trade, including trade in currencies.

Well, who is right in this debate? I think it is the radicals. The level of world trade today is much higher than it ever was before, and involves a much wider range of goods and services. But the biggest difference is in the level of finance and capital flows. Geared as it is to electronic money—money that exists only as digits in computers—the current world economy has no parallels in earlier times.

In the new global electronic economy, fund managers, banks, corporations, as well as millions of individual investors, can transfer vast amounts of capital from one side of the world to another at the click of a mouse. As they do so, they can destabilize what might have seemed rock-solid economies—as hap-pened in the events in Asia.

The volume of world financial transactions is usually measured in U.S. dol-lars. A million dollars is a lot of money for most people. Measured as a stack of hundred-dollar notes, it would be eight inches high. A billion dollars—in other words, a thousand million—would stand higher than St. Paul's Cathedral. A tril-lion dollars—a million million—would be over 120 miles high, 20 times higher than Mount Everest.

Yet far more than a trillion dollars is now turned over *each day* on global currency markets. This is a massive increase from only the late 1980s, let alone the more distant past. The value of whatever money we may have in our pockets, or our bank accounts, shifts from moment to moment according to fluctuations in such markets.

I would have no hesitation, therefore, in saying that globalisation, as we are experiencing it, is in many respects not only new, but also revolutionary. Yet I don't believe that either the sceptics or the radicals have properly understood either what it is or its implications for us. Both groups see the phenomenon almost solely in economic terms. This is a mistake. Globalisation is political, technological and cultural, as well as economic. It has been influenced above all by developments in systems of communication, dating back only to the late 1960s.

In the mid-nineteenth century, a Massachusetts portrait painter, Samuel Morse, transmitted the first message, "What hath God wrought?" by electric telegraph. In so doing, he initiated a new phase in world history. Never before could a message be sent without someone going somewhere to carry it. Yet the advent of satellite communications marks every bit as dramatic a break with the past. The first commercial satellite was launched only in 1969. Now there are more than 200 such satellites above the earth, each carrying a vast range of information. For the first time ever, instantaneous communication is possible from one side of the world to the other. Other types of electronic communication, more and more integrated with satellite transmission, have also accelerated over the past few years. No dedicated transatlantic or transpacific cables existed at all until the late 1950s. The first held fewer than 100 voice paths. Those of today carry more than a million.

On 1 February 1999, about 150 years after Morse invented his system of dots and dashes, Morse Code finally disappeared from the world stage. It was discontinued as a means of communication for the sea. In its place has come a system using satellite technology, whereby any ship in distress can be pinpointed immediately. Most countries prepared for the transition some while before. The French, for example, stopped using Morse Code in their local waters in 1997, signing off with a Gallic flourish: "Calling all. This our last cry before our eternal silence."

Instantaneous electronic communication isn't just a way in which news or information is conveyed more quickly. Its existence alters the very texture of our lives, rich and poor alike. When the image of Nelson Mandela may be more familiar to us than the face of our next-door neighbour, something has changed in the nature of our everyday experience.

Nelson Mandela is a global celebrity, and celebrity itself is largely a product of new communications technology. The reach of media technologies is growing with each wave of innovation. It took 40 years for radio in the United States to gain an audience of 50 million. The same number was using personal computers only 15 years after the personal computer was introduced. It needed a mere 4 years, after it was made available, for 50 million Americans to be regularly using the Internet.

It is wrong to think of globalisation as just concerning the big systems, like the world financial order. Globalisation isn't only about what is "out there," remote and far away from the individual. It is an "in here" phenomenon too, influencing intimate and personal aspects of our lives. The debate about family values, for example, that is going on in many countries might seem far removed from globalising influences. It isn't. Traditional family systems are becoming transformed, or are under strain, in many parts of the world, particularly as women stake claim to greater equality. There has never before been a society, so far as we know from the historical record, in which women have been even approximately equal to men. This is a truly global revolution in everyday life, whose consequences are being felt around the world in spheres from work to politics.

Globalisation thus is a complex set of processes, not a single one. And these operate in a contradictory or oppositional fashion. Most people think of globalisation as simply "pulling away" power or influence from local communities and nations into the global arena. And indeed this is one of its consequences. Nations do lose some of the economic power they once had. Yet it also has an opposite effect. Globalisation not only pulls upwards, but also pushes downwards, creating new pressures for local autonomy. The American sociologist Daniel Bell describes this very well when he says that the nation becomes not only too small to solve the big problems, but also too large to solve the small ones.

Globalisation is the reason for the revival of local cultural identities in different parts of the world. If one asks, for example, why the Scots want more independence in the U.K. or why there is a strong separatist movement in Quebec, the answer is not to be found only in their cultural history. Local nationalisms spring up as a response to globalising tendencies, as the hold of older nation-states weakens.

Globalisation also squeezes sideways. It creates new economic and cultural zones within and across nations. Examples are the Hong Kong region, northern Italy, and Silicon Valley in California. Or consider the Barcelona region. The area around Barcelona in northern Spain extends into France. Catalonia, where Barcelona is located, is closely integrated into the European Union. It is part of Spain, yet also looks outwards.

These changes are being propelled by a range of factors, some structural, others more specific and historical. Economic influences are certainly among the driving forces—especially the global financial system. Yet they aren't like forces of nature. They have been shaped by technology, and cultural diffusion, as well as by the decisions of governments to liberalise and deregulate their national economies.

The collapse of Soviet communism has added further weight to such developments, since no significant group of countries any longer stands outside. That collapse wasn't just something that just happened to occur. Globalisation explains both why and how Soviet communism met its end. The former Soviet Union and the East European countries were comparable to the West in terms of growth rates until somewhere around the early 1970s. After that point, they fell rapidly behind. Soviet communism, with its emphasis upon state-run enterprise and heavy industry, could not compete in the global electronic economy. The ideological and cultural control upon which communist political authority was based similarly could not survive in an era of global media.

The Soviet and the East European regimes were unable to prevent the reception of Western radio and television broadcasts. Television played a direct role in the 1989 revolutions, which have rightly been called the first "television revolution." Street protests taking place in one country were watched by television audiences in others, large numbers of whom then took to the streets themselves.

Globalisation, of course, isn't developing in an even-handed way, and is by no means wholly benign in its consequences. To many living outside Europe and North America, it looks uncomfortably like Westernisation—or, perhaps, Americanisation, since the U.S. is now the sole superpower, with a dominant economic, cultural and military position in the global order. Many of the most visible cultural expressions of globalisation are American—Coca-Cola, McDonald's, CNN.

Most of the giant multinational companies are based in the U.S. too. Those that aren't all come from the rich countries, not the poorer areas of the world. A pessimistic view of globalisation would consider it largely an affair of the industrial North, in which the developing societies of the South play little or no active part. It would see it as destroying local cultures, widening world inequalities and worsening the lot of the impoverished. Globalisation, some argue, creates a world of winners and losers, a few on the fast track to prosperity, the majority condemned to a life of misery and despair.

Indeed, the statistics are daunting. The share of the poorest fifth of the world's population in global income has dropped, from 2.3 per cent to 1.4 per cent between 1989 and 1998. The proportion taken by the richest fifth, on the other hand, has risen. In sub-Saharan Africa, 20 countries have lower incomes per head in real terms than they had in the late 1970s. In many less developed countries, safety and environmental regulations are low or virtually non-existent. Some transnational companies sell goods there that are controlled or banned in the industrial countries—poor-quality medical drugs, destructive pesticides or high tar and nicotine content cigarettes. Rather than a global village, one might say, this is more like global pillage.

Along with ecological risk, to which it is related, expanding inequality is the most serious problem facing world society. It will not do, however, merely to blame it on the wealthy. It is fundamental to my argument that globalisation today is only partly Westernisation. Of course the Western nations, and more generally the industrial countries, still have far more influence over world affairs than do the poorer states. But globalisation is becoming increasingly decentred—not under the control of any group of nations, and still less of the large corporations. Its effects are felt as much in Western countries as elsewhere.

This is true of the global financial system, and of changes affecting the nature of government itself. What one could call "reverse colonization" is becoming more and more common. Reverse colonization means that non-Western countries influence developments in the West. Examples abound—such as the latinising of Los Angeles, the emergence of a globally oriented high-tech sector in India, or the selling of Brazilian television programmes to Portugal.

Is globalisation a force promoting the general good? The question can't be answered in a simple way, given the complexity of the phenomenon. People who ask it, and who blame globalisation for deepening world inequalities, usually have in mind economic globalisation and, within that, free trade. Now, it is surely obvious that free trade is not an unalloyed benefit. This is especially so as concerns the less developed countries. Opening up a country, or regions within it, to free trade can undermine a local subsistence economy. An area that becomes dependent upon a few products sold on world markets is very vulnerable to shifts in prices as well as to technological change.

Trade always needs a framework of institutions, as do other forms of economic development. Markets cannot be created by purely economic means, and how far a given economy should be exposed to the world market-place must depend upon a range of criteria. Yet to oppose economic globalisation, and to opt for economic protectionism, would be a misplaced tactic for rich and poor nations alike. Protectionism may be a necessary strategy at some times and in some countries. In my view, for example, Malaysia was correct to introduce controls in 1998, to stem the flood of capital from the country. But more permanent forms of protectionism will not help the development of the poor countries, and among the rich would lead to warring trade blocs.

The debates about globalisation I mentioned at the beginning have concentrated mainly upon its implications for the nation-state. Are nation-states, and hence national political leaders, still powerful, or are they becoming largely irrelevant to the forces shaping the world? Nation-states are indeed still powerful and political leaders have a large role to play in the world. Yet at the same time the nation-state is being reshaped before our eyes. National economic policy can't be as effective as it once was. More importantly, nations have to rethink their identities now the older forms of geopolitics are becoming obsolete. Although this is a contentious point, I would say that, following the dissolving of the Cold War, most nations no longer have enemies. Who are the enemies of Britain, or France, or Brazil? The war in Kosovo didn't pit nation against nation. It was a conflict between old-style territorial nationalism and a new, ethically driven interventionalism.

Nations today face risks and dangers rather than enemies, a massive shift in their very nature. It isn't only of the nation that such comments could be made. Everywhere we look, we see institutions that appear the same as they used to be from the outside, and carry the same names, but inside have become quite different. We continue to talk of the nation, the family, work, tradition, nature, as if they were all the same as in the past. They are not. The outer shell remains, but inside they have changed—and this is happening not only in the U.S., Britain, or France, but almost everywhere. They are what I call "shell institutions." They are institutions that have become inadequate to the tasks they are called upon to perform.

As the changes I have described in this chapter gather weight, they are creating something that has never existed before, a global cosmopolitan society.

We are the first generation to live in this society, whose contours we can as yet only dimly see. It is shaking up our existing ways of life, no matter where we happen to be. This is not—at least at the moment—a global order driven by collective human will. Instead, it is emerging in an anarchic, haphazard fashion, carried along by a mixture of influences.

It is not settled or secure, but fraught with anxieties, as well as scarred by deep divisions. Many of us feel in the grip of forces over which we have no power. Can we reimpose our will upon them? I believe we can. The powerless-ness we experience is not a sign of personal failings, but reflects the incapacities of our institutions. We need to reconstruct those we have, or create new ones. For globalisation is not incidental to our lives today. It is a shift in our very life circumstances. It is the way we now live.

2

The World Is Flat

THOMAS L. FRIEDMAN

Friedman sets forth his thesis that the world is flat, a feature empowering indivi-duals to act globally.

Columbus reported to his king and queen that the world was round, and he went down in history as the man who first made this discovery. I returned home and shared my discovery only with my wife, and only in a whisper.

"Honey," I confided, "I think the world is flat."

How did I come to this conclusion? I guess you could say it all started in Nandan Nilekani's conference room at Infosys Technologies Limited. Infosys is one of the jewels of the Indian information technology world, and Nilekani, the company's CEO, is one of the most thoughtful and respected captains of Indian industry. I drove with the Discovery Times crew out to the Infosys campus, about forty minutes from the heart of Bangalore, to tour the facility and interview Nilekani. The Infosys campus is reached by a pockmarked road, with sacred cows, horse-drawn carts, and motorized rickshaws all jostling alongside our vans. Once you enter the gates of Infosys, though, you are in a different world. A massive resort-size swimming pool nestles amid boulders and manicured lawns, adjacent to a huge putting green. There are multiple restaurants and a fabulous health club. Glass-and-steel buildings seem to sprout up like weeds each week. In some of those buildings, Infosys employees are writing specific software programs for American or European companies; in others, they are running the back rooms of major American and European-based multinationals—everything from computer maintenance to specific research projects to answering customer calls routed there from all over the world. Security is tight, cameras monitor the doors, and if you are working for American Express, you cannot get into the building that is managing services and research for General Electric. Young Indian engineers, men and women, walk briskly from building to building, dangling ID badges. One looked like he could do my taxes. Another looked like she could take my computer apart. And a third looked like she designed it!

After sitting for an interview, Nilekani gave our TV crew a tour of Infosys's global conferencing center—ground zero of the Indian outsourcing industry. It was a cavernous wood-paneled room that looked like a tiered classroom from an

SOURCE: Excerpt from Thomas L. Friedman, *The World Is Flat*, New York: Farrar, Straus and Giroux, 2005, pp. 5–11.

Ivy League law school. On one end was a massive wall-size screen and overhead there were cameras in the ceiling for teleconferencing. "So this is our conference room, probably the largest screen in Asia—this is forty digital screens [put together]," Nilekani explained proudly, pointing to the biggest flat-screen TV I had ever seen. Infosys, he said, can hold a virtual meeting of the key players from its entire global supply chain for any project at any time on that supersize screen. So their American designers could be on the screen speaking with their Indian software writers and their Asian manufacturers all at once. "We could be sitting here, somebody from New York, London, Boston, San Francisco, all live. And maybe the implementation is in Singapore, so the Singapore person could also be live here.... That's globalization," said Nilekani. Above the screen there were eight clocks that pretty well summed up the Infosys workday: 24/7/365. The clocks were labeled US West, US East, GMT, India, Singapore, Hong Kong, Japan, Australia.

"Outsourcing is just one dimension of a much more fundamental thing happening today in the world," Nilekani explained. "What happened over the last [few] years is that there was a massive investment in technology, especially in the bubble era, when hundreds of millions of dollars were invested in putting broadband connectivity around the world, undersea cables, all those things." At the same time, he added, computers became cheaper and dispersed all over the world, and there was an explosion of software—e-mail, search engines like Google, and proprietary software that can chop up any piece of work and send one part to Boston, one part to Bangalore, and one part to Beijing, making it easy for anyone to do remote development. When all of these things suddenly came together around 2000, added Nilekani, they "created a platform where intellectual work, intellectual capital, could be delivered from anywhere. It could be disaggregated, delivered, distributed, produced, and put back together again—and this gave a whole new degree of freedom to the way we do work, especially work of an intellectual nature.... And what you are seeing in Bangalore today is really the culmination of all these things coming together."

We were sitting on the couch outside of Nilekani's office, waiting for the TV crew to set up its cameras. At one point, summing up the implications of all this, Nilekani uttered a phrase that rang in my ear. He said to me, "Tom, the playing field is being leveled." He meant that countries like India are now able to compete for global knowledge work as never before—and that America had better get ready for this. America was going to be challenged, but, he insisted, the challenge would be good for America because we are always at our best when we are being challenged. As I left the Infosys campus that evening and bounced along the road back to Bangalore, I kept chewing on that phrase: "The playing field is being leveled."

What Nandan is saying, I thought, is that the playing field is being flattened.... Flattened? Flattened? My God, he's telling me the world is flat!

Here I was in Bangalore—more than five hundred years after Columbus sailed over the horizon, using the rudimentary navigational technologies of his day, and returned safely to prove definitively that the world was round—and one of India's smartest engineers, trained at his country's top technical institute and backed by the most modern technologies of his day, was essentially telling me that the world was *flat*—as flat as that screen on which he can host a meeting of his whole global supply chain. Even more interesting, he was citing this development as a good thing, as a new milestone in human progress and a great opportunity for India and the world—the fact that we had made our world flat!

In the back of that van, I scribbled down four words in my notebook: "The world is flat." As soon as I wrote them, I realized that this was the underlying message of everything that I had seen and heard in Bangalore in two weeks of filming. The global competitive playing field was being leveled. The world was being flattened.

As I came to this realization, I was filled with both excitement and dread. The journalist in me was excited at having found a framework to better understand the morning headlines and to explain what was happening in the world today. Clearly, it is now possible for more people than ever to collaborate and compete in real time with more other people on more different kinds of work from more different corners of the planet and on a more equal footing than at any previous time in the history of the world—using computers, e-mail, networks, teleconferencing, and dynamic new software. That is what Nandan was telling me. That was what I discovered on my journey to India and beyond. And that is what this book is about. When you start to think of the world as flat, a lot of things make sense in ways they did not before. But I was also excited personally, because what the flattening of the world means is that we are now connecting all the knowledge centers on the planet together into a single global network, which—if politics and terrorism do not get in the way—could usher in an amazing era of prosperity and innovation.

But contemplating the flat world also left me filled with dread, professional and personal. My personal dread derived from the obvious fact that it's not only the software writers and computer geeks who get empowered to collaborate on work in a flat world. It's also al-Qaeda and other terrorist networks. The playing field is not being leveled only in ways that draw in and superempower a whole new group of innovators. It's being leveled in a way that draws in and superempowers a whole new group of angry, frustrated, and humiliated men and women.

Professionally, the recognition that the world was flat was unnerving because I realized that this flattening had been taking place while I was sleeping, and I had missed it. I wasn't really sleeping, but I was otherwise engaged. Before 9/11, I was focused on tracking globalization and exploring the tension between the "Lexus" forces of economic integration and the "Olive Tree" forces of identity and nationalism—hence my 1999 book, *The Lexus and the Olive Tree*. But after 9/11, the olive tree wars became all-consuming for me. I spent almost all my time traveling in the Arab and Muslim worlds. During those years I lost the trail of globalization.

I found that trail again on my journey to Bangalore in February 2004. Once I did, I realized that something really important had happened while I was fixated on the olive groves of Kabul and Baghdad. Globalization had gone to a whole new level. If you put *The Lexus and the Olive Tree* and this book together, the broad historical argument you end up with is that there have been three great eras of globalization. The first lasted from 1492—when Columbus set sail, opening trade between the Old World and the New World—until around 1800. I would call this era Globalization 1.0. It shrank the world from a size large to a size medium. Globalization 1.0 was about countries and muscles. That is, in Globalization 1.0 the key agent of change, the dynamic force driving the process of global integration was how much brawn—how much muscle, how much horsepower, wind power, or, later, steam power—your country had and how creatively you could deploy it. In this era, countries and governments (often inspired by religion or imperialism or a combination of both) led the way in breaking down walls and knitting the world together, driving global integration. In Globalization 1.0, the primary questions were: Where does my country fit into global competition and opportunities? How can I go global and collaborate with others through my country?

The second great era, Globalization 2.0, lasted roughly from 1800 to 2000, interrupted by the Great Depression and World Wars I and II. This era shrank the world from a size medium to a size small. In Globalization 2.0, the key agent of change, the dynamic force driving global integration, was multinational companies. These multinationals went global for markets and labor, spearheaded first by the expansion of the Dutch and English joint-stock companies and the Industrial Revolution. In the first half of this era, global integration was powered by falling transportation costs, thanks to the steam engine and the railroad, and in the second half by falling telecommunication costs—thanks to the diffusion of the telegraph, telephones, the PC, satellites, fiber-optic cable, and the early version of the World Wide Web. It was during this era that we really saw the birth and maturation of a global economy, in the sense that there was enough movement of goods and information from continent to continent for there to be a global market, with global arbitrage in products and labor. The dynamic forces behind this era of globalization were breakthroughs in hardware—from steamships and railroads in the beginning to telephones and mainframe computers toward the end. And the big questions in this era were: Where does my company fit into the global economy? How does it take advantage of the opportunities? How can I go global and collaborate with others through my company? *The Lexus and the Olive Tree* was primarily about the climax of this era, an era when the walls started falling all around the world, and integration, and the backlash to it, went to a whole new level. But even as the walls fell, there were still a lot of barriers to seamless global integration. Remember, when Bill Clinton was elected president in 1992, virtually no one outside of government and the academy had e-mail, and when I was writing *The Lexus and the Olive Tree* in 1998, the Internet and e-commerce were just taking off.

Well, they took off—along with a lot of other things that came together while I was sleeping. And that is why I argue in this book that around the year 2000 we entered a whole new era: Globalization 3.0. Globalization 3.0 is shrinking the world from a size small to a size tiny and flattening the playing field at the same time. And while the dynamic force in Globalization 1.0 was countries globalizing and the dynamic force in Globalization 2.0 was companies globalizing, the dynamic force in Globalization 3.0—the thing that gives it its unique character—is the newfound power for *individuals* to collaborate and compete globally. And the lever that is enabling individuals and groups to go global so easily and so seamlessly is not horsepower, and not hardware, but software—all sorts of new applications—in conjunction with the creation of a global fiber-optic network that has made us all next-door neighbors. Individuals must, and can, now ask, Where do *I* fit into the global competition and opportunities of the day, and how can *I,* on my own, collaborate with others globally?

But Globalization 3.0 not only differs from the previous eras in how it is shrinking and flattening the world and in how it is empowering individuals. It is different in that Globalization 1.0 and 2.0 were driven primarily by European and American individuals and businesses. Even though China actually had the biggest economy in the world in the eighteenth century, it was Western countries, companies, and explorers who were doing most of the globalizing and shaping of the system. But going forward, this will be less and less true. Because it is flattening and shrinking the world, Globalization 3.0 is going to be more and more driven not only by individuals but also by a much more diverse—non-Western, non-white—group of individuals. Individuals from every corner of the flat world are being empowered. Globalization 3.0 makes it possible for so many more people to plug and play, and you are going to see every color of the human rainbow take part.

(While this empowerment of individuals to act globally is the most important new feature of Globalization 3.0, companies—large and small—have been newly empowered in this era as well. I discuss both in detail later in the book.)

Needless to say, I had only the vaguest appreciation of all this as I left Nandan's office that day in Bangalore. But as I sat contemplating these changes on the balcony of my hotel room that evening, I did know one thing: I wanted to drop everything and write a book that would enable me to understand how this flattening process happened and what its implications might be for countries, companies, and individuals. So I picked up the phone and called my wife, Ann, and told her, "I am going to write a book called *The World Is Flat.*" She was both amused and curious—well, maybe *more* amused than curious! Eventually, I was able to bring her around, and I hope I will be able to do the same with you, dear reader. Let me start by taking you back to the beginning of my journey to India, and other points east, and share with you some of the encounters that led me to conclude the world was no longer round—but flat.

3

Why the World Isn't Flat

PANKAJ GHEMAWAT

Ghemawhat challenges Friedman's flat world argument by pointing out some of the ways in which the world lacks true global connections. Globalization has bound people, countries, and markets closer than ever, rendering national borders relics of a bygone era—or so we're told. But a close look at the data reveals a world that's just a fraction as integrated as the one we thought we knew. In fact, more than 90 percent of all phone calls, Web traffic, and investment is local. What's more, even this small level of globalization could still slip away.

Ideas will spread faster, leaping borders. Poor countries will have immediate access to information that was once restricted to the industrial world and traveled only slowly, if at all, beyond it. Entire electorates will learn things that once only a few bureaucrats knew. Small companies will offer services that previously only giants could provide. In all these ways, the communications revolution is profoundly democratic and liberating, leveling the imbalance between large and small, rich and poor. The global vision that Frances Cairncross predicted in her *Death of Distance* appears to be upon us. We seem to live in a world that is no longer a collection of isolated, "local" nations, effectively separated by high tariff walls, poor communications networks, and mutual suspicion. It's a world that, if you believe the most prominent proponents of globalization, is increasingly wired, informed, and, well, "flat."

It's an attractive idea. And if publishing trends are any indication, globalization is more than just a powerful economic and political transformation; it's a booming cottage industry. According to the U.S. Library of Congress's catalog, in the 1990s, about 500 books were published on globalization. Between 2000 and 2004, there were more than 4,000. In fact, between the mid-1990s and 2003, the rate of increase in globalization-related titles more than doubled every 18 months.

Amid all this clutter, several books on the subject have managed to attract significant attention. During a recent TV interview, the first question I was asked—quite earnestly—was why I still thought the world was round. The interviewer was referring of course to the thesis of *New York Times* columnist Thomas L. Friedman's bestselling book *The World Is Flat*. Friedman asserts that 10 forces—most of which enable connectivity and collaboration at a distance—are "flattening"

SOURCE: "Why the World Isn't Flat," by Pankaj Ghemawat, *Foreign Policy*, March/ April 2007, pp. 54–60. Used with permission.

the Earth and leveling a playing field of global competitiveness, the like of which the world has never before seen.

It sounds compelling enough. But Friedman's assertions are simply the latest in a series of exaggerated visions that also include the "end of history" and the "convergence of tastes." Some writers in this vein view globalization as a good thing—an escape from the ancient tribal rifts that have divided humans, or an opportunity to sell the same thing to everyone on Earth. Others lament its cancerous spread, a process at the end of which everyone will be eating the same fast food. Their arguments are mostly characterized by emotional rather than cerebral appeals, a reliance on prophecy, semiotic arousal (that is, treating everything as a sign), a focus on technology as the driver of change, an emphasis on education that creates "new" people, and perhaps above all, a clamor for attention. But they all have one thing in common. They're wrong.

In truth, the world is not nearly as connected as these writers would have us believe. Despite talk of a new, wired world where information, ideas, money, and people can move around the planet faster than ever before, just a fraction of what we consider globalization actually exists. The portrait that emerges from a hard look at the way companies, people, and states interact is a world that's only beginning to realize the potential of true global integration. And what these trend's backers won't tell you is that globalization's future is more fragile than you know.

THE 10 PERCENT PRESUMPTION

The few cities that dominate international financial activity—Frankfurt, Hong Kong, London, New York—are at the height of modern global integration; which is to say, they are all relatively well connected with one another. But when you examine the numbers, the picture is one of extreme connectivity at the local level, not a flat world. What do such statistics reveal? Most types of economic activity that could be conducted either within or across borders turn out to still be quite domestically concentrated.

One favorite mantra from globalization champions is how "investment knows no boundaries." But how much of all the capital being invested around the world is conducted by companies outside of their home countries? The fact is, the total amount of the world's capital formation that is generated from foreign direct investment (FDI) has been less than 10 percent for the last three years for which data are available (2003–05). In other words, more than 90 percent of the fixed investment around the world is still domestic. And though merger waves can push the ratio higher, it has never reached 20 percent. In a thoroughly globalized environment, one would expect this number to be much higher—about 90 percent, by my calculation. And FDI isn't an odd or unrepresentative example.

The levels of internationalization associated with cross-border migration, telephone calls, management research and education, private charitable giving, patenting, stock investment, and trade, as a fraction of gross domestic product (GDP); all stand much closer to 10 percent than 100 percent. The biggest exception in absolute terms—the trade-to-GDP ratio—recedes most of the way back down toward 20 percent if you adjust for certain kinds of double-counting. So if someone asked me to guess the internationalization level of some activity about which I had no particular information, I would guess it to be much closer to 10 percent—the average for the nine categories of data in the chart—than to 100 percent. I call this the "10 Percent Presumption."

More broadly, these and other data on cross-border integration suggest a semiglobalized world, in which neither the bridges nor the barriers between countries can be ignored. From this perspective, the most astonishing aspect of various writings on globalization is the extent of exaggeration involved. In short, the levels of internationalization in the world today are roughly an order of magnitude lower than those implied by globalization proponents.

A STRONG NATIONAL DEFENSE

If you buy into the more extreme views of the globalization triumphalists, you would expect to see a world where national borders are irrelevant, and where citizens increasingly view themselves as members of ever broader political entities. True, communications technologies have improved dramatically during the past 100 years. The cost of a three-minute telephone call from New York to London fell from $350 in 1930 to about 40 cents in 1999, and is now approaching zero for voice-over-Internet telephony. And the Internet itself is just one of many newer forms of connectivity that have progressed several times faster than plain old telephone service. This pace of improvement has inspired excited proclamations about the pace of global integration. But it's a huge leap to go from predicting such changes to asserting that declining communication costs will obliterate the effects of distance. Although the barriers at borders have declined significantly, they haven't disappeared.

To see why, consider the Indian software industry—a favorite of Friedman and others. Friedman cites Nandan Nilekani, the CEO of the second-largest such firm, Infosys, as his muse for the notion of a flat world. But what Nilekani has pointed out privately is that while Indian software programmers can now serve the United States from India, access is assured, in part, by U.S. capital being invested—quite literally—in that outcome. In other words, the success of the Indian IT industry is not exempt from political and geographic constraints. The country of origin matters—even for capital, which is often considered stateless.

Or consider the largest Indian software firm, Tata Consultancy Services (TCS). Friedman has written at least two columns in the *New York Times* on TCS's Latin American operations: "[I]n today's world, having an Indian company led by a Hungarian-Uruguayan servicing American banks with

Montevidean engineers managed by Indian technologists who have learned to eat Uruguayan veggie is just the new normal," Friedman writes. Perhaps. But the real question is why the company established those operations in the first place. Having worked as a strategy advisor to TCS since 2000, I can testify that reasons related to the tyranny of time zones, language, and the need for proximity to clients' local operations loomed large in that decision. This is a far cry from globalization proponents' oft-cited world in which geography, language, and distance don't matter.

Trade flows certainly bear that theory out. Consider Canadian–U.S. trade, the largest bilateral relationship of its kind in the world. In 1988, before the North American Free Trade Agreement (NAFTA) took effect, merchandise trade levels between Canadian provinces—that is, within the country—were estimated to be 20 times as large as their trade with similarly sized and similarly distant U.S. states. In other words, there was a built-in "home bias." Although NAFTA helped reduce this ratio of domestic to international trade—the home bias—to 10 to 1 by the mid-1990s, it still exceeds 5 to 1 today. And these ratios are just for merchandise; for services, the ratio is still several times larger. Clearly, the borders in our seemingly "borderless world" still matter to most people.

Geographical boundaries are so pervasive, they even extend to cyberspace. If there were one realm in which borders should be rendered meaningless and the globalization proponents should be correct in their overly optimistic models, it should be the Internet. Yet Web traffic within countries and regions has increased far faster than traffic between them. Just as in the real world, Internet links decay with distance. People across the world may be getting more connected, but they aren't connecting with each other. The average South Korean Web user may be spending several hours a day online—connected to the rest of the world in theory—but he is probably chatting with friends across town and e-mailing family across the country rather than meeting a fellow surfer in Los Angeles. We're more wired, but no more "global."

Just look at Google, which boasts of supporting more than 100 languages and, partly as a result, has recently been rated the most globalized Web site. But Google's operation in Russia (cofounder Sergey Brin's native country) reaches only 28 percent of the market there, versus 64 percent for the Russian market leader in search services, Yandex, and 53 percent for Rambler.

Indeed, these two local competitors account for 91 percent of the Russian market for online ads linked to Web searches. What has stymied Google's expansion into the Russian market? The biggest reason is the difficulty of designing a search engine to handle the linguistic complexities of the Russian language. In addition, these local competitors are more in tune with the Russian market, for example, developing payment methods through traditional banks to compensate for the dearth of credit cards. And, though Google has doubled its reach since 2003, it's had to set up a Moscow office in Russia and hire Russian software engineers, underlining the continued importance of physical location. Even now, borders between countries define—and constrain—our movements more than globalization breaks them down.

TURNING BACK THE CLOCK

If globalization is an inadequate term for the current state of integration, there's an obvious rejoinder: Even if the world isn't quite flat today, it will be tomorrow. To respond, we have to look at trends, rather than levels of integration at one point in time. The results are telling. Along a few dimensions, integration reached its all-time high many years ago. For example, rough calculations suggest that the number of long-term international migrants amounted to 3 percent of the world's population in 1900—the high-water mark of an earlier era of migration—versus 2.9 percent in 2005.

Along other dimensions, it's true that new records are being set. But this growth has happened only relatively recently, and only after long periods of stagnation and reversal. For example, FDI stocks divided by GDP peaked before World War I and didn't return to that level until the 1990s. Several economists have argued that the most remarkable development over the long term was the declining level of internationalization between the two World Wars. And despite the records being set, the current level of trade intensity falls far short of completeness, as the Canadian–U.S. trade data suggest. In fact, when trade economists look at these figures, they are amazed not at how much trade there is, but how little.

It's also useful to examine the considerable momentum that globalization proponents attribute to the constellation of policy changes that led many countries—particularly China, India, and the former Soviet Union—to engage more extensively with the international economy. One of the better-researched descriptions of these policy changes and their implications is provided by economists Jeffrey Sachs and Andrew Warner:

> The years between 1970 and 1995, and especially the last decade, have witnessed the most remarkable institutional harmonization and economic integration among nations in world history. While economic integration was increasing throughout the 1970s and 1980s, the extent of integration has come sharply into focus only since the collapse of communism in 1989. In 1995, one dominant global economic system is emerging.

Yes, such policy openings are important. But to paint them as a sea change is inaccurate at best. Remember the 10 Percent Presumption, and that integration is only beginning. The policies that we fickle humans enact are surprisingly reversible. Thus, Francis Fukuyama's *The End of History,* in which liberal democracy and technologically driven capitalism were supposed to have triumphed over other ideologies, seems quite quaint today. In the wake of Sept. 11, 2001, Samuel Huntington's *Clash of Civilizations* looks at least a bit more prescient. But even if you stay on the economic plane, as Sachs and Warner mostly do, you quickly see counterevidence to the supposed decisiveness of policy openings. The so-called Washington Consensus around market-friendly policies ran up against the 1997 Asian currency crisis and has since frayed substantially—for example, in the swing toward neopopulism across much of Latin America. In terms of economic outcomes, the number of countries—in Latin America, coastal Africa, and the former Soviet Union—that have dropped out of the "convergence club" (defined in terms of narrowing productivity and structural gaps vis-à-vis the advanced industrialized countries) is at least as impressive as the number of countries that have joined the club. At a multilateral level, the suspension of the Doha round of trade talks in the summer of 2006—prompting *The Economist* to run a cover titled "The Future of Globalization" and depicting a beached wreck—is no promising omen. In addition, the recent wave of cross-border mergers and acquisitions seems to be encountering more protectionism, in a broader range of countries, than did the previous wave in the late 1990s.

Of course, given that sentiments in these respects have shifted in the past 10 years or so, there is a fair chance that they may shift yet again in the next decade. The point is, it's not only possible to turn back the clock on globalization-friendly policies, it's relatively easy to imagine it happening. Specifically, we have to entertain the possibility that deep international economic integration may be inherently incompatible with national sovereignty—especially given the tendency of voters in many countries, including advanced ones, to support more protectionism, rather than less. As Jeff Immelt, CEO of GE, put it in late 2006, "If you put globalization to a popular vote in the U.S., it would lose." And even if cross-border integration continues on its upward path, the road from here to there is unlikely to be either smooth or straight. There will be shocks and cycles, in all likelihood, and maybe even another period of stagnation or reversal that will endure for decades. It wouldn't be unprecedented.

The champions of globalization are describing a world that doesn't exist. It's a fine strategy to sell books and even describe a potential environment that may someday exist. Because such episodes of mass delusion tend to be relatively short-lived even when they do achieve broad currency, one might simply be tempted to wait this one out as well. But the stakes are far too high for that. Governments that buy into the flat world are likely to pay too much attention to the "golden straitjacket" that Friedman emphasized in his earlier book, *The Lexus and the Olive Tree,* which is supposed to ensure that economics matters more and more and politics less and less. Buying into this version of an integrated world—or worse, using it as a basis for policymaking—is not only unproductive. It is dangerous.

4

Globalization and Its Specter

JEREMY BRECHER, TIM COSTELLO AND BRENDAN SMITH

This article offers a decidedly critical look at globalization and its contradictions.

GLOBALIZATION FROM ABOVE

Epochal changes can be difficult to grasp—especially when you are in their midst. Those who lived through the rise of capitalism or the Industrial Revolution knew something momentous was happening, but just what was new and what it meant were subjects of confusion and debate.

In a sense, there has been a global economy for 500 years. But the last quarter of the 20th century saw global economic integration take new forms. At first, globalization manifested itself as apparently separate and rather marginal phenomena: the emergence of the "Eurodollar market," "off-shore export platforms," and "supply-side economics," for example. It was easy to separate out one or another aspect of globalization—such as the growth of trade or of international economic institutions—and see it as an isolated phenomenon. These seemingly peripheral developments, however, gradually interacted in ways that changed virtually every aspect of life and defined globalization as a new global configuration.[1]

Globalization was not the result of a plot or even a plan. It was caused by people acting with intent—seeking new economic opportunities, creating new institutions, trying to outflank political and economic opponents. But it resulted not just from their intent, but also from unintended side effects of their actions and the consequences of unintended interactions.[2] Future historians will note at least the following aspects of the globalization process:

Production: In the 1970s, corporations began building factories and buying manufactured products in low-wage countries in the third world on a vastly expanded scale. Such off-shore production grew into today's "global assembly line," in which the components of a shirt or car may be made and assembled in a dozen or more different countries. Direct investment abroad by "American" companies has grown so rapidly that the value of the goods and services they produce and sell outside the United States is now three times the total value of all American exports.[3]

SOURCE: From *Globalization from Below: The Power of Solidarity* Cambridge, MA: South End Press, 2000, pp. 1–9. Used with permission.

Markets: Corporations came increasingly to view the entire world as a single market in which they buy and sell goods, services, and labor....

Finance: Starting with the rise of the Eurodollar market in the 1970s, international capital markets have globalized at an accelerating rate....

Technology: New information, communication, and transportation technologies—computers, satellite communications, containerized shipping, and, increasingly, the Internet—have reduced distance as a barrier to economic integration. Furthermore, the process of creating new technologies has itself become globalized.[4]

Global institutions: The World Trade Organization (WTO), the International Monetary Fund (IMF), the World Bank, and similar institutions at a regional level have developed far greater powers and have used them to accelerate the globalization process.

Corporate restructuring: While corporations have always operated internationally, starting in the 1980s they began to restructure in order to operate in a global economy. New corporate forms—strategic alliances, global outsourcing, captive suppliers, supplier chains, and, increasingly, transnational mergers—allowed for what the economist Bennett Harrison has called the "concentration of control [with] the decentralization of production."[5]

Changing structure of work: Globalization has been characterized by a "re-commodification of labor" in which workers have increasingly lost all rights except the right to sell their labor power. All over the world, employers have downsized, outsourced, and made permanent jobs into contingent ones. Employers have attacked job security requirements, work rules, worker representation, healthcare, pensions, and other social benefits, and anything else that defined workers as human beings and employers as partners in a social relationship, rather than simply as buyers and sellers of labor power.

Neoliberal ideology and policies: Starting with monetarism and supply-side economics, globalization has been accompanied—and accelerated—by an emerging ideology now generally known as neoliberalism or the Washington Consensus. It argues that markets are efficient and that government intervention in them is almost always bad. The policy implications—privatization, deregulation, open markets, balanced budgets, deflationary austerity, and dismantling of the welfare state—were accepted by or imposed on governments all over the world.

Changing role of the state: While some governments actively encouraged globalization and most acquiesced, globalization considerably reduced the power of nation states, particularly their power to serve the interests of their own people. Capital mobility undermined the power of national governments to pursue full employment policies or regulate corporations. International organizations and agreements increasingly restricted environmental and social protections. Neoliberal ideology reshaped beliefs about what government should do and what it is able to accomplish.

Neo-imperialism: Globalization reversed the post–World War II movement of third world countries out of colonialism toward economic independence. Globalization has restored much of the global dominance of the former imperialist powers, such as Western Europe, Japan, and, above all, the United States. With the collapse of Communism, that dominance has also spread to much of the formerly communist world. Globalization has taken from poor countries control of their own economic policies and concentrated their assets in the hands of first world investors. While it has enriched some third world elites, it has subordinated them to foreign corporations, international institutions, and dominant states. It has intensified economic rivalry among the rich powers.[6]

Movement of people: While people have always crossed national borders, the economic disruptions and reduction of national barriers caused by globalization are accelerating migration. International travel and tourism have become huge industries in their own right.

Cultural homogenization: Globalization has undermined the economic base of diverse local and indigenous communities all over the world. Growing domination of global media by a few countries and companies has led not to greater diversity, but to an increasingly uniform culture of corporate globalism.

As *New York Times* columnist and globalization advocate Thomas Friedman summed up, we are in a new international system: "Globalization is not just a trend, not just a phenomenon, not just an economic fad. It is the international system that has replaced the cold-war system." The driving force behind globalization is free-market capitalism: "Globalization means the spread of free-market capitalism to virtually every country in the world."[7]

THE CONTRADICTIONS OF GLOBALIZATION
FROM ABOVE

The proponents of globalization promised that it would benefit all: that it would "raise all the boats." Workers and communities around the globe were told that if they downsized, deregulated, eliminated social services, and generally became more competitive, the benefits of globalization would bless them. The poorest and most desperate were promised that they would see their standard of living increase if they accepted neoliberal austerity measures. They kept their end of the bargain, but globalization from above did not reciprocate. Instead, it is aggravating old and creating new problems for people and the environment.

Even conventional economic theory recognizes that the "hidden hand" of the market doesn't always work. Unregulated markets regularly produce unintended side effects or "externalities"—such as ecological pollution for which the producer doesn't have to pay, or the devastation of communities when corporations move away. Unregulated markets also produce unintended interaction effects, such as the downward spirals of depressions and trade wars. Unregulated markets do nothing to correct inequalities of wealth; indeed, they often intensify the concentration of wealth, leading to expanding gaps between rich and poor.[8] Globalization from above has globalized these problems, while dismantling at every level the non-market institutions that once addressed them.

Globalization promotes a destructive competition in which workers, communities, and entire countries are forced to cut labor, social, and environmental costs to attract mobile capital. When many countries each do so, the result is a disastrous "race to the bottom."

The race to the bottom occurs not just between developing and developed worlds, but increasingly among the countries of the third world. Consider the case of Argentina and Brazil. Early in 1999, Brazil devalued its currency by 40 percent. A *New York Times* reporter in Argentina found that "[a]bout 60 manufacturing companies have moved to Brazil in recent months, seeking lower labor costs and offers of tax breaks and other government subsidies." Companies closing Argentine factories to supply the Argentine market from Brazil included Tupperware, Goodyear, and Royal Philips Electronics. The Argentine auto and auto-parts industries suffered a 33 percent loss of production and a 59 percent fall in exports in 1999. "General Motors, Ford Motor and Fiat are all transferring production to Brazil."[9]

Argentine President Fernando de la Rua commented, "If you ask me what is my chief concern in a word, that word is 'competitiveness.'" The measures he has taken to become more "competitive" exemplify the race to the bottom. "The crown jewel of the De la Rua economic policy is his labor reform" intended to "reduce the bargaining power of labor unions and help businesses more easily hire and fire new workers."

But this gutting of labor rights was not enough to protect Argentine manufacturers against products from lower-wage countries. A shoe manufacturer who expected the new labor law to cut his labor costs by 10 percent "felt constrained because of the competitive disadvantage he continued to suffer in relation to Brazilian shoe producers who pay their workers one-third the wage an Argentine shoemaker earns." The director of a medical supply company who was considering closing his plant observed that it was impossible to compete with the flood of cheap Korean and Chinese syringes in recent years and that Brazilian officials were offering a package of tax breaks and subsidized loans to relocate to Brazil.

The role of international institutions in promoting the race to the bottom is illustrated by the fact that both Brazil and Argentina were shaping their economic policies in accord with loan agreements they had made with the IMF.

First world countries are also engaged in the race to the bottom. Over the past two decades, for example, the United States made huge cuts in corporate taxes while slashing federal funding for health, education, and community development. Canada, which did not make equivalent cuts, found that its tax structure was "making it difficult for companies to compete internationally. Many businesses have simply moved across the border to the U.S." In response, Canada decided in early 2000 to lower its corporate tax rate from 28 percent to 21 percent. In a fit of ingratitude, the Business Council on National Issues, representing Canada's 150 largest companies, condemned the cuts as "timid." The Business Council's president opined that "[t]he strategy should be to provide an environment more attractive than the U.S. now." The disappointed chief executive of an e-commerce services company said he had been planning to open offices in Calgary, Alberta, and Vancouver, British Columbia, but that after the inadequate tax cuts he was leaning toward Chicago or Minneapolis instead. The director of the Canadian Taxpayers Federation observed, "There's competition for tax cuts, just like everything else."[10]

The race to the bottom brings with it the dubious blessings of impoverishment, growing inequality, economic volatility, the degradation of democracy, and destruction of the environment.

Impoverishment: The past quarter-century of globalization has seen not a reduction but a vast increase in poverty. According to the 1999 U.N. *Human Development Report,* more than 80 countries have per capita incomes lower than they were a decade or more ago.[11] James Wolfensohn, president of the World Bank, says that, rather than improving, "global poverty is getting worse. Some 1.2 billion people now live in extreme poverty."[12] Global unemployment is approaching 1 billion.[13]

In the United States, the downward pressures of globalization are manifested in the stagnation of wages despite the longest period of economic growth in American history. Real average wages were $9 per hour in 1973; 25 years later, they were $8 per hour. The typical married-couple family worked 247 more hours per year in 1996 than in 1989—more than six weeks' worth of additional work each year.[14]

Inequality: Globalization has contributed to an enormous increase in the concentration of wealth and the growth of poverty both within countries and worldwide. Four hundred and forty-seven billionaires have wealth greater than the income of the poorest half of humanity. In the United States, the richest man has wealth equal to that of the poorest 40 percent of the American people.[15] The net worth of the world's 200 richest people increased from $440 billion to more than $1 trillion in just the four years from 1994 to 1998. The assets of the three richest people were more than the combined GNP of the 48 least developed countries.[16]

The downward pressures of globalization have been focused most intensively on discriminated-against groups that have the least power to resist, including women, racial and ethnic minorities, and indigenous peoples. Women have been the prime victims of exploitation in export industries and have suffered the brunt of cutbacks in public services and support for basic needs. Immigrants and racial and ethnic minorities in many parts of the world have not only been subject to exploitation, but have been abused as scapegoats for the economic troubles caused by globalization from above. Indigenous people have had their traditional ways of life disrupted and their economic resources plundered by global corporations and governments doing their bidding.

Volatility: Global financial deregulation has reduced barriers to the international flow of capital. More than $1.5 trillion now flows across international borders daily in the foreign currency market alone. These huge flows easily swamp national economies. The result is a world economy marked by dangerous and disruptive financial volatility.

In 1998, for example, an apparently local crisis in Thailand rapidly spread around the globe. In two years, Malaysia's economy shrunk by 25 percent, South Korea's by 45 percent, and Thailand's by 50 percent. Indonesia's economy shrunk by 80 percent; its per capita gross domestic product dropped from $3,500 to less than $750; and 100 million people—nearly half of the population—sank below the poverty line.[17] According to former World Bank chief economist Joseph Stiglitz,

> Capital market liberalization has not only not brought people the prosperity they were promised, but it has also brought these crises, with wages falling 20 or 30 percent, and unemployment going up by a factor of two, three, four or ten.[18]

Degradation of democracy: Globalization has reduced the power of individuals and peoples to shape their destinies through participation in democratic processes.

Of the 100 largest economies in the world, 51 today are corporations, not countries.[19] Globalization has greatly increased the power of global corporations relative to local, state, and national governments. The ability of governments to pursue development, full employment, or other national economic goals has been undermined by the growing ability of capital to simply pick up and leave.

There are few international equivalents to the anti-trust, consumer protection, and other laws that provide a degree of corporate accountability at the national level. As a result, corporations are able to dictate policy to governments, backed by the threat that they will relocate.

Governmental authority has been undermined by trade agreements such as NAFTA and WTO and by international financial institutions such as the IMF and World Bank, which restrict the power of national, state, and local governments to govern their own economies. These institutions are all too often themselves complicit in the denial of human rights. (At a time when 100 to 200 Algerians were having their throats cut every week, the IMF stated, "Directors agree that Algeria's exemplary adjustment and reform efforts deserve continued support of the international financial community.")[20] They make decisions affecting billions of people, but they are largely free of democratic control and accountability. As one unnamed WTO official was quoted by the *Financial Times,* "The WTO is the place where governments collude in private against their domestic pressure groups."[21]

Environmental destruction: Globalization is accelerating ecological catastrophe both globally and locally. Countries are forced to compete for investment by lowering environmental protections in an ecological race to the bottom. (Seventy countries have rewritten their mining codes in recent years to encourage investment.[22]) Neoliberal policies imposed by international institutions or voluntarily accepted by national governments restrict environmental regulation. Worldwide, corporations promote untested technologies, such as pesticides and genetic engineering, turning the planet into a testing lab and its people into guinea pigs. Growing poverty leads to desperate overharvesting of natural resources.

Global corporations' oil refineries, chemical plants, steel mills, and other factories are the main source of greenhouse gases, ozone-depleting chemicals, and toxic pollutants. Overfishing of the world's waters, overcutting of forests, and abuse of agricultural land result from the search for higher corporate profits, the drive to increase exports, and the increase in poverty.

Globally, environmental destruction is changing the basic balances on which life depends. Carbon dioxide has reached record levels in the atmosphere.[23] Global warming is already resulting in the melting of glaciers, the dying of coral reefs, climate instability, and "a disturbing change in disease patterns."[24] An estimated one-quarter of the world's mammal species and 13 percent of plant species are threatened with extinction in the worst period of mass extinction of species in 65 million years.[25]

ENDNOTES

1. Some analysts still debate to what extent globalization is genuine or significant. See, for example, Doug Henwood, "What Is Globalization Anyway?" ZNet Commentary, November 26, 1999, maintaining that globalization is exaggerated (http://zmag.org/ZSustamers/ZDaily/1999-ll/26henwood.htm), and Richard Du Boff and Edward Herman, "Questioning Henwood on Globalization," ZNet Commentary, December 1, 1999, finding Henwood's arguments "incomplete and unconvincing" (http://zmag.org/ZSustainers/ZDaily/1999-12/01herman.htm).

2. For a more detailed analysis of the first phase of globalization, see Brecher and Costello, *Global Village or Global Pillage*, and works cited there. For a review of developments up to 1998, see the Introduction to the second edition.

3. John Tagliabue, "For Americans, an Indirect Route to the Party," *New York Times*, June 14, 1998, p. 3, 4, citing James E. Carlson, economist at Merrill Lynch in New York.

4. For the globalization of the process of technological change, see Peter Dorman, "Actually Existing Globalization," prepared for "Globalization and Its Discontents," Michigan State University, April 3, 1998, pp. 4–7.

5. Bennett Harrison, *Lean and Mean: The Changing Landscape of Corporate Power in the Age of Flexibility* (New York: Basic Books, 1994), pp. 9, 171. For further discussion of changing corporate and work structures and their implications for labor organization, see Jeremy Brecher and Tim Costello, "Labor and the Dis-Integrated Corporation," *New Labor Forum* 2 (Spring/Summer 1998) 5ff.

6. For a view emphasizing the similarity between contemporary globalization and previous periods of shift in global hegemony, see Giovanni Arrighi and Beverly J. Silver, *Chaos and Governance in the Modern World System* (Minneapolis: University of Minnesota Press, 1999). For a view that globalization is replacing traditional national imperialism with a universal, non-national system of empire, see Michael Hardt and Antonio Negri, *Empire* (Cambridge: Harvard UP, 2000).

7. *New York Times Magazine*, March 28, 1999.

8. Economists often discuss such failings under the heading "market failures." For further discussion of market failures and "political failures," see Jeremy Brecher, "*Can'st Thou Draw Out Leviathan with a Fishhook'?*" (Washington: Grassroots Policy Project, 1995), and Charles E. Lindblom, *Politics and Markets* (New York: Basic Books, 1977), Chapters 5 and 6.

9. Clifford Krauss, "Injecting Change Into Argentina: New President Tries to Keep Industry from Leaving the Country," *New York Times*, March 8, 2000, p. CI. See also Craig Torres and Matt Moffett, "Neighbor-Bashing: Argentina Cries Foul as Choice Employers Beat a Path Next Door," *Wall Street Journal*, May 2, 2000, p. 1.

10. "Canada's Tax Cut Underwhelms Businesses: CEOs Fault Slowness of Phase-In and Look Abroad for Expansion," *Wall Street Journal*, May 2, 2000, p. A23.

11. United Nations Development Program (UNDP), *Human Development Report 1999* (New York: Oxford UP, 1999).

12. James D. Wolfensohn, "Let's Hear Everyone and Get on with Imaginative Solutions." *International Herald Tribune*, January 28, 2000 (http://www.iht.com/).

13. *World Employment Report, 1996/97* (Geneva: ILO, 1996).

14. Lawrence Mishel, Jared Bernstein, and John Schmitt, *The State of Working America 1998–99* (Ithaca: Economic Policy Institute/Cornell UP, 1999).

15. Anderson et al., *Field Guide to the Global Economy*, p. 53. *BusinessWeek*, April 20, 1998.

16. UNDP, *Human Development Report 1999*, p. 37.

17. Fareed Zakaria, "Will Asia Turn Against the West?" *New York Times*, July 10, 1998, p. A15.

18. Reuters, "Worker Rights Key to Development," January 8, 2000.

19. Anderson et al., *Field Guide to the Global Economy*, p. 69.

20. Jan Pronk, "Globalization: A Developmental Approach," in Jan Nederveen Pieterse, ed., *Global Futures: Shaping Globalization* (London: Zed Books, 2000), p. 48.

21. "Network Guerrillas," *Financial Times*, April 30, 1998, p. 20.

22. Hilary French, *Vanishing Borders: Protecting the Planet in the Age of Globalization* (New York: Norton, 2000), p. 27.

23. Hilary French, *Vanishing Borders*, p. 8.

24. "There are strong indications that a disturbing change in disease patterns has begun and that global warming is contributing to them," according to Paul Epstein, associate director of Harvard Medical School's Center for Health and the Global Environment. Quoted in Hilary French, *Vanishing Borders*, p. 46.

25. Hilary French, *Vanishing Borders*, pp. 8–9.

REFLECTION QUESTIONS FOR CHAPTER 2

1. Are the arguments in this section more persuasive for the position that "globalization is good" or the position that "globalization is bad"? Why? Continually revisit this question as you read the other chapters in this book.

2. How do the various dimensions of globalization affect your life? Consider, for example, job opportunities, stock market fluctuations, diseases (AIDS, SARS, West Nile virus), music, food, and fashion.

3. What additional evidence can you provide for Freidman's "flat world" thesis, especially in the non-Western world?

4. The readings in this section focus on the consequences of globalization for societies, institutions within societies (democracy), and opportunities for individuals (jobs, wages). Ghemawat disagrees with Friedman's view of information technology in the global age as profoundly liberating. What do you think of the reasons and evidence he gives to make the argument that the world's playing field is not leveling?

Chapter 3

Transnational Migration

Introduction

In 2009, about 200 million people worldwide (about 3 percent of the world's population) lived outside the country of their birth for at least a year. They left their homelands for a variety of reasons, such as escaping political tyranny, fleeing the dangers of war, seeking better jobs and a better life, or going to school. Other migrants were forced to move across political borders because their livelihoods were in jeopardy (for example, due to the depletion of fish in an area), because of drought or rising sea levels resulting from climate change, or because they were taken captive and trafficked as slaves (as described in Chapter 8).

Nations typically limit the number of immigrants allowed to settle within their borders. In the 1920s, for example, the United States placed limits on the number of immigrants it would accept, the operating principle being that the new immigrants should resemble the old ones, thus severely limiting the acceptance of Eastern Europeans and the total denial of Asians (Eitzen, Baca Zinn, and Smith, 2012, Chapter 5). The United States is currently trying to stifle the illegal entry of immigrants with a fence along part of its 2,000 mile shared border with Mexico, as well as through numerous government border guards and private vigilante groups. Nevertheless, some 11 to 12 million undocumented immigrants, mostly Latinos and Asians, reside in the United States. These immigrants are changing the racial and ethnic landscape of the U.S., as 12.5 percent of the nation's population are now foreign born. This is also the trend in Europe and Scandinavia, especially in the Netherlands and France, where significant numbers of Muslims have settled.

There are several reasons for temporary migration. About 900 million people travel annually across national boundaries as tourists (and this number is expected to double by 2020). These transient migrants travel for pleasure (seeing the sights or sex tourism, for example), adventure (such as climbing Mount Everest or scuba diving at the Great Barrier Reef off the coast of Australia), knowledge (for example, through ecotourism or visiting the Galapagos Islands), and health (medical tourism). Regarding medical tourism, many Americans travel outside the country for cheaper health care. For instance, the small town of Los Algodones in northern Mexico has 160 dental clinics and about 350 dentists that serve mostly Americans in search of cheap root canals and other dental work [Adams, 2009], and Americans travel across the Canadian and Mexican borders to purchase less-expensive pharmaceuticals).

This chapter contains five perspectives on transnational migration. The first, "Why Migration Matters" by Khalid Koser, provides an excellent introduction to the scope and consequences of migration. Khoser argues that one of the most important results of such migration is increasing diversity. He writes, people "who speak different languages and practice different customs and religions [...] are coming into unprecedented contact with each other. For some this is a threat, for others opportunity."

In the next reading, William Robinson shows how globalization both creates and requires pools of immigrant labor that global elites and transnational capital then shape for their own purposes. Immigrant Latinos, for example, are often exploited with low wages and difficult working conditions. The immigrant rights movement is the leading edge of the social struggle against oppressive class relations in global capitalism.

The third selection goes a step further, looking at foreign workers who are indentured—that is, who arrive in the U.S. under a guest-worker program. These workers are bound to the companies that requested them. They remain in the U.S. at the pleasure of their employers, and they cannot change employers. In this selection John Bowe, the author of *Nobodies* (2007), an exposé of modern slave labor, describes the contours of this exploitation of foreign workers.

In the fourth selection John Gibler discusses the implications of the fact that "Mexico has become the world's largest exporter of its people." As large numbers of Mexicans migrate to the United States, they leave behind ghost towns. The Mexican state of Zacatecas, for example, has lost over half of its population—about 1.8 million people—who now live and work in the United States. Those leaving Mexico provide the goods and services in the United States that their hometowns lack, resulting in shrinking, and ultimately disappearing, local Mexican economies.

The final selection focuses on migration resulting from climate change. The authors, Michael Werz and Kari Manlove, predict that by 2050 some 200 million people will be climate migrants due to the effects of global warming. Extreme weather events (such as hurricanes, droughts, and tsunamis) and resource shortages such as insecure food and water supplies will force many people, especially in Africa, Australia, and Latin America, to move to more favorable climates.

REFERENCES

Adams, Bobby Neel. 2009. "Border Crossing for a Root Canal." *Utne Reader*, #156 (November/December):30–32.

Bowe, John. 2007. *Nobodies: Modern American Slave Labor and the Dark Side of the New Global Economy*. New York: Random House.

Eitzen, D. Stanley, Maxine Baca Zinn, and Kelly Eitzen Smith. 2012. *Social Problems*, 12th ed. Boston: Allyn and Bacon.

5

Why Migration Matters

KHALID KOSER

Koser provides a historical overview of migration, showing that while migration has always mattered, its consequences are greater today.

Migration has always mattered—but today it matters more than ever before. The increasing importance of migration derives from its growing scale and its widening global reach, but also from a number of new dynamics. These include the feminization of migration, the growth of so-called irregular migration, and migration's inextricable linkages with globalization in terms of economic growth, development, and security. Climate change, moreover, is certain to raise migration still higher on nations' and international institutions' policy agendas.

The history of migration begins with humanity's very origins in the Rift Valley of Africa. It was from there that Homo sapiens emerged about 120,000 years ago, subsequently migrating across Africa, through the Middle East to Europe and Central and South Asia, and finally to the New World, reaching the Bering Strait about 20,000 years ago. Then, in the ancient world, Greek colonization and Roman expansion depended in migration; significant movements of people were also associated with the Mesopotamian, Incan, Indus, and Zhou empires. Later we see major migrations such as those involving the Vikings along the shorelines of the Atlantic and the North Sea, and the Crusaders to the Holy Land.

In more recent history—in other words, in the past two or three centuries—it is possible to discern, according to migration historian Robin Cohen, a series of major migration periods or events. In the eighteenth and nineteenth centuries, one of the most prominent migration events was the forced transportation of slaves. About 12 million people were taken, mainly from West Africa, to the New World (and also, in lesser numbers, across the Indian Ocean and the Mediterranean Sea). One of the reasons this migration is considered so important, other than its scale, is that it still resonates for descendants of slaves and for African Americans in particular. After slavery's collapse, indentured laborers from China, India, and Japan moved overseas in significant numbers—1.5 million from India alone—to work the plantations of the European powers.

SOURCE: From Khalid Koser, "Why Migration Matters," *Current History* 108 (April 2009), pp. 147–153. Reprinted with permission, Current History, Inc.

European expansion, especially during the nineteenth century, brought about large-scale voluntary migration away from Europe, particularly to the colonies of settlement, dominions, and the Americas. The great mercantile powers—Britain, the Netherlands, Spain, and France—all promoted settlement of their nationals abroad, not just workers but also peasants, dissident soldiers, convicts, and orphans. Migration associated with expansion largely came to an end with the rise of anti-colonial movements toward the end of the nineteenth century, and indeed over the next decades some significant reverse flows back to Europe occurred, for example of the so-called *pieds noirs* to France.

The next period of migration was marked by the rise of the United States as an industrial power. Between the 1850s and the Great Depression of the 1930s, millions of workers fled the stagnant economies and repressive political regimes of northern, southern, and eastern Europe and moved to the United States. (Many fled the Irish famine as well.) Some 12 million of these migrants landed at Ellis Island in New York Harbor. Opportunities for work in the United States also attracted large numbers of Chinese migrants in the first wave of the so-called Chinese diaspora, during the last 50 years of the nineteenth century.

The next major period of migration came after World War II, when the booming postwar economies in Europe, North America, and Australia needed labor. This was the era when, for example, many Turkish migrants arrived to work in Germany and many North Africans went to France and Belgium. It was also the period when, between 1945 and 1972, about one million Britons migrated to Australia as so-called "Ten Pound Poms" under an assisted passage scheme. During the same era but in other parts of the world, decolonization continued to have an impact on migration, most significantly in the movement of millions of Hindus and Muslims after the partition of India in 1947, and of Jews and Palestinians after the creation of Israel.

By the late 1970s, and in part as a consequence of the 1973 oil crisis, the international migrant labor boom had ended in Europe, though in the United States it continued into the 1990s. Now, with the global economy's momentum shifting decisively to Asia, labor migration on that continent has grown heavily, and it is still growing. How much longer this will be true, given the current global financial crisis, is a matter open to debate.

MORE AND MORE

As even this (inevitably selective) overview of international migration's history should make clear, large movements of people have always been associated with significant global events like revolutions, wars, and the rise and fall of empires; with epochal changes like economic expansion; and with enduring challenges like conflict, persecution, and dispossession. Nevertheless, one reason to argue that migration matters more today than ever before is sheer numbers. If we define an international migrant as a person who stays outside his usual country of residence for at least one year, there are about 200 million such migrants

worldwide. This is roughly equivalent to the population of the fifth-most popu-
lous country on earth, Brazil. In fact, one in every 35 people in the world today
is an international migrant.

Of course, a less dramatic way to express this statistic is to say that only
3 percent of the world's population is composed of international migrants. (In
migration, statistics are often used to alarm rather than to inform.) And it is also
worth noting that internal migration is a far more significant phenomenon than is
international migration (China alone has at least 130 million internal migrants). Still,
the world total of international migrants has more than doubled in just 25 years;
about 25 million were added in just the first 5 years of the twenty-first century.

And international migration affects many more people than just those who
migrate. According to Stephen Castles and Mark Miller, authors of the influential volume *The Age of Migration,* "There can be few people in either industrialized or less developed countries today who do not have personal experience of
migration and its effects; this universal experience has become the hallmark of
the age of migration." In host countries, migrants' contributions are felt keenly
in social, cultural, and economic spheres. Throughout the world, people of different national origins, who speak different languages and practice different customs and religions, are coming into unprecedented contact with each other. For
some this is a threat, for others opportunity.

Migration is also a far more global process than ever before, as migrants
today travel both from and to all of the world's regions. In 2005 (the most recent
year for which global data are available) there were about 60 million international migrants in Europe, 44 million in Asia, 41 million in North America,
16 million in Africa, and 6 million each in Latin America and Australia. A significant portion of the world's migrants—about 35 million—lived in the United
States. The Russian Federation was the second-largest host country for migrants,
with about 13 million living there. Following in the ranking were Germany,
Ukraine, and India, each with between 6 million and 7 million migrants.

It is much harder to say which countries migrants come from, largely
because origin countries tend not to keep count of how many of their nationals
are living abroad. It has been estimated that at least 35 million Chinese currently
live outside their country, along with 20 million Indians and 8 million Filipinos.
But in fact the traditional distinctions among migrants' countries of origin, transit, and destination have become increasingly blurred. Today almost every country in the world fulfills all three roles—migrants leave them, pass through them,
and head for them.

A WORLD OF REASONS

The reasons for the recent rise in international migration and its widening global
reach are complex. The factors include growing global disparities in development,
democracy, and demography; in some parts of the world, job shortages that will be
exacerbated by the current economic downturn; the segmentation of labor markets

in high-income economies, a situation that attracts migrant workers to so-called "3D" jobs (dirty, difficult, or dangerous); revolutions in communications and transportation, which result in more people than ever before knowing about life elsewhere and having the ability to travel there; migration networks that allow existing migrant and ethnic communities to act as magnets for future migration; and a robust migration industry, including migrant smugglers and human traffickers, that profits from international migration.

In addition to being bigger, international migration today is also a more complex phenomenon than it has been in the past, as people of all ages and types move for a wide variety of reasons. For example, child migration appears to be on the increase around the world. Migrants with few skills working "3D" jobs make important contributions to the global economy, but so do highly skilled migrants and students. Some people move away from their home countries permanently, but an increasing proportion moves only temporarily, or circulates between countries. And though an important legal distinction can be made between people who move for work purposes and those who flee conflict and persecution, members of the two groups sometimes move together in so-called "mixed flows."

One trend of particular note is that women's representation among migrants has increased rapidly, starting in the 1960s and accelerating in the 1990s. Very nearly half the world's authorized migrants in 2005 were women, and more female than male authorized migrants resided in Europe, North America, Latin America and the Caribbean, the states of the former Soviet Union, and Oceania. What is more, whereas women have traditionally migrated to join their partners, an increasing proportion who migrate today do so independently. Indeed, they are often primary breadwinners for families that they leave behind.

A number of reasons help explain why women comprise an increasing proportion of the world's migrants. One is that global demand for foreign labor, especially in more developed countries, is becoming increasingly gender-selective. That is, more jobs are available in the fields typically staffed by women—services, health care, and entertainment. Second, an increasing number of countries have extended the right of family reunion to migrants, allowing them to be joined by their spouses and children. Third, in some countries of origin, changes in gender relations mean that women enjoy more freedom than previously to migrate independently. Finally, in trends especially evident in Asia, there has been growth in migration of women for domestic work (this is sometimes called the "maid trade"); in organized migration for marriage (with the women sometimes referred to as "mail-order brides"); and in the trafficking of women, above all into the sex industry.

MOST IRREGULAR

Another defining characteristic of the new global migration is the growth of irregular migration and the rapid rise of this phenomenon in policy agents. Indeed, of all the categories of international migrants, none attracts as much attention or divides opinion as consistently as irregular migrants—people often described as "illegal," "undocumented," or "unauthorized."

Almost by definition, irregular migration defies enumeration (although most commentators believe that its scale is increasing). A commonly cited estimate holds that there are around 40 million irregular migrants worldwide, of whom perhaps one-third are in the United States. There are between 3.5 million and 5 million irregular migrants in the Russian Federation, and perhaps 5 million in Europe. Each year, an estimated 2.5 million to 4 million migrants are thought to cross international borders without authorization.

One reason that it is difficult to count irregular migrants is that even this single category covers people in a range of different situations. It includes migrants who enter or remain in a country without authorization; those who are smuggled or trafficked across an international border; those who seek asylum, are not granted it, and then fail to observe a deportation order; and people who circumvent immigration controls, for example by arranging bogus marriages or fake adoptions.

What is more, an individual migrant's status can change—often rapidly. A migrant can enter a country in an irregular fashion but then regularize her status, for example by applying for asylum or entering a regulation program. Conversely, a migrant can enter regularly then become irregular by working without a permit or overstaying a visa. In Australia, for example, British citizens who have stayed beyond the expiration of their visas account for by far the largest number of irregular migrants.

THE RICH GET RICHER

International migration matters more today than ever because of its new dimensions and dynamics, but even more because of its increased impact—on the global economy, on international politics, and on society. Three impacts are particularly worth noting: international migration's contribution to the global economy; the significance of migration and security.

Kodak, Atlantic Records, RCA, NBC, Google, Intel, Hotmail, Sun, Microsoft, Yahoo, eBay—all these U.S. firms were founded or cofounded by migrants. It has been estimated that international migrants make a net contribution to the U.S. economy of $60 billion, and that half of the scientists, engineers, and holders of Ph.D. degrees in the United States were born overseas. It is often suggested (though this is hard to substantiate) that migrants are worth more to the British economy than is North Sea oil. Worldwide migrant labor is thought to earn at least $20 trillion. In some of the Gulf States, migrants comprise 90 percent of the labor force.

Such a selection of facts and figures can suggest a number of conclusions about international migration's significance for the global economy. First, migrants are often among the most dynamic and entrepreneurial members of society. This has always been the case. In many ways the history of U.S.

economic growth is the history of migrants: Andrew Carnegie (steel), Adolphus Busch (beer), Samuel Goldwyn (movies), and Helena Rubenstein (cosmetics) were all migrants. Second, migrants fill labor market gaps both at the top end and the bottom end—a notion commonly captured in the phrase "migrants do the work that natives are either unable or unwilling to do." Third, the significance of migrant labor varies across countries but more importantly across economic sectors. In the majority of advanced economies, migrant workers are overrepresented in agriculture, construction, heavy industry, manufacturing, and services—especially food, hospitality, and domestic services. (It is precisely these sectors that the global financial crisis is currently hitting hardest.) Finally, migrant workers contribute significantly more to national economies than they take away (through, for example, pensions and welfare benefits). That is, migrants tend to be young and they tend to work.

This last conclusion explains why migration is increasingly considered one possible response to the demographic crisis that affects increasing numbers of advanced economies (though it does not affect the United States yet). In a number of wealthy countries, a diminishing workforce supports an expanding retired population, and a mismatch results between taxes that are paid into pension and related programs and the payments that those programs must make. Importing youthful workers in the form of migrants appears at first to be a solution—but for two reasons, it turns out to be only a short-term response. First, migrants themselves age and eventually retire. Second, recent research indicates that, within a generation, migrants adapt their fertility rates to those that prevail in the countries where they settle. In other words, it would not take long for migrants to exacerbate rather than relieve a demographic crisis.

THE POOR GET RICHER

International migration does not affect only the economies of countries to which migrants travel—it also strongly affects the economies of countries from which migrants depart, especially in the realm of development in poorer countries. The World Bank estimates that each year migrants worldwide send home about $300 billion. This amounts to triple the value of official development assistance, and is the second-largest source of external funding for developing countries after foreign direct investment. The most important recipient countries for remittances are India ($27 billion), China ($26 billion), Mexico ($25 billion), and the Philippines ($17 billion). The top countries from which remittances are dispatched are the United States ($42 billion), Saudi Arabia ($16 billion), Switzerland ($14 billion), and Germany ($12 billion).

The impact of remittances on development is hotly debated, and to an extent the impact depends on who receives the money and how it is spent. It is indisputable that remittances can lift individuals and families out of poverty: annual household incomes in Somaliland are doubled by remittances. Where remittances are spent on community projects such as wells and schools, as is often the case in Mexico, they also have a wider benefit. And remittances make a significant contribution to gross domestic product (GDP) at the national level, comprising for example 37 percent of GDP in Tonga and 27 percent in both Jordan and Lesotho.

Most experts emphasize that remittances should not be viewed as a substitute for official development assistance. One reason is that remittances are private monies, and thus it is difficult to influence how they are spent or invested. Also, remittances fluctuate over time, as is now becoming apparent in the context of the global economic crisis. Finally, it has been suggested that remittances can generate a "culture of migration," encouraging further migration, and even provide a disincentive to work where families come to expect money from abroad. It has to be said, even so, that the net impact of remittances in developing countries is positive.

International migration, moreover, can contribute to development through other means than remittances. For example, it can relieve pressure on the labor markets in countries from which migrants originate, reducing competition and unemployment. Indonesia and the Philippines are examples of countries that deliberately export labor for this reason (as well as to obtain remittance income). In addition, migrants can contribute to their home countries when they return by using their savings and the new skills they have acquired—although the impact they can have really depends on the extent to which necessary infrastructure is in place for them to realize their potential.

At the same time, however, international migration can undermine development through so-called "brain drain." This term describes a situation in which skills that are already in short supply in a country depart that country through migration. Brain drain is a particular problem in sub-Saharan Africa's health sector, as significant numbers of African doctors and nurses work in the United Kingdom and elsewhere in Europe. Not only does brain drain deprive a country of skills that are in high demand—it also undermines that country's investment in the education and training of its own nationals.

SAFETY FIRST

A third impact of international migration—one that perhaps more than any other explains why it has risen toward the top of policy agendas—is the perception that migration constitutes a heightened security issue in the era after 9/11. Discussions of this issue often revolve around irregular migration—which, in public and policy discourses, is frequently associated with the risk of terrorism, the spread of infectious diseases, and criminality.

Such associations are certainly fair in some cases. A strong link, for example, has been established between irregular migrants from Morocco, Algeria, and Syria and the Madrid bombings of March 2004. For the vast majority of irregular migrants, however, the associations are not fair. Irregular migrants are often assigned bad intentions without any substantiation. Misrepresenting evidence can criminalize and demonize all irregular migrants, encourage them to remain underground—and divert attention from those irregular migrants who actually are criminals and should be prosecuted, as well as those who are suffering from disease and should receive treatment.

Irregular migration is indeed associated with risks, but not with the risks most commonly identified. One legitimate risk is irregular migration's threat to the exercise of sovereignty. States have a sovereign right to control who crosses their borders and remains on their territory, and irregular migration challenges this right. Where irregular migration involves corruption and organized crime, it can also become a threat to public security. This is particularly the case when illegal entry is facilitated by migrant smugglers and human traffickers, or when criminal gangs compete for control of migrants' labor after they have arrived.

When irregular migration results in competition for scarce jobs, this can generate xenophobic sentiments within host populations. Importantly, these sentiments are often directed not only at migrants with irregular status but also at established migrants, refugees, and ethnic minorities. When irregular migration receives a great deal of media attention, it can also undermine public confidence in the integrity and effectiveness of a state's migration and asylum policies.

In addition, irregular migration can undermine the "human security" of migrants themselves. The harm done to migrants by irregular migration is often underestimated—in fact, irregular migration can be very dangerous. A large number of people die each year trying to cross land and sea borders while avoiding detection by the authorities. It has been estimated, for example, that as many as 2,000 migrants die each year trying to cross the Mediterranean from Africa to Europe, and that about 400 Mexicans die annually trying to cross the border into the United States.

People who enter a country or remain in it without authorization are often at risk of exploitation by employers and landlords. Female migrants with irregular status, because they are confronted with gender-based discrimination, are often obliged to accept the most menial jobs in the informal sector, and they may face specific health-related risks, including exposure to HIV/AIDS. Such can be the level of human rights abuses involved in contemporary human trafficking that some commentators have compared it to the slave trade.

Migrants with irregular status are often unwilling to seek redress from authorities because they fear arrest and deportation. For the same reason, they do not always make use of public services to which they are entitled, such as emergency health care. In most countries, they are also barred from using the full range of services available both to citizens and to migrants with regular status. In such situations, already hard-pressed nongovernmental organizations, religious bodies, and other civil society institutions are obliged to provide assistance, at times compromising their own legality.

IN HARD TIMES

What might the future of international migration look like? Tentatively at least, the implications of the current global economic crisis for migration are beginning to emerge. Already a slowdown in the movement of people at a worldwide level has been reported, albeit with significant regional and national variations, and this appears to be largely a result of declining job opportunities in destination countries. The economic sectors in which migrants tend to be overrepresented have been hit first; as a consequence migrant workers around the world are being laid off in substantial numbers.

Interestingly, it appears that most workers are nevertheless not returning home, choosing instead to stay and look for new jobs. Those entitled to draw on social welfare systems can be expected to do so, thus reducing their net positive impact on national economies. (It remains to be seen whether national economic stimulus packages, such as the one recently enacted in the United States, will help migrant workers get back to work.) Scattered cases of xenophobia have been reported around the world, as anxious natives increasingly fear labor competition from migrant workers.

In the last quarter of 2008 remittances slowed down. Some project that in 2009 remittances, for the first time in decades, may shrink. Moreover, changes in exchange rates mean that even if the volume of remittances remains stable, their net value to recipients may decrease. These looming trends hold worrying implications for households, communities, and even national economies in poor countries.

Our experience of previous economic downturns and financial crises—including the Great Depression, the oil crisis of the early 1970s, and the Asian, Russian, and Latin American financial crises between 1997 and 2000—tells us that such crises' impact on international migration is relatively short-lived and that migration trends soon rebound. Few experts are predicting that the current economic crisis will fundamentally alter overall trends toward increased international migration and its growing global reach.

HOT IN HERE

In the longer term, what will affect migration patterns and processes far more than any financial crisis is climate change. One commonly cited prediction holds that 200 million people will be forced to move as a result of climate change by 2050, although other projections range from 50 million to a startling 1 billion people moving during this century.

The relationship that will develop between climate change and migration appears complex and unpredictable. One type of variable is in climate change events themselves—a distinction is usually made between slow-onset events like rising sea levels and rapid-onset events like hurricanes and tsunamis. In addition, migration is only one of a number of possible responses to most climate change events. Protective measures such as erecting sea walls may reduce the impact. Societies throughout history have adapted to climate change by altering their agricultural and settlement practices.

Global warming, moreover, will make some places better able to support larger populations, as growing seasons are extended, frost risks reduced, and new crops sustained. Where migration does take place, it is difficult to predict whether the movement will mainly be internal or cross-border, or temporary or permanent. And finally, the relationship between climate change and migration may turn out to be indirect. For example, people may flee conflicts that arise over scarce resources in arid areas, rather than flee desertification itself.

Notwithstanding the considerable uncertainty, a consensus has emerged that, within the next 10 years, climate-related international migration will become observably more frequent, and the scale of overall international migration will increase significantly. Such migration will add still further complexity to the migration situation, as the new migrants will largely defy current classifications.

One immediately contentious issue is whether people who cross borders as a result of the effects of climate change should be defined as "climate refugees" or "climate migrants." The former conveys the fact that at least some people will literally need to seek refuge from the impacts of climate change, will find themselves in situations as desperate as those of other refugees, and will deserve international assistance and protection. But the current definition of a refugee in international law does not extend to people fleeing environmental pressures, and few states are willing to amend the law. Equally, the description "climate migrant" underestimates the involuntariness of the movement, and opens up the possibility for such people to be labeled and dealt with as irregular migrants.

Another legal challenge arises with the prospect of the total submergence by rising sea levels of low-lying island states such as the Maldives—namely, how to categorize people who no longer have a state. Will their national flags be lowered outside UN headquarters in New York, and will they be granted citizenship in another country?

The complexities of responding to climate-related movements of people illustrate a more general point, that new responses are required to international migration as it grows in scale and complexity. Most of the legal frameworks and international institutions established to govern migration were established at the end of World War II, in response to a migration reality very different from that existing today, and as a result new categories of migrants are falling into gaps in protection. New actors have also emerged in international migration, including most importantly the corporate sector, and they have very little representation in migration policy decisions at the moment.

Perhaps most fundamentally, a shift in attitude is required, away from the notion that migration can be controlled, focusing instead on trying to manage migration and maximize its benefits.

6

Globalization and the Struggle for Immigrant Rights in the United States

WILLIAM ROBINSON

Robinson makes a case for the centrality of race in global capitalism, focusing on how globalization both creates and requires pools of immigrant labor. Latinos in the United States are, in his words, the superexploited sector.

It is an honor and a privilege to be here with you today, with the leaders and organizers of one of the most vital, just and cutting-edge struggles of our time. I am very grateful to Javier Rodriguez and the other conveners of this conference for inviting me to participate. I want to start by highlighting three things that are unprecedented and interconnected, three current "upsurges" [....]

The first is an upsurge in Latino immigration to the United States. Officially, there are 34 million immigrants in the U.S., 12 to 15 million of them undocumented, although we know that these are underestimates. Migration levels in recent years have surpassed those of the turn of the 19th century. Of these 34 million, 18 to 20 million are from Latin America, the majority from Mexico, but also from Central America, the Dominican Republic, Peru, Ecuador, Colombia, Brazil, Argentina, and elsewhere.

The second is the upsurge of repression, racism, and discrimination against immigrants—the minutemen, the denial of driver's licenses, attacks, evictions, escalating raids, public segregation and anti-immigrant Jim Crow, and so on. We are witnessing the criminalization of immigrants and the militarization of their control by the state.

Third is an unprecedented mass immigrant rights movement. We saw last Spring the largest demonstration in U.S. history. They had the powers that be quite frightened. This is what *poder popular* looks like; what "power of the people" means.

What is the larger context and backdrop of anti-immigrant politics and immigrant struggle? In an attempt to answer this question I would like to put forward 10 points for analysis and discussion.

SOURCE: William Robinson, "Globalization and the Struggle for Immigrant Rights in the United States." Keynote Presentation for "El Gran Paro Americano II" Immigrant Rights Conference, Feb 3–4, 2007, Los Angeles, CA. Used with permission.

1. These upsurges are situated in an age of globalization, this new system of global capitalism we face.

   This system entered a new stage that began in the late 1970s and 1980s, a transnational phase. The capitalist system began a dramatic expansion worldwide, and that is the structural underpinning of the immigrant issue and ultimately those are the structures we need to address. Capital has become truly transnational, with newfound global mobility, and new powers to reorganize the whole world. In this new phase, the system depends on new methods of control over workers worldwide, and relies much more heavily on migrant workers who can be denied their rights and superexploited. This denial of rights and superexploitation requires, in turn, new levels of control and repression.

   Since the 1980s, global capital has been waging a worldwide offensive against global labor with the objective of capturing natural resources, markets, and labor pools around the world. There is a new global production and financial system into which every country is being integrated. There are new levels of global social and economic integration and new webs of interdependence. A few thousand of the most powerful transnational corporations dominate the process.

   Capitalist globalization constitutes a war of global rich against global poor. There has been a massive transfer of wealth from the poor to the rich. In the United States, relative wages have declined steadily since 1973 and we are witness to unprecedented global inequalities. Currently just a little more than 10 percent of the world's population consumes 85 percent of the world's wealth, while the rest have to make do with just 15 percent of that wealth—wealth which the world's workers generate but do not receive. This is the new global social apartheid. Hundreds of millions of people worldwide have been displaced—turned into workers for global economy and thrown into a new global labor market that global elites and transnational capital have been able to shape for their own purposes.

2. In Latin America every country has been violently integrated into global capitalism through free trade agreements, privatizations, deregulation, and neo-liberal social and economic policies.

   In Mexico, this process began under De la Madrid in 1982 and accelerated under Salinas de Gortari starting in 1988. But it really took off when NAFTA went into effect in 1994. Under Calderón the process will continue and even deepen. The PAN is now the party of global capital in Mexico. But this process has taken place all over Latin America. Throughout the continent millions of campesinos have been displaced, indigenous communities have been broken up, whole countries have become deindustrialized, millions of public sector workers fired, small businesses forced out by the onslaught of transnational corporations, such as Wal-Mart in Mexico (there are over 700 Wal-Marts now in Mexico and they are the country's biggest retailer and employer), systematic and ongoing austerity, the dismantling of social welfare programs, and so on. Hundreds of millions have been thrown into poverty, unemployment, and dispossession. As a result, political and military conflict has spread.

These policies of neo-liberalism, free trade, and capitalist globalization have been imposed by global elites and their local allies, and especially by the U.S. government, against the wishes of the great majority. They have created social disasters of unprecedented magnitude in Latin America and generated a crisis of survival for hundreds of millions.

This is the backdrop for transnational migration.

In more academic terms, we can say that the transnational circulation of capital and the disruption and deprivation it causes in turn generate the transnational circulation of labor. In other words, global capitalism creates immigrant workers. The wave of outmigration from socially and economically devastated communities in Latin America began in the 1980s and accelerated in the 1990s and into the new century, coinciding with globalization and neo-liberalism. In a sense, this must be seen as a coerced or forced migration, since global capitalism exerts a structural violence over whole populations and makes it impossible for them to survive in their homeland.

Yet, while transnational capital is free to move about the world, to reshape the world in its interests, transnational labor is subject to ever tighter and more repressive controls. Nine-eleven gave the Bush regime a pretext to step up the war against immigrant rights, side by side with its war in Iraq, and to militarize society and create the beginnings of a police state that is wielded against immigrants. The war in Iraq reflects the war on immigrants.

We should acknowledge that borders are not in our interest. Borders are instruments of dominant groups, of powerful economic groups, of capital, not labor. They are functional to the system as mechanisms of transnational control.

3. In this system, the U.S. and the global economy are increasingly dependent on immigrant labor that can be superexploited and super-controlled.

We have the following data for the percentage of the workforce in different categories in California that were immigrants in 1980 and in 1990:

TABLE 3.1

	1980	1990
Construction worker	20	64
Janitor	26	49
Farm worker	58	91
Maid	34	76
Electronics worker	37	60
Child care provider	20	58
Restaurant worker	29	69
Gardener	37	66
Drywall installer	9	48

Since 1990 there has been further tremendous increase in employer dependence on immigrant labor. The U.S. and the global economy would grind to a halt without immigrant labor. Employers don't want expensive labor, labor with citizenship rights. They are seeking cheap, superexploited labor, super-controllable labor. Moreover, the 20 percent of the population that is affluent or well off wants cheap "throw-away labor." They want to be able to maintain their privileges by drawing on an army of maids, nannies, gardeners, *jornaleros*, and so on. Elites, employers, and the more affluent wish to maintain a reserve army of immigrant labor.

4.   Sustaining such an immigrant labor force means creating—and reproducing— the division of workers into immigrants and citizens.

This is a new axis of inequality worldwide, between citizen and non- citizen. This axis is racialized. We are talking about racialized class relations worldwide, meaning that these are class relations of exploitation, but they are also racist relations of racial/ethnic oppression and discrimination against Latinos/as and other immigrants.

5.   The phenomenon of a superexploited and super-controlled Latino/a immi- grant workforce is part of a larger global phenomenon, that of transnational migration flows worldwide and the creation of immigrant labor pools. We have:

- Latinos/as and other immigrants in North America;

- Turkish, Eastern European, North African, and Asian labor in Europe;

- Indian and Pakistani workers in Middle East oil-producing countries;

- Central and Southern African immigrants in South Africa;

- Nicaraguans in Costa Rica, Peruvians in Chile, and Bolivians in Argentina;

- Asian immigrants in Australia;

- Thai, Korean, and others in Japan, etc....

In all these cases, repressive state control over transnational labor creates conditions for "immigrant labor" as a distinct category of workers in relation to capital. The creation of these distinct categories—"immigrant labor" groups all over the world—becomes central to the whole global capitalist economy. What we are seeing is the rise of a transnational capitalist class that draws on immigrant labor pools around the world for its own use. And alongside this transnational capitalist class we see the rise of a transnational, or global, working class, but one split between immigrant and native workers. In this situation, borders and nationality are used by capital, the powerful, and the privileged, to exploit, control, and dominate this global working class.

6.   In all these cases—but returning the focus to the United States—the system needs this immigrant labor; it can't function without it, without this reserve army of immigrant labor. But—and this is the crux of the matter—the system needs it to remain just that, immigrant labor: vulnerable, undocumented,

without citizenship and civil, political, and labor rights, deportable … in a word—controllable. The aim of the powers that be is not to do away with Latino/a (and other) immigrants, but to exercise repressive control over immigrants. It is this condition of "deportable" that they wish to create or preserve, since that condition assures the ability to superexploit with impunity, and dispose of without consequences should this labor organize, demanding its rights and its dignity. Latino/a immigrant labor is the new superexploited sector of the labor force. We need here to focus on black-brown unity. African Americans used to constitute the superexploited segment of the working class outside of the Southwest (where Chicanos/as played this role). But in the 1960s and 1970s African Americans organized to win their civil rights. They fought also for full social and labor rights, launched the movement for black liberation and also became the most militant group within the trade unions. Since they have citizenship rights they cannot be deported. So capital decided that African Americans were not desirable workers since they were not as vulnerable and were too militant and organized. Employers turned massively in the 1980s and onward to using immigrant workers and to structurally marginalizing black workers. So while African Americans are increasingly the structurally unemployed and marginalized sector of the working class—subject to hostility, neglect, and incarceration—Latinos/as are now increasingly the superexploited sector. Black-brown unity is crucial. We cannot let the system pit African Americans and immigrants against each other.

7. This is therefore a contradictory situation. From the viewpoint of dominant groups, this situation presents a dilemma: How to have their cake and eat it too? How to superexploit the Latino/a immigrant population, yet simultaneously assure that it is super-controllable and super-controlled? Hence the dual emphasis on guest-worker programs alongside heightened criminalization, enforcement, and militarization.

8. We need to deepen a working-class focus! The immigrant issue is a labor issue, one in which we see how race and class come together.

   Let us recall that this is about transnational immigrants as workers for global capitalism, and Latino/a immigrants as immigrant workers. Transnational capital wants a class of workers, and the twin instruments in this endeavor become:

   a. the division of the working class into immigrant and citizen, and
   b. the racialization of the former.

   The struggle of immigrant labor is the struggle of all working and poor people. Any improvement of status of immigrant labor, any advance in immigrant rights, is in the interests of all workers.

   But there is a technical point we need to stress because it is so important. Global elites and dominant groups around the world have imposed new capital-labor relations on all workers based on oppressive new systems of labor control and the cheapening of labor. These involve diverse contingent categories of devalued labor, including subcontracted, outsourced, and flexibilized work; deunionization, casualization, informalization, part-time, temp, and contract work replacing steady full-time jobs; and the loss of benefits, the erosion of wages, longer hours, and so on. This is what we could call the "Walmartization" of labor. It is not just immigrant workers, but all workers, immigrant and citizen alike, who are increasingly subject to these new capital-labor relations, in the United States and all around the world.

Here's the key point. An immigrant workforce reflects the new global class relations; from the viewpoint of dominant groups, they are the perfect workforce for global capitalism. Latino/immigrant workers are reduced to nothing but a commodity, a flexibilized and expendable input into the global capitalist economy, a transnationally mobile commodity deployed when and where capital needs them throughout North America, and utterly dehumanized in the process.

9. Why has there been increasing hostility against and oppression of the Latino/a community, not just from the state and the right wing, but in the mass media, among the general public, and so on?

This system needs Latino/a immigrant labor, yet the presence of that labor scares both dominant groups and privileged—generally white/native—strata. Dominant groups and privileged strata fear that a rising tide of Latino immigrants will lead to a loss of cultural and political control, so the dynamic becomes racialized hostility towards Latinos/as, and the problem becomes how to control them. Thus we have a rising tide of xenophobia and nativism, escalating racism, and the minutemen.

Really, what this amounts to is the beginnings of fascism, of a twenty-first-century fascism. This neo-fascist movement is led and manipulated by elites, but its base is drawn from those displaced from previously privileged positions. White working- and middle-class sectors who face downward mobility and insecurities brought about by capitalist globalization are particularly prone to being organized into racist anti-immigrant politics by right-wing forces. The loss of caste privileges for these white sectors of the working class is problematic for political elites and the state, since legitimation and domination in the United States have historically been constructed through the white racial hegemonic bloc. Therefore, anti-immigrant forces try to draw in white workers with appeals to racial solidarity and to xenophobia, and through scapegoating immigrant communities.

10. Conclusions. Some have called this the "new civil rights movement." It is, but this is about much more than "civil rights." This is fundamentally about human rights, about what kind of a world we are going to live in. No one can be left out in this struggle. There is no room for compromise—full legalization for all. But also freedom from all forms of repression and persecution, and full labor, social, cultural, and human rights for all.

The immigrant rights movement is the leading edge of popular struggle in the United States. In the larger picture, beyond its immediate demands, the movement for immigrant rights challenges the oppressive and exploitative class relations that are at the very core of global capitalism. Bound up with immigrant debates in the United States is the entire political economy of global capitalism, with all its injustices and inequalities. This is the same political economy that is now being sharply contested throughout Latin America by the upsurge in mass, popular, and democratic struggles.

The movement for immigrant rights in the United States is part and parcel of this larger Latin American—and worldwide—struggle for social justice and human dignity. We are integral to that worldwide struggle, on its cutting edge. We need to see our struggle as part of a broader transnational movement, to develop transnational links with other immigrant movements around the world and also with social movements of the poor and indigenous and workers in Latin America and elsewhere.

Please allow me by way of conclusion to humbly express my opinion on one matter.

The powers that be in the United States were terrified by the mass mobilizations of Spring 2006, and as we know they tried to intimidate movement by unleashing a wave of shameless repression that is still continuing. Some have pointed out that we don't have the organizational capacity to defend all the victims of this repression. That is very true. But it is also true that the only real defense from this repression is not to back off, but to push forward, to step up and intensify, the mass struggle. Unfortunately, there is no change without sacrifice. Backing down or holding back only makes it easier for the state and the right wing to retake the initiative and carry out repression. When you have seized the initiative—as we did last Spring—when your enemy is on the defensive, it is not the time to back down or demobilize but to sustain and deepen the offensive.

Let me close by quot[ing] Che Guevara, something he said that is written on a poster that I saw on my way in here, and that is fitting for the moment: "Seamos realistas; soñamos lo imposible."

Thank you.

7

Bound for America

JOHN BOWE

Many of the services we receive and the food we eat are provided by exploited immigrants who experience a new kind of indentured servitude.

In the spring of 2004, Nikhom Intajak, a 35-year-old rice farmer in Thailand's Lampang province, met a labor recruiter who made him an attractive offer: a contract to do farm labor in the United States. He'd work for three years and earn the minimum wage of $7 to $10 an hour, depending on where he was deployed; best of all, he'd be a legal temporary worker, protected by American laws.

Intajak, who weighs 139 pounds and stands 5 feet 4 inches tall in a baseball cap, had worked overseas before, spending a total of about seven years in chemicals, electronics, and luggage plants in Taiwan. The money he'd sent home helped build a new house and pay school fees for two daughters. For each of his stints abroad, Intajak had paid a recruiting fee somewhat higher than the Thai legal maximum (currently about $2,000), and so he wasn't surprised when the new recruiter, Pochanee Sinchai, asked for one as well. He was, however, taken aback by the size of her demand: the job in America would cost him $11,700 up front.

Intajak's home, a hamlet called Banh Santicome, is poor, but not destitute. The climate is suitable for growing rice and produce, and earning opportunities range from farming garlic to foraging for mushrooms, bamboo, and wood. A formal job, if one can be found, might pay $2,000 a year. Three years of work in America at $7 an hour would come out to about $50,000. If one-fifth of that went to Sinchai, Intajak figured, then so be it. He asked his mother to put up her new house as collateral to borrow the money from a bank at 15 percent interest. Then he traveled to the Bangkok office of the recruiting firm that hired Sinchai, AACO International Recruitment, where he signed a number of documents, including several written in English, and also some blank pieces of paper.

Intajak (who asked me not to use his real name for fear of retribution) landed in Seattle on the Fourth of July. He was met by an employee of Global Horizons, the American company for which Sinchai and AACO had and a vanload of new arrivals from Thailand to an isolated Yakima Valley apple grower

SOURCE: From John Bowe, "Bound for America," *Mother Jones* 35 (May/June 2010): 61–65. Copyright © 2010, Foundation for National Progress. Used with permission.

named Green Acre Farms, where he confiscated their passports. Global Horizons agents stayed in the barracks and came to work in the orchards, Intajak says, to make sure the Thais didn't run away.

Intajak worked there for about three months. The pay, $8.53 per hour, was reasonable enough, he told me, but the work was so unsteady that he earned far less than he had been promised. Some days there might be eight hours of work, other days four—or none. After witnessing 30 or so coworkers get sent home after only a few months' work, Intajak began to realize that the contract he had signed back in Bangkok guaranteed nothing like three years of steady employment. Rather, he was eligible to work as many hours as Global saw fit to give him, for *up* to three years—as long as Global chose to renew his visa. If it didn't, if the work ran out, or if he did anything to displease his bosses, he'd have no way to pay off the $11,700 he'd borrowed. Ever.

Last year, some 60,000 workers arrived in the US under the federal H-2A guestworker program, which allows agribusinesses to bring in foreign labor for jobs they say are hard to fill at minimum wage. Similar temp-worker programs in industries like seafood processing, tree planting, and hotel maintenance brought in an additional 59,000 workers, and 60,000 more came in through temporary programs for professionals in fields deemed to have labor shortages—teachers, nurses, computer programmers.

These men and women are bound to the companies that requested them. They remain on American soil at the pleasure of their employers, who can send them home at any time. As Mary Bauer, an expert on temporary-worker programs at the Southern Poverty Law Center, has written: "These workers are not treated like 'guests'.... Unlike US citizens, guestworkers do not enjoy the most fundamental protection of a competitive labor market—the ability to change jobs if they are mistreated."

Many, like Intajak, arrive with crushing debt from recruiting fees. I reviewed the cases of dozens of Thai workers employed by Global Horizons who had paid between $11,000 and $21,000 in recruiting fees, money they had borrowed from banks or relatives, often with family or communal property as collateral. In theory, they were free to leave their job anytime. In practice, they were modern-day indentured servants.

Global Horizons, which brought in more than 1,000 Thais in 2004 and 2005 but was banned from recruiting guest workers in 2006, is now being investigated by the Department of Justice for human trafficking, according to Susan French, a long-term prosecutor in the DOJ's civil rights division. If a charge is brought, it could be the largest human trafficking case in U.S. history.

Yet Global is neither an isolated case, nor—except in terms of size—is it especially egregious. Every year, several companies are prosecuted for keeping H-2A workers in near-slavery conditions, and it is safe to assume that those cases are the tip of the iceberg, given federal and state authorities' minimal capacity to oversee the programs. To monitor the job sites of all 137 million U.S. workers, the Department of Labor's Wage and Hour Division has 953 staffers—a number that has fallen by 14 percent since 1973, while the number of workers under their purview has increased by half. Exactly two of them speak Thai.

Despite the obvious flaws of H-2A, the program assuages concerns of all interested parties. Pro-immigrant liberals can feel good about bringing foreign workers into the "light of legality." Ag employers, sick of regulations and politicized uneven enforcement, are freed from dealing with recruiting, housing, and supervision of workers. Anti-immigration activists find relief in the fact that, in theory, these temporary workers will all be sent home when their job is done. Which is why bringing in more temporary foreign workers is the one thing almost everyone in the immigration debate can agree on. In 2007, a proposal known as the Agricultural Job Opportunities, Benefits, and Security Act (AgJOBS) garnered the support of everyone from the late Sen. Ted Kennedy to President Bush and the U.S. Chamber of Commerce, and it became the cornerstone of the immigration reform bill that almost passed the Senate. It would have increased the number of H-2A jobs more than eightfold, from 60,000 to half a million. With the Obama administration under pressure to take a run at immigration reform by next year, it is widely expected that AgJOBS—or something like it—will be at the core of whatever legislation finally emerges.

Three months after Intajak arrived in America, in October 2004, Global Horizons sent him to Hawaii to work for the Maui Pineapple Co. Here, the pay was better than in Yakima—$9.50 an hour—but the conditions were worse. One Global Horizons agent, Intajak and other workers told me, was in the habit of carrying a knife, a gun, or a baseball bat, and of threatening workers with "deportation" if they didn't behave or meet their quotas. Just four days in, Intajak says, he watched the man beat a coworker.

The Maui Pineapple Co.'s land is nestled among gorgeous foothills, shrouded in mist and covered with volcanic soil the color of dark coffee. The now-defunct company was part of Maui Land & Pineapple Co., whose majority owner is Steve Case, cofounder of AOL; another primary shareholder is eBay founder Pierre Omidyar, a generous benefactor of anti-slavery organizations. The terrain is so lovely that *Martha Stewart Living* featured the Maui Pineapple Co. and a smattering of pineapple recipes in its January 2007 issue.

I visited Maui in January 2008, meeting with Intajak and taking a tour of the plantation. He took me to a one-room barracks where he says he stayed with 17 other workers. He pointed to a muddy spot near the parking lot, saying this was where he and others slept on the ground to make sure they were chosen to get work when the van came at 4:30 a.m. Here, he indicated, pointing to a chain-link fence, was where they snuck out to run to a store for Ramen noodles because their food rations were too small, or too disgusting. Finally, he pointed to the bush whose leaves they'd boil when they couldn't afford Ramen noodles. "*Khom!*" he said. "Very bitter." (A Maui Land & Pineapple spokesman says Maui Pineapple Co. was not aware of the workers' allegations at the time, but terminated its contract with Global after learning of them in 2006.)

Intajak was asking a lot of questions and his supervisor began threatening to send him home. On September 12, 2005, Intajak took stock of his predicament. After 14 months of work, his visa had two days remaining. He had no idea whether Global planned to renew it, and he still owed $6,000 on his recruiting fee. Being sent home would mean losing his home and his land, which had belonged to his family for generations. It would mean homelessness for his wife, daughters, grandmother, and aunt. He could see only one way out.

Teaming up with a friend who'd also decided to run, Intajak threw his backpack out a window, then snuck out of the pineapple compound. Within moments, another friend called his cell phone to tell him the guards were coming after him.

Intajak ran into the cane fields for cover, snaking his way in shorts and flip-flops through the twisting, two-inch-thick, 12-foot-high stalks. "I was sweating like crazy, and it was muddy and slippery," he says. "I really had no idea what was going to happen, or if I'd make it, or what would happen if I got caught." Listening for cars, trying to stay close to the road, Intajak headed down the mountain, toward the ocean. After an hour in the cane, he found his friend, and together, they walked into Paia, a surfer town.

Disoriented in a world of dreadlocked, smoothie-sipping *falangs,* or white people, among stores called OmZone and Drums & Tings, Intajak and his friend made their way to a grocery where, by chance, their conversation was overheard by a Thai employee. She and her husband agreed to help the runaways. They put them up overnight and, for $500 each, bought them one-way tickets to Los Angeles. Only after they boarded the plane to safety did Intajak and his friend notice that the tickets had cost $175 apiece.

Several weeks after arriving in Los Angeles, Intajak made his way to the offices of the Thai Community Development Center, a nonprofit serving the area's burgeoning Thai population. According to Chanchanit Martorell, the center's director, he was the first of many Thai workers who'd escaped Global Horizons' employ. The center was getting reports from worker advocates and legal-aid attorneys in Washington state, Utah, Pennsylvania, Colorado, and Hawaii. The staff posted a U.S. map on the wall, using pushpins to keep track of Global's sprawling empire. As the number of pins began to run into the hundreds, they realized they'd never seen anything of this magnitude.

"In the past, we had always dealt with small operations that were highly localized and had a fairly limited reach," says Martorell. "Up until that point, our biggest case involved 72 workers living under one roof and working in the same factory. All of a sudden, we were looking at over 1,000 workers in over a dozen states, working for hundreds of different establishments. Trafficking on this scale could not have taken place in the typical underground fashion. It was almost as though trafficking had gone mainstream and everybody was doing it."

Martorell uses the term "trafficking" deliberately. Some years ago, she worked on a case involving Thai welders brought in on temporary visas to help rebuild San Francisco's Bay Bridge. Prosecutors, she recalls, were leery of bringing the workers to the stand: unlike with female sex-trafficking victims, they feared, juries would find it hard to associate able-bodied male construction workers with trafficking. (In the Global case, the DOJ has actually helped a number of the workers obtain special visas designed to protect victims of trafficking who could aid an investigation.)

Since his escape more than four years ago, Intajak has lived underground in Los Angeles, working without papers six-and-a-half days a week as a cook in a

Thai restaurant. (He has also filed for a trafficking visa.) He makes $100 a day. Although he's free from Global's direct control, he says his family is still being threatened by Sinchai, the recruiter back in Lampang province. Intajak says Sinchai has told him that one of the pieces of paper he signed promised a "guarantee fee" of nearly $5,000 if he ever "ran away" from Global's employ—and that's on top of the thousands he's already repaid. Sinchai has filed court proceedings and taken title to a portion of his property, and she is pushing to assume the rest of it. Intajak's wife and children live under the threat of losing everything. It's been more than five years since Intajak has seen them.

The president and founder of Global Horizons is a 44-year-old Israeli named Mordechai "Motty" Orian. I interviewed him in Global's office, a vibrant 3,000-square-foot chunk of an office tower in West Los Angeles, rising above Santa Monica Boulevard. The floors are blond wood, and the *esprit d'office* is urbane and upbeat.

Orian entered the manpower business in Israel in 1989, during his compulsory national military service. Jobs that could no longer be trusted to Palestinians needed to be filled, and Orian was tasked by his commanders with finding replacements. He experimented with Eastern Europeans, then branched out, discovering along the way that each population had its own workforce peculiarities: Romanians were clumsy with tiles, marble, and mosaic, but great with plastering, concrete, and brickwork. Chinese did wonderful things with marble but had zero interest in farmwork, even if they'd been farmers at home. Thais he found to be sensitive to others' religious preferences and, moreover, highly suitable cooks and farmworkers.

After leaving the military, Orian took his manpower operation private, expanding in 1992 into the U.S. and European markets. Y2K was a huge boon, creating a seller's market in computer programmers. Nurses were good for a while, but not so much after 9/11, when visa requirements became more stringent.

For H-2A farmworker contracts, Global charged growers a percentage over and above each worker's wages—somewhere between 45 to 80 percent, Orian told me. In return, the company handled transportation, housing, food, payroll, workers' comp, and health care. Besides these conveniences, a key reason farmers would pay a premium for bringing in H-2A workers may have to do with control. Orian recalled a North Carolina client who complained, "If I bring 200 Mexicans from Mexico, I know 100 will run away. Then I apply for another 200 so I can have another 100 stay with me." Orian laughed. "Mexicans run away. Right away! After one week! Because somebody down the road was offering them 50 cents more!" Imported Thais, isolated by debt, distance, and an absence of cultural or community links, were simply more stable.

When I asked Orian about the debt his workers took on, he offered a variety of responses. One was that the workers were lying about the size of the fees they'd paid. Another was to scoff at the idea that anyone could be stupid enough to sign blank pieces of paper. A third was to blame the system. "The problem," he said, "is that the workers go though sub-sub-sub-subagents." Each subagent makes extravagant promises and extracts their cut. But what can you do?

"Middlemen are always going to seek an incentive," Orian said. "And Third World governments are always going to be corrupt."

Orian stressed that in the messy, imprecise, red-tape-filled business of labor contracting, he had tried everything to keep his operation as clean as possible. Still, by 2004, he found himself in hot water with agencies in several states over housing and tax violations. In 2006, after finding that Global "knowingly gave false information" to applicants, the Department of Labor banned Orian from bringing in more foreign workers.

Several times during our conversations, Orian launched into cogent diatribes detailing the shortcomings of American immigration policy. The system was broken. America had become the world's largest prison camp. No one wants to do farmwork in *any* country. "You know how much I pay when I came to this country?" he said. "You know how much I spend on immigration until now? For my own paperwork? Over $25,000. From visa fee, embassy fee, government fee, lawyers—25,000 goddamn dollars." When I suggested that this hardly compared to loss of family land and home, he scoffed. "Come on. Come *on!*" Everyone knew that poor workers will say anything to stay and work in the US. Where was the proof that these Thai workers were really losing their homes? "When it comes to money," he shrugged, "people will do crazy stuff. You cannot stop it and come to blame me."

When I mentioned that most people with whom I'd discussed the case felt that his workers had been not just exploited, but trafficked, he dismissed the idea with a jerk of the head. "Let me tell you something," he said. "Every day, I take my kids to school. Sometimes, I get into a traffic jam. That's the only trafficking I do."

Delta egg farms is two hours south of Salt Lake City, 14 miles from the town of Delta (population 3,200). It's surrounded by scrub desert and improbable agriculture: a stockyard here, a peach plantation there. The egg farm is a behemoth of industrial agriculture—the processing facility is next to a series of 600-foot-long "layer houses" housing 1.5 million chickens. Another Delta egg facility—another few hundred thousand chickens—lay half a mile to the right. At sunset, the alien isolation of the landscape seemed both bleak and beautiful. The Intermountain power plant, about a mile behind the egg plant, shot indigo steam into the pale evening light. If I had just arrived here from Thailand, I might wonder if I'd landed on the moon.

In the town, Alfredo Laguna, an outreach coordinator with Utah Legal Services, showed me some two-bedroom homes where Global's Thais had stayed between shifts at the egg factory. A few blocks away stood a run-down hotel that looked like an immense horse stable. Empty now, it sometimes housed farmworkers too. "The Mexicans stayed on that side, and the Peruvians stayed on the other," Alfredo said. The Peruvians had paid about $5,000 each for their jobs. The Mexicans, who knew? A bit farther down the road, there was a mushroom plant staffed by Laotians. Everyone in their own shadow.

Later, I met a Thai worker who'd been sent to Utah by Global. In his trafficking visa application, he stated, "We never really knew where we were." The worker, who had done two previous stints working outside Thailand, testified that "in Israel and Singapore, they would give us maps of where we were along with bus routes so that we could get around. Here we would just be dropped off in the middle of nowhere. We would have to wait until a supervisor would take us to Wal-Mart or something like that to check our accounts. We were forbidden from running errands on our own."

Another worker I met in Salt Lake City recalled his time at Green Acre Farms, where Intajak had worked. "I felt I had to be very careful about what I said and what I did," he said. "The group before us had been deported. We felt we had no control over our lives or our pay."

Over the last 200 years, writes Cindy Hahamovich, a history professor at the College of William and Mary who has researched farm labor and guest-worker programs, these schemes have represented "an uncomfortable marriage between those who desired and those who resented foreign workers." In a journal article comparing guest-worker programs in South Africa's diamond mines, Germany's World War II munitions factories, Japan's pre–WWII buildup, and America's rush to build the trans-Pacific railroad with Chinese laborers, she found that temporary labor schemes consistently represented "state-brokered compromises designed to maintain high levels of migration while placating anti-immigrant movements. They offered employers foreign workers who could still be bound like indentured servants but who could also be disciplined by the threat of deportation. They placated trade unionists who feared foreign competition by promising to restrict guest workers to the most onerous work and to expel them during economic downturns. And they assuaged nativists by isolating guest workers from the general population." As "the perfect immigrants," guest workers serve to please employers whose problem was "not so much a shortage of labor as it was a shortage of tractable labor."

The experiences of Intajak and his coworkers are the increasingly common outcomes of these pressures. A 2007 report from the Southern Poverty Law Center notes that H-2A workers have so few rights that abuse of the system is not limited to a few "bad apple employers," but systematic and predictable. Recruiting fees, legal or otherwise, offer "a powerful incentive" to import as many workers as possible, for as long as possible, even when there is little work. The report cites ongoing legal cases involving Peruvian, Dominican, and Bolivian workers who arrived in the U.S. owing enormous debts from their recruiting fees; when work and pay in America were not as promised, they and their families were bankrupted.

Several recent court cases document how easily guest-worker status devolves into forced labor. In one 2009 case, *U.S. v. Sou,* three Hawaii growers were indicted for bringing in 44 Thai workers, pocketing a portion of their recruitment fees, then "maintaining their labor at the farm through threats of serious economic harm," according to the Justice Department. In another case, *Asanok v. Million Express Manpower,* Thai and Indonesian workers alleged that they had been promised well-paid, steady farmwork in North Carolina, only to find themselves housed in a Katrina-damaged New Orleans hotel, demolishing the building by day and sleeping in what remained at night, going so hungry they sometimes trapped pigeons for dinner. The list could go on, with several cases filed each year for as long as the U.S. has deployed guest-worker schemes.

Proponents of expanding guest-worker programs say there are ways to safeguard against these problems—giving workers legal recourse, increasing enforcement, rewarding employers for treating guest workers well. Last year's reauthorization of the Trafficking Victims Protection Act added a new penalty specifically aimed at labor recruiters found guilty of defrauding foreign workers. But no one has devised a way to deal with the workers' indebtedness to whoever helps them find work in America—or suggested giving them, as European guest-worker programs do, the free-market right to change employers.

Intajak's extended family lives in a pair of teak homes on a family plot perhaps a quarter-acre in size. a path from the houses leads past herb and vegetable gardens and a 10-foot-square concrete frog tank before opening onto a communal rice paddy. i traveled there with a researcher helping the advocates at the thai community development center document the hardship among the families affected by global's hiring practices. she had expected three or four families and was surprised to see a group of 45 to 50 men and women sitting on carpets thrown outside the house.

Each worker had a similar tale to tell: they had been sent home from the U.S. 5, 7, 13, 14 months into their contract with Global. Most looked ashamed as they spoke of their financial calamities. One after another showed me documents: a signed contract with Global Horizons; a Thai passport bearing U.S. entry and departure stamps; a series of bank withdrawals and loans—$3,000, $5,000, $11,000. Some had papers indicating direct payments and debts to Sinchai, the subrecruiter. Whoever they were paying, it was clear they had collectively spent hundreds of thousands of dollars for the privilege of performing a few months' farmwork in America. Each of them said that since signing with Global, they and their families had been subjected to relentless pressure from courts, judges, agents, and lawyers. When I asked one worker why they didn't organize and demand justice, either by approaching local authorities or by hiring a lawyer, he held up his pinkie finger and shrugged, "We're just small strings!" Referring to Sinchai, he complained, "She's a big rope. She knows everyone in the government." (Sinchai could not be reached for comment; the number on the business card she gave Intajak has been disconnected.)

Another worker was almost dapper in a cream polo shirt and crewneck sweater, and cracked jokes about his own haplessness even as he described Sinchai's ongoing attempts to wrest his property from him. He had worked at Maui Pineapple and confirmed the conditions Intajak had told me about. From the first day he'd arrived in Maui, he said, "Everybody was saying, 'We're screwed.'" He recited a song he and his friends had made up to make light of their situation.

But just beneath the jokes and the struggle to save face the desperation was palpable. Besides Sinchai, the workers had borrowed from aunts, uncles, and grandparents, who now expected repayment. In a town where I met seven sisters (all of them grandmothers) living in the same family compound, where 10 miles away was described as "far from here," where family was the primary source of identification, families were breaking apart. I spoke to a woman who told me that her husband, like Intajak, was working underground as a cook in Los

Angeles. She said that they had already paid $12,300 to Sinchai, but the recruiter said they still owed $11,600. Her husband was remitting about $900 dollars a month–a lot of money in Thailand, but nowhere near enough to meet the 20 percent monthly interest demanded by Sinchai.

I visited this woman in her home, which was perched atop round log pillars 10 feet high. We sat on the floor with a dozen or so women in similar circumstances. It was her job to care for seven people, to work, to cook, to raise the kids, to keep things going. "This is our life now," she said. "Our husbands work 13-hour days, and then they call us at midnight or 1 a.m. I've been so stressed by these debts that I can't sleep at night." In Thai, she sighed, "*Mot nua, mot dua,*" a saying meaning, "Out of flesh, out of body. There's nothing left."

8

Mexico's Ghost Towns

JOHN GIBLER

Typically, the focus in immigration debates is on immigrants who come to the United States (legally and illegally) to work. For Mexicans, the lure of significantly greater wages in the north can have a downside, however, leaving many Mexican towns relatively empty. According to Gibler, the emigrants are generally younger and stronger than those who remain in Mexico. As a result some Mexican communities are faltering, while U.S. communities with an influx of workers are flourishing.

Zacatecas, mexico—cerrito del agua, population 3,000, has no paved roads—either leading to it or within it. no restaurants, no movie theaters, no shopping malls. in fact, the small town located in the central mexican state of zacatecas has no middle schools, high schools or colleges; no cell phone service, no hospital. its surrounding fields are dry and untended. the streets are empty.

The explosion of emigration to the United States over the past 15 years has emptied much of central Mexico, even reaching into southernmost states like Chiapas and Yucatan. But it has simply devastated Zacatecas, a dry, rolling agricultural region located about 400 miles northwest of Mexico City.

A little more than half of Zacatecas' population—about 1.8 million people—now live in the United States, especially in the areas surrounding Atlanta, Chicago, and Los Angeles. Between 2000 and 2005, three out of its four municipalities registered a negative population growth. A 2004 state law created two new state legislative posts for migrants living in the United States. In 2006, depopulation cost the state one of its five congressional districts.

"Well, you've seen what this place is like," says Dr. Manuel Valadez Lopez, gesturing out the door of his small private clinic, when I ask him how emigration has affected the town. "There has not been even minimal development here. There is not a single yard of pavement. The few people who have sidewalks in front of their houses built them themselves. Most people defecate outdoors."

Lopez, 40, a native of Cerrito del Agua, is one of the few to leave the town and return. All six of his brothers now live and work in the United States. All four of his sisters married men who left to work in the United States.

SOURCE: From John Gibler, "Mexico's Ghost Towns," *In These Times* 32 (June 2008), pp. 28–30. This article is reprinted with permission from *In These Times*, June 2008, and is available at www.inthesetimes.com

In his teens, Lopez himself had moved to Guadalajara (about a five-hour drive southwest of Zacatecas) to attend high school and university, then stayed on to study medicine and receive a specialist's training in gynecology. He later returned to Cerrito del Agua for a visit and realized "there was so much work to do here that I stayed," he says.

That was eight years ago.

"The whole culture now is that people grow up and go to the U.S.—their parents, their uncles, their brothers and sisters, everyone goes," Lopez says. "The kids who are strong and smart, they all go to the U.S. There are no basic services here; the government has not carried out a single project."

The situation has been so dire, he says, that the staff at the clinic had to install its own sewage system. "There is running water, but it's not clean," he continues. "People get all sorts of infections, a typical Third-World situation."

Worst of all, says Lopez, is that "people who could possibly stay here and do something, they all go."

THE NEW U.S. COLONY

A January [2008] report by Richard Nadler, president of the conservative Americas Majority Foundation, found that the strongest state economies in the United States are those with high numbers of migrant workers. Nadler writes: "An analysis of data from 50 states and the District of Columbia demonstrates that a high resident population and/or inflow of immigrants is associated with elevated *levels* and *growth* in gross state product, personal income, per capita personal income, disposable income, per capita disposable income, median household income and median per capita income."

Those who are leaving Mexico—those whose land goes unplanted, whose roads remain unpaved—are laboring in the United States, building shopping malls and factories, washing dishes in restaurants and cafes, picking grapes and pulling lettuce.

They are creating within the U.S. economy precisely the goods and services that their hometowns lack. At the same time, their anemic home economies falter on the brink of collapse.

"I think that the U.S.'s plan is to make Mexico into a kind of colony," says Lopez, with a half smile. "People go to the U.S. to work and earn dollars. They come back to Mexico and spend their dollars on American products. It's a nice, round business." He continues: "Everyone here depends on the U.S. If this isn't a colony, then how do you define colony?"

CONDEMNED TO DISAPPEAR

In the heated debates over U.S. immigration policy, the pressing questions seem to be "How many immigrants should be allowed in, if any?" and "How should they be processed into the system?" But rarely considered is what this massive influx is doing to Mexico.

With nearly half a million Mexicans crossing the U.S.–Mexico border every year to look for work, Mexico has become the world's largest exporter of its people. More people flee destitution in Mexico than in China or India—each with populations 10 times larger than Mexico's.

Their remittances—the money Mexican immigrants in the United States save and send back to their families—equaled $24 billion last year, and made up the third-largest source of revenue for the Mexican economy (after illegal drugs and oil).

Theories of migration always show the interests of the North," says Raul Delgado Wise, director of the Graduate School of Development Studies at the Autonomous University of Zacatecas and an expert on migration. He says migrants born in Mexico contribute 8 percent of the U.S. gross domestic product (GDP)—about $900 billion—which is more than Mexico's entire GDP.

Wise is one of several researchers studying Mexican migration at the University of Zacatecas. Together they publish an international journal called *Migration and Development* and are laying the groundwork for an alternative think tank to the World Bank, which will be called the Consortium for Critical Development Studies.

"With all of this, we need to see really how much it is costing Mexico, how much Mexico is losing," Wise says.

He says that the mass migration from Mexico to the United States cannot be fully understood without considering U.S.–Mexico economic integration. Begun in the '80s, this integration reached its maximum expression with the North American Free Trade Agreement (NAFTA), which took effect Jan. 1, 1994.

What Mexico really exports, Wise argues, is labor.

The supposed growth in Mexico's manufacturing sector is a "smokescreen," he wrote in a 2005 article in *Latin American Perspectives,* a scholarly journal. Almost half of all manufacturing exports come from the *maquiladora* assembly plants (foreign-owned factories in Mexico) that import production materials and export their final products—and their profits.

Mexico adds only the labor.

Neoliberal policies—first implemented in the '80s, and later through NAFTA—cut government investment in public works and agriculture, privatized key state enterprises, and created low interest rates that attracted foreign capital. These policies opened the way for a 25-fold increase in *maquiladora* sales between 1982 and 2003 (though that growth peaked in 2000 and has since fallen as *maquiladora* owners seek ever lower wages and looser environmental regulations to compete with China's abundant labor supply).

From 1994 to 2002, Mexico lost more than 1 million agricultural jobs. And from 1980 to 2002—the same period *maquiladora* sales soared—migration from Mexico to the United States grew by 452 percent, with more than 400,000 people crossing each year, on average.

"In Mexico, we have exported the factory of migrants," says Rodolfo Garcia Zamora, an economics professor who also teaches at the Graduate School of

Development Studies. Zamora, author of *Migration, Remittances and Local Development,* says Mexico "is mortgaging its future" with migration and remittances. In the 10 Mexican states with the longest migration histories, he says, 65 percent of municipalities have a negative population growth. "This means that in the future," says Zamora, "these communities will not be able to reproduce, neither economically nor socially, because the demographics of migration have condemned them to disappear."

NO ESCAPE?

"The United States economy demands cheap labor. Mexico has an excess of laborers. We complement each other," says Fernando Robledo, director of the Zacatecas State Migration Institute, a government office that administers development projects in conjunction with several U.S. migrant organizations.

He dismisses talk of depopulation and an abandoned countryside as "fatalism." "Zacatecas has a 120-year history of migration," he says. "Migration is historical."

Robledo describes the state government's development priorities as variations on the "three for one" program—where local, state, and federal governments match each dollar provided by U.S. migrant organizations for use in local development projects, such as building interstate highways heading north and constructing greenhouses for growing export crops.

"If you had $50 million in the budget," Robledo says, "would you use that to increase production in the countryside or to build an interstate highway? It is a political and economic decision."

Robledo puts priority on the highway.

But doesn't building super-highways toward the Mexico–U.S. border and changing agriculture to a cash-crop export reproduce the very neoliberal policies that dispossess migrants in the first place?

"We do not live in a socialist country," he responds, "where the government controls every aspect of the economy. We are in a neoliberal country. We cannot escape from neoliberal economics."

Garcia Zamora, who helped write the Zacatecas state development plan, is unconvinced. The main problem, he says, is the lack of real political alternatives to neoliberalism. According to Zamora, "there is only one political party in Mexico—the PRI," referring to Mexico's notoriously corrupt Institutional Revolutionary Party, which ruled the country from 1929 to 2000. "The PRD government in Zacatecas now acts just like a PRI government," Zamora says, this time referencing the Party of the Democratic Revolution, the opposition party to the PRI. "The same lack of planning and nepotism. It spends its time mainly implementing federal programs. They drafted a good development plan, but they ... have never carried out a serious regional economic development policy that seeks to diminish the massive exodus of 40,000 Zacatecas residents who abandon Mexico every year."

ABANDONED BY MIGRATION

A few years ago, Mario Gardia left Zacatecas to work in construction in Southern California, but after about five months he decided to return to El Cargadero, a tiny town about 50 miles west of the city of Zacatecas, the state capital.

"I thought, 'In Mexico, if you work a couple of shifts, you can live OK,'" he says. "Without so many luxuries and freeways, but you can live a more peaceful life."

Garcia, in his early 40s, is a small farmer and municipal delegate. His wife and three daughters live in El Cargadero. All nine of his brothers and sisters, and more than 50 cousins, live in the United States.

El Cargadero, with a population of about 350, and a population in the United States of more than 1,000, is supposed to be a success story. Most of its roads are freshly paved, and residents have electricity and potable water, thanks to remittances and the "three for one" program.

"There are many points of view, but as you can see here, this is a community abandoned by migration," Garcia says. "The government should work to keep people in the country, to find jobs, better living conditions. Here we have pretty streets, but where are the people?"

Driving from the city of Zacatecas to El Cargadero, mile after mile of empty fields, closed restaurants and boarded-up houses span the countryside. Jose Manuel, a taxi driver, who worked in California for four years, washing dishes and making salads, accompanies me on the drive. He says he remembers when these roads weren't paved yet, but the fields were full of corn and beans. It is now vast emptiness.

"Nobody works most of this land anymore," he says. "The owners went to the U.S. and left the land behind."

This is precisely what brought Mario Garcia back. "The countryside is broken," he says. "The rural economy needs to be reactivated. But we export one of the most valuable things: our workers. And now we don't produce anything."

The legalization debate is misguided, he says, because it focuses, always, on the U.S. economy: how many immigrants to allow in and how to stamp their passports. That focus needs to shift to include Mexico.

"Mexico does not need an open border with the U.S. that invites Mexicans to go work there. People always talk about legalization, but no, what needs to be legalized is the Mexicans' ability to stay [home] so that Mexico can grow and produce."

9

Climate Change on the Move

MICHAEL WERZ AND KARI MANLOVE

Global warming exacerbates extreme weather events and contributes to resource shortages. Thus, the number of climate refugees grows as agricultural and other economic activities in the affected areas decline. The expanding climate migration will have consequences for the world's security.

Fast forward to the year 2050. The world's population will be up to 9 billion people according to the United Nations—an increase of one-third. More than 90 percent of this growth will take place in developing countries. Estimates also predict than 200 million people will be newly mobilized as climate migrants by 2050 due to global warming's effects. This increased migration will very likely affect global security, which makes it imperative for the United States and other nations to begin formulating responses to climate migration now.

As Thomas Friedman so bluntly writes, the world in 2050 will be crowded and it will be hot. Even if industrial and emerging societies were to reduce their greenhouse gas emissions tomorrow and reach instant carbon neutrality, existing pollution has locked into the atmosphere at least some unavoidable warming. No matter what steps the global community takes to mitigate emissions, we will still be forced to adapt to a warmer climate.

Global warming's consequences will be felt much earlier than 2050, too. Climate scientists argue that extreme weather events and resource shortages will affect millions of people in Africa, Australia, and Latin America by 2050. In Asia, warming will shrink freshwater resources from large river basins that could adversely affect one billion people. Parts of Africa could see rain-fed agricultural yields fall by much as 50 percent from today's output, threatening food insecurity on top of water insecurity. Melting snowcaps in the Andean region will harm important agricultural regions in Latin America.

CAP President and CEO John Podesta and his former CAP colleague Peter Ogden framed the stark reality of adaptation to a warmer climate two years ago: "Science only tells part of the story. The geopolitical consequences of climate change are determined by local political, social, and economic factors as much as by the magnitude of the climate shift itself."

SOURCE: Michael Werz and Kari Manlove, "Climate Change on the Move," *Center for American Progress* (December 2009). www.americanprogress.org. Used by permission.

And it's inevitable that as global warming intensifies hurricanes and droughts and adds to resource shortages, we will need to prepare for extreme circumstances, including human migration. In 1991, Tropical Cyclone Gorky hit the Chittagong district of Southeastern Bangladesh and a 20-foot storm surge made landfall, killing approximately 138,000 people and leaving 10 million homeless. Refugees from natural disasters usually can return home over time—as in this case—but future climate migrants could be permanently forced to leave. Climate migration is often a result of natural disasters, but resource scarcity, food security, and water shortages will also be important drivers of voluntary and long-term climate migration in the 21st century.

Worldwide estimates suggest that as many as 200 million people could become climate migrants by 2050. Today there are roughly 214 million migrants globally, meaning if climate migration projections come true, they will double the total level of migration worldwide. In some cases, climate migration hot spots overlap with already volatile and unstable regions, where substantial migration could easily give rise to border conflicts and national security concerns.

These factors will undoubtedly affect twenty-first-century migration, and the United States and other nations would be wise to factor these new forms of human mobility into long-term policy strategies and security assessments now. We'll offer a framework that can be used to address climate migration and its effects on security below. But first, we'd like to outline a handful of specific regions that could see migration and conflict due to a changing climate. Any assessment of climate migration and its security impacts should pay close attention to these areas.

NORTHERN AFRICA AND THE MEDITERRANEAN

Africa will be one of the continents hardest hit by global warming. In the west and northwest, drought and desertification will intensify and threaten the livelihoods of local habitants. Sir Nicholas Stern, chair of the Grantham Research Institute on Climate Change and the Environment at the London School of Economics, considers both the humanitarian and economic impacts on Africa in his groundbreaking climate change analysis. He cites Mali in the 1970s and 1980s as a precedent for the looming potential conflict that water shortages and drought can cause. In that case, the native Tuareg were so devastated by drought that they were driven to other countries to seek sustainable livelihoods. Their migration and eventual return caused much unrest and rebellion.

This example illustrates how northwestern Africa can become a battlefront, but other factors contribute to risk in the region: Sub Saharan migrants enter the European Union through the region, and Islamist rebels threaten the governments of Morocco, Algeria, and Tunisia, where Al Qaeda Maghreb's presence has steadily grown in recent years. Resource allocation policies, drought, and water shortages are also factors within migration hubs like Morocco, Algeria, and Egypt, which all face their own environmental challenges.

The European Union is keeping a close eye on these developments and how they could be affected by climate change. For example, the E.U. report "Climate Change and International Security" discusses mass migration and political destabilization that "puts the multilateral system at risk." While this is a dark assessment, it appropriately calls attention to the issue.

In response to growing migration pressures in Northern Africa, the European Union and especially the Spanish government have tightened border controls. The Spanish government, for example, has gone so far as to set up operations, such as sending officials of the Interior Ministry to countries like Senegal, Guinea-Conakry, Mali, Mauritania, and Cape Verde, often discouraging potential migrants from leaving. FRONTEX, a cooperative European effort on border security, has invested 24 million euros to control migratory routes toward Spain and to control the coastal waters of Cape Verde, Morocco, Mauritania, and Senegal.

Both the E.U. analysis and Spain's actions reflect the understanding that increased climate migration amplifies existing security and humanitarian threats, which increases pressure on weak states along the North African coastline, and that these developments will affect Mediterranean security in a much broader sense.

BANGLADESH AND INDIA

Bangladesh is particularly vulnerable from both a climate and a security perspective. Using scientific modeling, the World Bank estimates that a 1.5-yard rise in sea level would flood 18 percent of Bangladesh, affecting large parts of its population of 162 million. To make matters worse, the Intergovernmental Panel on Climate Change's climate modeling predicts that global warming could cause Bangladesh's rice and wheat production to fall anywhere from 8 to 32 percent by 2050, given increased warming and water stress.

Experts expect that the combination of this change in food resources and sea level rise will cause major migratory movements from Bangladesh into neighboring India. For example, the National Defense University in Washington, D.C., ran an exercise in 2008 that explored a severe flood's impact in Bangladesh. The result was hundreds of thousands of refugees taking shelter in India. Given the already tense conditions at the border, such a situation could easily result in increased religious conflict and potentially foster the spread of contagious disease.

Within the next few decades India's role as a strategic partner to the United States, as well as its role as a regional anchor of political stability, will continue to grow. But India is not in a position to absorb many climate-induced pressures or large number of climate migrants. By 2050 it will have contributed 22 percent of global population growth and will have close to 1.6 billion inhabitants.

Bangladeshi migration to India's northeast region of Assam has already incited social friction and conflict. Roughly a dozen ethnic insurgencies reside in the northeastern region's seven states, motivated by causes ranging from greater autonomy within India to complete independence.

The Pentagon's Deputy Assistant Secretary of Defense for Strategy Amanda J. Dory has stated that things in the region "get real complicated real quickly." And because of climate change's impact on migration and conflict the Pentagon is incorporating climate change into the national security strategy planning. The 2008 National Defense Authorization Act required that the Pentagon do this assessment as a part of its Quadrennial Defense Review, which is expected in February of 2010.

CHINA

China will also be a country to watch. Across its vast territory China will experience the full spectrum of climate consequences that have the potential to drive migration, most likely internally. Consequences include "water stress; increased droughts, flooding, and more severe natural disasters; increased coastal erosion and saltwater inundation; glacial melt in the Himalayas that could affect hundreds of millions; and shifting agricultural zones" that will affect food supplies.

SUSTAINABLE SECURITY STRATEGIES OFFER A NEW PERSPECTIVE ON CLIMATE MIGRATION

When natural disasters occur or humanitarian crises break out, the United States usually has been and is likely to be a first responder, particularly with military assistance and operations. In 2004 an earthquake shook the Indian Ocean, sending one of the deadliest tsunamis—and natural disasters—in recent memory to the shores of Indonesia. The U.S. military's role in the response to distribute aid and provide assistance in the cleanup was monumental in dealing with the consequences.

Our traditional responses to these disasters have worked, but climate migration has multiple humanitarian, security, and legal implications and is a more complex issue than we've previously faced. Traditional methods of response may prove insufficient.

Military and national security experts describe climate change as a "threat multiplier" and Department of Defense officials use the term "instability accelerant" because it stands to affect communities already at risk and especially sensitive to even the smallest changes. An example is competition over shrinking or less-reliable natural resources, which under the worst circumstances can incite violent conflict. In already volatile regions, fluid populations can radicalize more easily and take up a myriad of transnational concerns—trafficking, pandemics, terrorism, weapon smuggling, or drug trade.

Recognizing climate change's potential threat to security, the Center for American Progress and the Center for a New American Security ran an extensive Climate War Game in Washington, D.C., in the summer of 2008. International teams reacted to simulations and projections on climate change and extreme weather events for the years 2015 and 2050. The war game was based on extensive research and sophisticated modeling by the Pew Center on Global Climate Change, the Sustainability Institute, and the Oak Ridge National Laboratory team.

What we found in running the Climate War Game was that traditional frameworks for understanding global security threats are insufficient to deal with the looming specter of climate change. To approach the emerging challenges with traditional means such as aid or military force alone and independently from each other is insufficient for the complexity they pose.

Many policymakers also have little understanding of what to expect and how to prepare for small- or large-scale climate migration. For this reason, Susan Martin at Georgetown University's Institute for the Study of International Migration calls for new frameworks to manage climate-induced migration. She notes that "to date, there are no examples of legislation or policies that address migration of persons from gradual climate changes that may destroy habitats or livelihoods in the future."

The Center for American Progress has developed concepts such as integrated power and sustainable security to establish broad frameworks combining political, economic, and security assets to adequately address complex challenges such as climate migration. This includes thinking about new mechanisms and inter-agency solutions that incorporate economic development diplomacy, aid, and security. The goal is to muster effective responses, realizing that it is critical to set climate migration and international security agendas in the near future.

Climate change is in essence an attack on the shared interests or collective security of the world, and both climate change and climate migration assault the well-being and safety of people, or human security. It will therefore test the ability of countries to preserve natural resources and protect people. Since we are entering unknown territory we must expect the unexpected and prepare for worst-case scenarios.

In response to these challenges the Center for American Progress is bringing together our energy and national security teams to launch a project focused on the intersections between global warming, human migration, and national security. Our work will focus on better understanding the climate challenges at hand and articulating a set of progressive policy recommendations aimed at addressing these challenges. Ultimately, the proper response is likely to require new governance and management structures that can deal with the fallout at different levels and combine humanitarian and developmental policies along with public diplomacy and military assets.

REFLECTION QUESTIONS FOR CHAPTER 3

1. How does transnational migration affect you? Your community? The nation?
2. Why do we have an open border with Canada but not with Mexico? Why do we have open borders for commerce but not for people?

3. Should recent immigrants be allowed to retain elements of their culture, such as the hijab worn by traditional Muslim women?

4. Over the next 50 years, climate change will cause significant numbers of people to migrate not only across national boundaries but within nations. In the United States, for example, the best agriculture will move north. Desert areas will be deprived of sufficient water. And low-lying areas will experience a greater likelihood of floods and hurricane damage. What do you think your life will be like in 50 years as a result of climate change?

Chapter 4

Economic Globalization

At the heart of the globalization phenomenon is a complex web of transnational economic interconnections that is created through the flow of goods, services, and capital. While trade between countries has existed for centuries, the current era "is the first in which the international economic system has become truly *interdependent*" (O'Meara, et al., 2000: 215)[1]. The goods we consume and the prices we pay for them are determined by transnational corporations, and by the wages these corporations pay the workers who produce their products. Jobs move across borders to low-wage economies. Workers move across these same borders to find more profitable work. Trillions of dollars move across borders instantaneously. The effects of a financial crisis in one country or region can ripple across the globe, causing havoc in stock, bond, and commodity prices, as well as defaults on huge loans and the devaluation of currencies. Transnational corporations have incredible power. As Barbara Ehrenreich writess:

> There are 193 nations in the world, many of them ostensibly democratic, but most of them are dwarfed by the corporations that alone decide what will be produced, and where, and how much people will be paid to do the work. In effect, these multinational enterprises have become a kind of covert world government—motivated solely by profit and unaccountable to any citizenry. Only a small group of humans on the planet, roughly overlapping the world's 475-member billionaire's club, rule the global economy. And wherever globalization impinges, inequality deepens (Ehrenreich, 2000: x)[2].

The first selection in this chapter, by the *Dollars & Sense* collective, provides an overview of the global economy, with an emphasis on the international institutions supporting free trade and capitalism—the World Bank, the International

Monetary Fund, the World Trade Organization, and the International Standards Organization. The authors make clear the bias of these organizations against the world's working class and poor, and toward corporations.

The second selection shows how Wal-Mart, the biggest private employer in the world, drives down not only the wages of its own workers, but also those of other workers in the U.S., as well as those of workers in the low-wage countries that supply Wal-Mart.

The subject of the third selection is the globalization of beer. Over half of the world's beer is produced by just four giant transnational corporations. This global concentration of corporate wealth tends to result in a homogenization of culture and a concentration of power, but local initiatives do rise up to challenge corporate hegemony.

The fourth selection deals with two major shocks to the United States, the terrorist attack in 2001 and the Great Recession of 2008. The first changed the way the United States views the world. The second changed the way the world works, functioning as a true global watershed that called into question the fundamental premise of capitalism—that it is self-correcting and serves the public interest. This unraveling of the economy thus undercut the operating principle of the World Bank and the International Monetary Fund.

The fifth selection, by Nobel laureate Joseph E. Stiglitz, also addresses the terrorist attack and subsequent financial failure. One of the lessons of September 11, Stiglitz points out, was how easily bad as well as good things travel across borders in the age of globalization. The global economic downturn of 2008 reinforced this fact. Stiglitz assesses the world's response to the global crisis, identifying ways in which that response must be strengthened if we are to repair the world's economy and guard against future disasters.

ENDNOTES

1. O'Meara, Patrick, Howard D. Mehlinger, and Matthew Krain, *Globalization and the Challenges of a New Century* (Bloomington: University of Indiana Press, 2000), p. 215.

2. Ehrenreich, Barbara, "Foreword" to Sarah Anderson, and John Cavanagh, *Field Guide to the Global Economy* (New York: The New Press, 2000), p. x.

10

The ABCs of the Global Economy

THE *DOLLARS & SENSE* COLLECTIVE

This collection of essays by Dollars & Sense *offers a primer on the global economy. In particular it provides an understanding of important global economic organizations, such as the World Bank, the International Monetary Fund, the Multinational Agreement on Investment, the World Trade Organization, and others. In each instance the reader is provided answers to two questions: What is this organization up to? and Why should you care?*

In the 1960s, U.S. corporations changed the way they went after profits in the international economy. Instead of producing goods in the U.S. to export, they moved more and more toward producing goods overseas to sell to consumers in those countries and at home. They had done some of this in the 1950s, but really sped up the process in the '60s. Before the mid-1960s, free trade probably helped workers and consumers in the United States while hurting workers in poorer countries. Exporters invested their profits at home in the United States, creating new jobs and boosting incomes. The AFL–CIO thought this was a good deal and backed free trade.

But when corporations changed strategies, they changed the alliances. By the late 1960s, the AFL–CIO began opposing free trade as they watched jobs go overseas. But unionists did not see that they had to start building alliances internationally. The union federation continued to take money secretly from the U.S. government to help break up red unions abroad, not a good tactic for producing solidarity. It took until the 1990s for the AFL–CIO to reduce (though not eliminate) its alliance with the U.S. State Department. In the 1990s, unions also forged their alliance with the environmental movement to oppose free trade.

But corporations were not standing still; in the 1980s and 1990s they were working to shift the architecture of international institutions created after World War II to work more effectively in the new global economy they were creating. More and more of their profits were coming from overseas—by the 1990s, 30% of U.S. corporate profits came from their direct investments overseas, up from 13% in the 1960s. This includes money made from the operations of their subsidiaries abroad. But the share of corporate profits earned overseas is even higher than

SOURCE: RepFrom The *Dollars & Sense* Collective, "The ABCs of the Global Economy," March/April 2000. Reprinted by permission of *Dollars & Sense,* a progressive economics magazine, www.dollarsandsense.org.

that because the 30% figure doesn't include the interest companies earn on money they loan abroad. And the financial sector is an increasingly important player in the global economy.

Financial institutions and other global corporations without national ties now use governments to dissolve any national restraints on their activities. They are global, so they want their government to be global too. And while trade used to be taken care of through its own organization (GATT) and money vaguely managed through another organization (the International Monetary Fund), the new World Trade Organization erases the divide between trade and investment in its efforts to deregulate investment worldwide.

In helping design some of the global institutions after World War II, John Maynard Keynes assumed companies and economics would operate within national bounds, with the IMF and others regulating exchanges across those borders. The instability created by ruptured borders is made worse by the deregulation sought by corporations, and especially the financial sector. The most powerful governments of the world seem oblivious to this threat in giving them what they want.

This is a world-historical moment in which it is possible to stop the corporate offensive, a moment when the ruling partnership composed of the United States, Europe, and to a lesser extent Japan is fracturing, as the European Union reaches its limit on the amount of deregulation it will take and Japan's economy is in turmoil. This may allow those opposing the ruling bloc—Third World governments (which may be conservative), labor, and environmentalists worldwide—to build alliances of convenience with sympathetic elements within the E.U. to guide the reshaping of the global institutions in a liberatory manner.

What follows is a primer on the most important of those institutions. We hope in the near future to publish primers on other aspects of the global economy: regional trade agreements and alternative visions of how to regulate it. Stay tuned.

—Abby Scher

THE WORLD BANK AND
INTERNATIONAL MONETARY FUND

Where Did They Come From?

The basic institutions of the postwar international capitalist economy were framed in 1944, at an international conference in the town of Bretton Woods, New Hampshire. Among the institutions coming out of the conference were the World Bank and the International Monetary Fund (IMF). These two are often discussed together because they were founded together, because countries must be members of the IMF before they can become members of the World Bank, and because both practice what is known as "structural adjustment" (where borrower countries unable to obtain credit from other sources must change government policies before loans are released).

At both the World Bank and IMF, the number of votes a country receives is based on how much capital it gives the institution, so rich countries like the United States enjoy disproportionate voting power. In both, five powerful countries (the United States, Great Britain, France, Germany, and Japan) get to appoint their own representatives to the institution's executive board (with 19 other directors elected by the rest of the 150-odd member countries). The president of the World Bank is elected by the Board of Executive Directors, and traditionally nominated by the U.S. representative. The managing director of the IMF, meanwhile, is traditionally a European. The governments of a few rich countries, obviously, call the shots in both institutions.

Why Should You Care?

Just after World War II, the World Bank mostly loaned money to Western European governments to help rebuild their countries. It was during the long tenure (1968–1981) of former U.S. Defense Secretary Robert S. McNamara as president that the bank turned towards "development" loans to Third World countries. McNamara brought the same philosophy to "development" that he had used in war—more is better. Ever since, the Bank's approach has drawn persistent criticism for favoring large, expensive projects regardless of their appropriateness to local conditions. Critics have argued that the bank pays little heed to the social and environmental impact of the projects it finances, and that it often works through dictatorial elites that channel benefits to themselves rather than those who need them (and leave the poor to foot the bill later).

The most important function of the IMF is as a "lender of last resort" to member countries that cannot borrow money from other sources. The loans are usually given to prevent a country from defaulting on previous loans from private banks. Funds are available from the IMF, on the condition that the country implement what is formally known as a "structural adjustment program" (SAP), but more often referred to as an "austerity plan." Typically, a government is told to eliminate price controls or subsidies, devalue its currency, or eliminate labor regulations like minimum wage laws—all actions whose costs are born by the working class and the poor whose incomes are cut.

The conditions imposed by the IMF and the World Bank, which places similar conditions on "structural adjustment" loans, are motivated by an extraordinary devotion to the free-market model. As Colin Stoneman, an expert on Zimbabwe, put it, the World Bank's prescriptions for that country during the 1980s were "exactly those which someone with no knowledge of Zimbabwe, but familiarity with the World Bank, would have predicted."

The IMF and World Bank wield power disproportionate to the size of the loans they give out because private lenders take their lead in deciding which countries are credit-worthy. Both institutions have taken advantage of this leverage, and of debt crises in Latin America, Africa, and now Asia, to impose their cookie-cutter model (against varying levels of resistance from governments and people) on poor countries around the world.

—*Alejandro Reuss*

THE MULTILATERAL AGREEMENT ON INVESTMENT (MAI), TRADE-RELATED INVESTMENT MEASURES (TRIMS), AND THE INTERNATIONAL MOVEMENT OF CAPITAL

Where Did They Come From?

You're probably not the sort of person who would own a chemical plant or luxury hotel, but imagine you were. Imagine you built a chemical plant or luxury hotel in a foreign country, only to see a labor-friendly government take power and threaten your profits. This is the scenario that makes the CEOs of footloose global corporations wake up in the middle of the night in a cold sweat. To avert such threats, ministers of the richest countries met secretly at the Organization for Economic Cooperation and Development (OECD) in Paris in 1997 and tried to hammer out a bill of rights for international investors, the Multilateral Agreement on Investment (MAI).

When protests against the MAI broke out in the streets and the halls of government alike in 1998 and 1999, scuttling the agreement in that form, corporations turned to the World Trade Organization to achieve their goal. (See "Rage Against the Machine" by Chantell Taylor, *Dollars & Sense*, September/ October 1998.)

What Are They Up To?

Both the MAI and Trade Related Investment Measures (or TRIMs, the name of the WTO version) would force governments to compensate companies for any losses (or reductions in profits) they might suffer because of changes in public policy. Governments would be compelled to tax, regulate, and subsidize foreign business exactly as they do local businesses. Policies designed to protect fledgling national industries (a staple of industrial development strategies from the United States and Germany in the 19th century to Japan and Korea in the 20th) would be ruled out.

TRIMs would also be a crowning blow to the control of governments over the movement of capital into or out of their countries. Until fairly recently, most governments imposed controls on the buying and selling of their currencies for purposes other than trade. Known as capital controls, these curbs significantly impeded the mobility of capital. By simply outlawing conversion, governments could trap investors into keeping their holdings in the local currency. But since the 1980s, the IMF and the U.S. Treasury have pressured governments to lift these controls so that international companies can more easily move money around the globe. Corporations and wealthy individuals can now credibly threaten to pull liquid capital out of any country whose policies displease them.

Malaysia successfully imposed controls during the Asian crisis of 1997 and 1998, spurring broad interest among developing countries. The United States wants to establish a new international discussion group—the Group of 20 (G-20), consisting of ministers from 20 developing countries handpicked by the U.S.—to consider reforms. Meanwhile, it continues to push for the MAI-style liberation of capital from any control whatsoever.

Why Should You Care?

It is sometimes said that the widening chasm between the rich and poor is due to the fact that capital is so easily shifted around the globe while labor, bound to family and place, is not. But there is nothing natural in this. Human beings, after all, have wandered the earth for millennia—traversing oceans and continents, in search of food, land, and adventure—whereas a factory, shipyard, or office building, once built, is almost impossible to move in a cost-effective way. Even liquid capital (money) is less mobile than it seems. To be sure, a Mexican can fill a suitcase with pesos, hop a plane and fly to California, but once she disembarks, who's to say what the pesos will be worth, or whether they'll be worth anything at all? For most of this century, however, capitalist governments have curbed labor's natural mobility through passports, migration laws, border checkpoints, and armed border patrols, while capital has been rendered movable by treaties and laws that harmonize the treatment of wealth around the world. The past two decades especially have seen a vast expansion in the legal rights of capital across borders. In other words, labor fights with the cuffs on, while capital takes the gloves off.

WORLD INTELLECTUAL PROPERTY
ORGANIZATION (WIPO) AND TRADE-RELATED
ASPECTS OF INTELLECTUAL PROPERTY
RIGHTS (TRIPS)

What Are They Up To?

One of the less familiar members of the "alphabet soup" of international economic institutions, the World Intellectual Property Organization (WIPO) has governed "intellectual property" issues since its founding in 1970 (though it oversees treaties and conventions dating from as early as 1883). Companies are finding it harder to control intellectual property in two new fields—computer software and biotechnology—because it is so cheap and easy to reproduce electronic information and genetic material in virtually unlimited quantities. This is what makes software, music and video "piracy" widespread.

In the old days, "intellectual property" only covered property rights over inventions, industrial designs, trademarks, and artistic and literary works. Now it covers computer programs, electronic images and recordings, and even biological processes and genetic codes.

WIPO has been busy staking out a brave new world of property rights in the electronic domain. A 1996 WIPO treaty, which now faces ratification battles around the world, would outlaw the "circumvention" of electronic security measures. It would be illegal, for example, to sidestep the security measures on a website (such as those requiring that users register or send payment in exchange for access). The treaty, if ratified, would also prevent programmers from cracking open commercial software to view the underlying code. This could prevent programmers from crafting their own programs so that they are compatible with existing software, and prevent innovation in the form of "reengineering"—drawing on one design as the basis of another. Reengineering has been at the heart of many [a] country's economic development—not just Taiwan but also the United States. Lowell, Massachusetts, textile manufacturers built their looms based on English designs.

WIPO now faces a turf war over the intellectual property issue with none other than the World Trade Organization (WTO). Wealthy countries are attempting an end run around WIPO because it lacks enforcement power and less developed countries have resisted its agenda. But the mass-media, information-technology, and biotechnology industries in wealthy countries stand to lose the most from "piracy" and to gain the most in fees and royalties if given more extensive property rights. So they introduced, under the name "Trade-Related Aspects of Intellectual Property Rights" (TRIPS), extensive provisions on intellectual property into the most recent round of WTO negotiations.

TRIPS would put the muscle of trade sanctions behind intellectual property rights. It would also stake out new intellectual property rights over plant, animal, and even human genetic codes. The governments of some developing countries have objected, warning that private companies based in rich countries will declare ownership over the genetic codes of plants long used for healing or crops within their countries. By manipulating just one gene of a living organism, a company can be declared the sole owner of an entire plant variety.

Why Should You Care?

These proposals may seem like a new frontier of property rights, but except for the defense of ownership over life forms, TRIPS are actually a defense of the old regime of property rights. It is because current computer and bio-technology make virtually unlimited production and free distribution possible that the fight for private property has become so extreme. By extending private property to previously unimagined horizons, we are reminded of the form of power used to defend it.

—Alejandro Reuss

THE WORLD TRADE ORGANIZATION (WTO)

Where Did It Come From?

Since the 1950s, government officials from around the world have met irregularly to hammer out the rules of a global trading system. Known as the General Agreements on Trade and Tariffs (GATT), these negotiations covered, in excruciating detail, such matters as what level of taxation Japan would impose on foreign rice, how many American automobiles Brazil would allow into its market, and how large a subsidy France could give its vineyards. Every clause was carefully crafted, with constant input from business representatives who hoped to profit from expanded international trade.

The GATT process, however, was slow, cumbersome, and difficult to monitor. As corporations expanded more rapidly into global markets they pushed governments to create a more powerful and permanent international body that could speed up trade negotiations as well as oversee and enforce provisions of the GATT. The result is the World Trade Organization, formed out of the ashes of GATT in 1994.

What Is It up To?

The WTO functions as a sort of international court for adjudicating trade disputes. Each of its 135 member countries has one representative, who participates in negotiations over trade rules. The heart of the WTO, however, is not its delegates, but its dispute resolution system. With the establishment of the WTO, corporations now have a place to complain to when they want trade barriers—or domestic regulations that limit their freedom to buy and sell—overturned.

Though corporations have no standing in the WTO—the organization is, officially, open only to its member countries—the numerous advisory bodies that provide technical expertise to delegates are overflowing with corporate representation. The delegates themselves are drawn from trade ministries and confer regularly with the corporate lobbyists and advisors who swarm the streets and offices of Geneva, where the organization is headquartered. As a result, the WTO has become, as an anonymous delegate told the *Financial Times*, "a place where governments can collude against their citizens."

Lori Wallach and Michelle Sforza, in their new book *The WTO: Five Years of Reasons to Resist Corporate Globalization*, point out that large corporations are essentially "renting" governments to bring cases before the WTO, and in this way, to win in the WTO battles they have lost in the political arena at home. Large shrimping corporations, for example, got India to dispute the U.S. ban on shrimp catches that were not sea-turtle safe. Once such a case is raised, the resolution process violates most democratic notions of due process and openness. Cases are heard before a tribunal of "trade experts," generally lawyers, who, under WTO rules, are required to make their ruling with a presumption in favor of free trade. The WTO puts the burden squarely on governments to justify any restriction of what it considers the natural order of things. There are no amicus briefs (statements or legal opinion filed with a court by outside parties), no observers, and no public record of the deliberations.

The WTO's rule is not restricted to such matters as tariff barriers. When the organization was formed, environmental and labor groups warned that the WTO would soon be rendering decisions on essential matters of public policy. This has proven absolutely correct. Currently, the WTO is considering whether "selective purchasing" laws—like a Massachusetts law barring state agencies and local governments from buying products made in Burma and intended to withdraw an economic lifeline to that country's dictatorship—are a violation of "free trade." It is feared that the WTO will rule out these kinds of political motives from government policy making. The organization has already ruled against Europe for banning hormone-treated beef and against Japan for prohibiting pesticide-laden apples.

Why Should You Care?

At stake is a fundamental issue of popular sovereignty—the rights of the people to regulate economic life, whether at the level of the city, state, or nation. Certainly, the current structure of institutions like the WTO allows for little if any expression of the popular will. Can a city, state, or country insist that goods sold in its markets meet labor and environmental standards determined in a democratic forum by its citizens? What if the U.S., for example, insisted that clothing manufactured for the Gap by child laborers not be permitted for sale here? The U.S. does not allow businesses operating within its borders to produce goods with child labor, so why should we allow those same businesses—Disney, Gap, or Wal-Mart—to produce their goods with child labor in Haiti and sell the goods here?

—Ellen Frank

INTERNATIONAL STANDARDS
ORGANIZATION (ISO)

There's at least one global institution shaping commerce that corporations control completely, with no pretense of public involvement. That is the International Standards Organization (ISO).

It was founded in 1947 (around the same time as the International Monetary Fund, World Bank, and GATT), with the aim of easing trade by standardizing the dimensions of industrial products. Most famously, it set the dimensions of screw threads so that an auto manufacturer in the United States can be confident that screws it buys in China can be used in its cars. More recently, the ISO trumpets its success in standardizing ATM and credit card dimensions so they can be used in machines worldwide.

Without set standards, buyers cannot roam the world in search of the cheapest deal; the dissimilar products thus act as a "technical barrier to trade." Not surprisingly, the ISO, although privately run, is intimately linked to the World Trade Organization, with whom it says it is creating "a strategic partnership."

"The political agreements reached within the framework of the WTO require underpinning by technical agreements" devised by the ISO, according to the ISO.

"From an environmental perspective, the ISO isn't ideal because it's captured by industry," says trade lawyer Stephen Porter of the Washington, D.C., Center for International and Environmental law. Companies send their expert reps to national standards organizations, which in turn send reps to the ISO.

That might not be a problem if the ISO stuck to screws, but in the 1990s it expanded its scope to setting environmental standards, including the process used for producing organic agricultural products.

"The part that's most troublesome is when an ISO standard becomes a default standard under the WTO rules," says Porter. "Does it become impossible to go beyond that in a practical matter if Austria wants to set an environmental standard that is 130% of the ISO standard?" And once ISO standards become part of the WTO, what was a voluntary system receives the force of law, without public involvement.

—Abby Scher

THE INTERNATIONAL LABOR
ORGANIZATION (ILO)

Every year it is becoming more obvious that the global economy needs global regulation to protect the interests of workers and their communities. This was a central demand of some WTO protesters in Seattle. But who can regulate at a global level, and how can this regulation be made democratically accountable? There are no easy answers to these questions, but we can learn a lot by studying the successes and shortcomings of the International Labor Organization.

Where Did It Come From?

The ILO was established in 1919 in the wake of World War I, the Bolshevik revolution in Russia, and the founding of the Third (Communist) International, a world federation of revolutionary socialist political parties. Idealistic motives mingled with the goal of business and political elites to offer workers an alternative to revolution, and the result was an international treaty organization (established by agreement between governments) whose main job was to promulgate codes of practice in work and employment.

After World War II the ILO was grafted onto the UN structure, and it now serves a wide range of purposes: drafting conventions on labor standards (182 so far), monitoring their implementation, publishing analyses of labor conditions around the world, and providing technical assistance to national governments.

Why Should You Care?

The ILO's conventions set high standards in such areas as health and safety, freedom to organize unions, social insurance, and ending abuses like workplace discrimination and child labor. It convenes panels to investigate whether countries are upholding their legal commitment to enforce these standards, and by general agreement their reports are accurate and fair. ILO publications, like its flagship journal, *The International Labour Review*, its World Labor and Employment Reports, and its special studies, are of very high quality. Its staff, which is headquartered in Geneva and numbers 1,900, has many talented and idealistic members. The ILO's technical assistance program is minuscule in comparison to the need, but it has changed the lives of many workers. (You can find out more about the ILO at its website: www.ilo.org.)

As a rule, international organizations are reflections of the policies of their member governments, particularly the ones with the most clout, such as the United States. Since governments are almost always biased toward business and against labor, we shouldn't expect to see much pro-labor activism in official circles. The ILO provides a partial exception to this rule, and it is worth considering why. There are probably four main reasons:

- The ILO's mission explicitly calls for improvements in the conditions of work, and the organization attracts people who believe in this cause. Compare this to the mission of the IMF (to promote the ability of countries to repay their international debts) or the WTO (to expand trade), for instance.

- Governments send their labor ministers (in the U.S., the Secretary of Labor) to represent them at the ILO. Labor ministers usually specialize in social protection issues and often serve as liaisons to labor unions. A roomful of labor ministers will generally be more progressive than a similar gaggle of finance (IMF) or trade (WTO) ministers.

- The ILO's governing body is based on tripartite principles: representatives from unions, employers, and governments all have a seat at the table. By institutionalizing a role for nongovernmental organizations, the ILO achieves a greater degree of openness and accountability.

- Cynics would add that the ILO can afford to be progressive because it is largely powerless. It has no enforcement mechanism for its conventions, and some of the countries that are quickest to ratify have the worst records of living up to them.

On Balance?

The ILO has significant shortcomings as an organization. Perhaps the most important is its cumbersome, bureaucratic nature: it can take forever for the apparatus to make a decision and carry it out. (Of course, that beats the IMF's approach: decisive, reactionary, and authoritarian.) The experience of the ILO tells us that creating a force capable of governing the global economy will be extremely difficult, and that there are hard tradeoffs between democracy, power, and administrative effectiveness. But it also demonstrates that reforming international organizations—changing their missions and governance systems—is worth the effort, especially if it brings nongovernmental activists into the picture.

—Peter Dorman

Resources: Arthur MacEwan, "Markets Unbound: The Heavy Price of Globalization," *Real World International (Dollars & Sense,* 1999); David Mermelstein, ed., *The Economic Crisis Reader* (Vintage, 1975); Susan George and Fabrizio Sabelli, *Faith and Credit: The World Bank's Secular Empire* (Penguin Books, 1994); Hans-Albrecht Schraepler, *Directory of International Economic Organizations* (Georgetown University Press, 1997); Jayati Ghosh, *Lectures on the History of the World Economy,* (Tufts University Press, 1995); S. W. Black, "International Monetary Institutions," *The New Palgrave: A Dictionary of Economics,* John Eatwell, Murray Milgate, and Peter Newman, eds. (The Macmillan Press Limited, 1987).

11

Why the Bosses Need Wal-Mart

FRED GOLDSTEIN

Fred Goldstein's thesis is that Wal-Mart's low wages and low prices for its goods serve the entire capitalist class by driving down the cost of the means of subsistence and thus making it easier for all the bosses to lower wages.

Not only does Wal-Mart pay low wages, but its low prices allow the entire capitalist class to pay low wages to a large section of the working class.

Wal-Mart has been considered one of the engines driving the process of low-wage capitalism. The company in 2006 employed 1.2 million workers in the United States and 400,000 workers abroad, making it the biggest private employer in the world. Its rise to power has been on the basis of high technology, low prices, and low wages. Wal-Mart is the pre-eminent low-wage company in the U.S. capitalist establishment. Its wages have been declining since 1970, driving down wages throughout the retail industry.[1]

But Wal-Mart not only drives down the wages of its own workers. By using its leverage as the world's largest retailer, it pressures its suppliers to lower costs. This can only be done by increasing the exploitation of the suppliers' own workers or by offshoring to low-wage countries.

Wal-Mart spreads its monopoly power by destroying its competition with its famous "lowest prices." But it achieves these low prices by impoverishing its own workers and the workers of its suppliers and by forcing its suppliers to super-exploit low-wage labor across the globe. Wal-Mart and its apologists claim that these low prices are saving people hundreds of billions of dollars a year. But the truth is that Wal-Mart's low prices serve the entire capitalist class in the United States. They drive down the cost of the means of subsistence and thus make it easier for all the bosses to lower wages.

Wages at Wal-Mart have been variously estimated at anywhere from $8 to $10 per hour. It considers a full-time job to be thirty-four hours a week. But even at $10 an hour with a forty-hour week, a Wal-Mart worker would take home $1,280 a month after taxes. No one can support a family on Wal-Mart pay; even a single worker can barely survive on these wages. By 2005 the company faced forty different lawsuits filed by workers all across the country for making them work off the clock, work through scheduled breaks, or punch

SOURCE: Fred Goldstein, *Low-Wage Capitalism*. New York: World View Forum, 2008, pp. 127–132. Copyright © 2008 Workers World.

out and continue to work for no pay. Wal-Mart has been sued for sex discrimination (two-thirds of its workers are women). It has abused undocumented
workers. In fact, Wal-Mart workers are so poor that many of them have to use
Medicaid, food stamps, and other forms of government assistance just to survive.

The company is fiercely anti-union. In 2000, when eleven butchers at a
Wal-Mart in Jacksonville, Texas, voted to join the United Food and Commercial Workers (UFCW), Wal-Mart announced that henceforth it would buy only
pre-wrapped meat. It then eliminated butchers in all of its hundreds of supercenters.[2] When the UFCW, after a long and successful organizing drive, opened
up bargaining at a store in Jonquière, Quebec, Wal-Mart shut the store down
after five months at the negotiating table.[3] It is constantly on the lookout for
any signs of union sympathy among its workers and fires them on trumped-up
charges.

When Wal-Mart moves into an area, it puts unionized supermarkets out of
business or gives the bosses of its competitors leverage at the bargaining table to
demand concessions based on not being able to compete with Wal-Mart. The
important grocery workers' strike in southern California of 2003–2004, waged
by the UFCW, was triggered by Wal-Mart's threat to put forty supercenters in
the state and drive wages down in the entire area.

In 2004 Wal-Mart accounted for 2.3 percent of the U.S. gross domestic
product (GDP). It sold 14 percent of the groceries in the country and 20 percent
of the toys. It racks up $300 billion in sales, running neck-and-neck with Exxon-
Mobil for first place. Wal-Mart has close to 4,000 stores in the U.S. and
hundreds abroad. It is the largest retailer in Mexico and Canada. It is the
second-largest grocer in England.[4] A typical Wal-Mart has 60,000 different items
on the shelf. A typical supercenter carries 120,000.[5] With this kind of leverage,
suppliers don't tell Wal-Mart what price an item should sell for; Wal-Mart tells
them. Even the largest monopolies, such as Procter & Gamble and Levi's, have
lost out in the struggle over pricing with Wal-Mart. Furthermore, Wal-Mart is
famous for telling its suppliers to cut prices by 5 percent year over year.[6]

Wal-Mart drives down prices by driving down wages at its 60,000 suppliers
in the United States—and also in China, Singapore, Mexico, Indonesia, and Sri
Lanka. It puts suppliers in all these countries in competition with each other to
get their products onto Wal-Mart's shelves. Wal-Mart has such dominance in
some industries that it can play a major role in establishing sweatshop wages
that affect entire countries or regions. By 2003 Wal-Mart had over 3,000 factory
suppliers in southern China at low wages. Wal-Mart is Bangladesh's most important customer. Bangladeshi sweatshop workers, most of them women, supply
clothing to Wal-Mart.

The cost of Wal-Mart's low prices is illustrated by the plight of workers at
the Western Dresses factory in Dhaka, Bangladesh. In 2003 a sixteen-year-old
junior sewing machine operator, Robina Akther—whose job was to sew flaps
on the back pockets of pants destined for Wal-Mart—worked for thirteen cents
an hour, fourteen hours a day, making $26.98 a month. If she did not sew the
required 120 pairs of pants per hour she was beaten. "They slapped you and
lashed you hard on the face with the pants. This happens very often. It is no

joke."[7] The work went on from 8:00 a.m. to 10:00 or 11:00 p.m., seven days a week, with ten days off in the whole year. Charles Fishman, the author of this account, calculated that it would take half a century for Akther to earn $16,200, while in 2003 Wal-Mart's profits were $19,597 a minute!

Akther brought a lawsuit against Wal-Mart in the United States for failure to provide basic wages, overtime pay, and protection from physical abuse that Bangladeshi law provides. Fourteen other workers were plaintiffs in the lawsuit—from China, Indonesia, Swaziland, and Nicaragua. According to Fishman, "all make merchandise for Wal-Mart, and all have nearly identical claims."[8]

On the home front, Wal-Mart tracks every item rung up on every cash register by every cashier. It has central communications and its loading workers must wear headsets for perpetual monitoring. It owns the largest private satellite communications system in the United States and links every store location to its central office. Combining communications technology with software, Wal-Mart tracks every item sold. It compels many of its suppliers to adopt similar technology in order to speed up workers—what the bosses call "achieving efficiency"—so the suppliers can come in with the lowest prices.

Wal-Mart and its capitalist admirers continually wax eloquent about the company's low prices, as if it were giving away money to the masses. The estimates of how much money consumers save goes all the way from $100 billion a year to $300 billion a year, depending upon which authority is used. All the bourgeois experts say it's a tradeoff, low wages for low prices. Wal-Mart CEO H. Lee Scott bragged, "In effect it gives them a raise every time they shop with us."[9]

But this argument flies in the face of intuition. After all, if every time workers shopped at Wal-Mart they were getting a raise and saving so much money, then why are the vast majority of the workers who shop there living from paycheck to paycheck? Why are so many of the customers of Wal-Mart (and other low-price discount retailers) in personal debt? In fact, CEO Scott contradicted himself in virtually the same sentence when he declared how much of a benefactor his company was to the working class.

"These savings are a lifeline for millions of middle- and lower-income families who live from payday to payday," he declared. Customers shopped at Wal-Mart 7.2 billion different times in the year 2006.[10] The masses should be rolling in wealth if they got a raise every time they went there. The fact is, however, that the average annual income of the people who shop at Wal-Mart is $35,000 a year. An article in the *Washington Post* that reported this figure called Wal-Mart a "force for poverty relief," saying its "$200 billion–plus assistance to consumers may rival many federal programs."[11]

Of course workers can benefit from lower prices for the necessities of life, but only if their wages do not go down at the same time. But workers in the U.S. are getting poorer. The truth is that tens of millions of workers who shop at Wal-Mart cannot pay higher prices at the declining wage levels. Life is getting harder. Yet Wal-Mart's low prices are "saving money." Where is all that money going, if everyone is still poor?

Marx's analysis of wages explains that the real beneficiary of Wal-Mart's low prices is the capitalist class as a whole.

In Marx's explanation of the bosses' drive to increase their surplus value, i.e., to increase the unpaid labor of the workers and thus increase profits, he showed the different ways the capitalists go about it. One way to get more profits from the workers is to simply make them work longer hours without increasing their pay. This elongation of the workday Marx called absolute surplus value. (Wal-Mart used this method by making workers work off the clock.)

But another way the bosses can get more surplus value, without making the workday longer, is to lower wages. However, wages have to be high enough to keep workers alive, so there is a limit to how low the bosses can push them and still retain the labor force without a mass rebellion. The limit of how low wages for the majority of workers can go is what Marx terms the price of necessary labor.

Necessary labor time is the hours it takes a worker to earn the wages necessary to sustain herself or himself and family. Let's assume the worker is paid by the week. As soon as the worker puts in the hours on the job needed to produce a value equivalent in money to his or her weekly wage, then the rest of the time worked, and the value the worker adds to the product or service during that time, the boss gets for free. This is the source of all profit. The value added to the product, or to the service rendered, during this time is unpaid labor-time.

For example, suppose a worker on a production line producing shoes adds $500 a week in new value to the shoes on which he or she works. That comes to $100 a day for five days. And further suppose that the worker receives $300 a week in wages to live on. But the worker has added labor to the shoes worth $300 in value in just three days. The money equivalent of value produced for the rest of the week—$200 for the last two days—goes into the boss's pocket once the shoes are sold. The worker has given two free days of labor to the boss.

The wage form of payment conceals the fact that only part of the workday represents paid labor; the rest is unpaid. That is the secret, uncovered by Marx, that has concealed the real nature of capitalist exploitation. Because of the form of capitalist production and the wage system—whether the pay or salary is calculated hourly, weekly, or monthly—workers are led to believe that they are being paid for the entire time that they work. They may get high wages, middle wages, or low wages, but whatever their pay, the capitalist myth is that the workers are getting paid for the entire day, week, or month, as the case may be.

Most auto workers know, because they see the prices of the cars they work on and they know how many come off the assembly line, that their labor produces vehicles worth hundreds of times what they get paid. In the case of automobiles, because they are so key to the entire capitalist economy, these figures are public and constantly in the press. But because of capitalist secrecy, most workers have to guess at the relationship between their wages and the wealth that they produce. Every worker needs to know that their wages come to only a fraction of the value they add to products with their labor. Otherwise, there would be no profit. That fraction of the total value of the product, representing the time it took to earn their wages, is the paid labor time; the other fraction is the unpaid labor time taken by the capitalist. If wages can be reduced because prices for the means of subsistence go down, then the boss gains the difference.

The bottom line is that Wal-Mart's low prices are at the expense of the low wages of their 1.2 million workers, as well as at the expense of the millions of workers in the United States and in sweatshops around the world who work for Wal-Mart's suppliers. In fact, these low prices have made it possible for the capitalist class to lower wages without driving all the workers to absolute hunger and rebellion. Lowering the price for the necessities of life has the objective economic effort of cheapening labor-power and reducing necessary labor time. When this happens, the bosses get more unpaid labor-time, more surplus value.

Only a class struggle that stops the slide in wages can allow workers to take advantage of lower prices. Otherwise, that $200 billion in "savings" goes into the pockets of the bosses, because the lower prices allow bosses all over the United States to pay lower wages.

ENDNOTES

1. Nelson Lichtenstein, ed., *Wal-Mart: The Face of Twenty-First Century Capitalism* (New York: New Press Books, 2006), pp. 15 and 148.

2. Ibid., p. 268, Wade Rathke; "A Wal-Mart Workers' Association?: An Organizing Plan."

3. Ibid., pp. 268-269.

4. Charles Fishman, *The Wal-Mart Effect* (New York: Penguin Books, 2006), p. 235.

5. Ibid., p.15.

6. Lichtenstein, *Wal-Mart*, p. 176.

7. Fishman, *The Wal-Mart Effect*, p. 185.

8. Ibid., p. 186.

9. Lichtenstein, *Wal-Mart*, p. 26.

10. Fishman, *The Wal-Mart Effect*, p. 6.

11. Sebastian Mallaby, "Progressive Wal-Mart. Really," *Washington Post*, Nov. 28, 2005.

12

Beer Globalization in Latin America

BENJAMIN DANGL

Mexico is known for its great beer, which is mostly produced by FEMSA, the company that brews Dos Equis, Tecate, and Sol, among others. Now the Dutch brewing giant Heineken has bought FEMSA, and the majority of the world's beer production now rests with a mere four gigantic transnational corporations. This global concentration of a commodity results in the homogenization of culture. But there are local challenges to the corporate globalization of beer, with homemade alcoholic drinks flourishing in places like Colombia and Bolivia.

On a pleasant autumn day in 1890 the Cuauhtémoc brewery was founded in Monterrey, Mexico. This brewery, which also specialized in ice production, went on to become Mexican Economic Development Inc. (FEMSA), brewing such beers as Dos Equis, Tecate and Sol. Recently the Dutch brewing giant Heineken bought FEMSA, bringing over half of the world's beer production into the hands of just four mega-corporations. One Mexican columnist wrote of the merger in *La Jornada*, "Just a bit more globalization and we will all be lost."

The concentration of beer production into the hands of a few brewers is reflective of what is happening in economies across the globe. Homogenization of culture and the centralization of wealth and power naturally follow corporate globalization. Though the recent merger in Mexico is emblematic of this profit-driven trend, homegrown examples of grassroots alternatives have emerged in the kitchens and coca fields of Colombia and Bolivia.

RIDING THE BEER WAVE OF CONSOLIDATION

A number of major beer company mergers have taken place in recent years, the largest being Belgian—Brazilian InBev's purchase of Anheuser-Busch for $52 billion in 2008. According to the *Wall Street Journal*, this "wave of consolidation in the global beer market" has "put pressure on smaller brewers to find larger homes." Subsequently, Heineken purchased FEMSA, the brewer of just under half of Mexico's beer, for $5.7 billion.

SOURCE: From Benjamin Dangl, "Beer Globalization in Latin America," *Counterpunch* (February 23, 2010), Used with permission.

Jean-François van Boxmeer, the chairman and chief executive of Heineken, said this purchase will help his company become a "more competitive player in Latin America, one of the world's most profitable and fastest-growing beer markets."

Business Times reporter Chew Xiang met Van Boxmeer at the company's Singapore office on a rainy day in December. Though cold beer was available at an office bar, Van Boxmeer chose to drink coffee instead as he was recovering from a long flight. The executive told Xiang, "In the niche of premium international brands, we want to stay number one in the world. And in each market, we want to be number one or number two."

THE DEVIL FERNÁNDEZ AND
BOLIVIAN COCA LEAVES

Jose Antonio Fernández, who became chairman of Mexico's FEMSA in 1995, is still known by a nickname from his youth: "the Devil Fernández." From the beginning of his work for FEMSA, Fernández focused on other aspects of his business besides beer, namely the company's convenience stores and bottling operations.

"The Devil has a love affair with his two new babies, Coca-Cola and Oxxo [convenience stores]," said Monterrey corporate lawyer Ernesto Canales. Since 1993, when the company became partners with Coca-Cola, FEMSA has become the second largest Coca-Cola bottler in the world. Following the sale to Heineken, Fernández said he planned to focus on expanding his Coca-Cola bottling business. He is optimistic, and prepared to "go hunting" for new business ventures.

However, some new competition for Coca-Cola—and therefore, the Devil Fernández—has emerged in Bolivia. In January, Bolivian president, and former coca farmer, Evo Morales announced plans for the production of a soft drink called Coca Colla. The name's play on words is a nod to indigenous culture in Bolivia, where approximately 60% of the country self-identifies as indigenous; the term "colla" refers to indigenous people living in the western highlands of Bolivia. The drink would be based on actual coca leaves produced legally by coca farmers in the country. Coca has been widely used for medicinal and cultural purposes throughout the Andes for centuries.

The drink may be produced exclusively by the state or in partnership with coca growers, and would have a label similar to that of Coca-Cola. Other Bolivian products like tea, candy, shampoo, cookies, and liquor are made with coca. The Vice-Minister of Coca and Integrated Development Jerónimo Meneses said, "Coca Colla will be part of the industrialization of coca production."

The AFP pointed out that Bolivia's new constitution, passed in January of last year, officially states that coca is not a drug in its natural form, and recognizes the leaf as a "cultural heritage, a natural and renewable resource of biodiversity in Bolivia and a factor of social cohesion."

CHICHA RESURGENCE IN COLOMBIA

Another development in Latin America also challenges the type of corporate globalization Heineken and the Devil Fernández are pushing. Chicha, a home-made alcoholic drink typically made from corn for centuries throughout the Americas, is enjoying a comeback in major Colombian cities, according to Inter Press Service reporter Helda Martínez. Many people, particularly students, are drawn to chicha as an alternative to beer from massive commercial producers.

In Bogotá, Gloria Cecilia Delgado sells chicha that she prepares based on a recipe from her grandmother and advice on the brewing process provided by an indigenous man who visited her shop years ago. Delgado stirs the ingredients in a ceramic pot, and believes that the mood of the brewer, particularly "the love they put into it," has an impact on the flavor, according to Martínez.

Throughout the 20th century, campaigns and bans were initiated against chicha, in part due to pressure from competing beer and soda companies in Colombia. In her IPS article, Martínez cited historians Oscar Iván Calvo and Marta Saade, who wrote of the bans on chicha during the 1940s in Colombia.

Part of the desire for prohibition was based on the fact that local indigenous people and workers would gather in chicha bars to complain about their work-ing conditions and low salaries. Calvo and Saade's words underline the urgency for contemporary alternatives to capitalist monopolies which squash local culture and diversity: "The elites feared the existence of recreational spaces where the popular social classes, discontented with their poverty, came together."

In contrast, Heineken's purchase of FEMSA in Mexico signals a global move not just toward the consolidation of beer brewing, but of political and social power as well. Abraham Nuncio, in the *La Jornada* column on the beer merger, lamented the corporate drive in Mexico toward a national economy that "falls into the hands of the monopolies that control the global market and loot the riches and the sovereignty of the country."

13

Two Septembers

DAVID ROTHKOPF

Journalist David Rothkopf describes two momentous events: the terrorist attack in 2001 and the Great Recession of 2008. The first changed the way the United States views the world, while the second changed the way the world views us and it will change how the world works. The economic shock is a true watershed. The World Bank and the International Monetary Fund, for example, loan money to struggling economies based on capitalist principles. But the economic chaos of 2008 that affected the U.S. and other economies, called into question the fundamental premise of capitalism that it is self-correcting and serves the public interest.

Two September shocks will define the presidency of George W. Bush. Stunningly enough, it already seems clear that the second—the financial crisis that has only begun to unfold—may well have far greater and more lasting ramifications than the terrorist attacks of Sept. 11, 2001.

That's because while 9/11 changed the way we view the world, the current financial crisis has changed the way the world views us. And it will also change, in some very fundamental ways, the way the world works.

Of course, the Sept. 11 attacks left a deep scar on the soul of the country and caused immense tragedy. Beyond human losses, they also revealed that being the sole superpower did not make us safe. But the attacks themselves were not, in a real sense, as significant a turning point in world history as they may have seemed at the time. (Remember, it was actually Bush's father who had first been put in charge of an American "war on terror" during the 1980s when he was Ronald Reagan's vice president.)

The current economic debacle is far more likely to be seen by historians as a true global watershed: the end of one period and the beginning of another. The financial chaos has brought down the curtain on a wide range of basic and enduring tenets also closely linked with the Reagan era, those associated with neoliberal economics, the system that the Nobel Prize–winning economist Joseph Stiglitz has called "that grab-bag of ideas based on the fundamentalist notion that markets are self-correcting, allocate resources efficiently and serve the public interest well." Already this crisis has seen not just our enemies but

SOURCE: From David Rothkopf, "Two Septembers: The Terrorist Attacks of 2001 Were Enormous, But the Economic Shock of 2008 Is Even Bigger," *The Washington Post National Weekly Edition* (October 13–19, 2008), p. 26. Used by permission of the author.

even some of our closest allies wondering whether we are at the beginning of the end of both American-style capitalism and of American supremacy.

On Sept. 16, 2001, President Bush addressed the nation to express his faith in the American people and "the resiliency" of the U.S. economy. Seven years later, the president again spoke to a country in crisis, using eerily similar language to try to shore up concerns about the market. This time, however, he felt compelled to go further. During a prime-time broadcast to the nation, the president of the richest and most powerful nation on Earth felt compelled to offer a defense of the free-market capitalism whose final and enduring triumph we had been celebrating only a few years earlier after the fall of our Soviet foes. "Despite corrections in the marketplace and instances of abuse," Bush said, "democratic capitalism is the best system ever devised."

To many around the world, however, the president's words were not so reassuring. Not only did his argument ring hollow to those who felt anxiety and rage over Wall Street's ineptitude and greed, its attempt to buoy American capitalism by lashing it to the virtues of democracy contrasted uncomfortably with a chorus of critical assessments from leaders in democracies worldwide.

French President Nicolas Sarkozy concluded recently that the world has seen the end to free-market economies. "Laissez-faire, it's finished. The all-powerful market that is always right, it's finished," he said. We would, he added, need "to rebuild the entire global financial and monetary system from the bottom up, the way it was done at Bretton Woods after World War II." Germany's finance minister offered a similar perspective in remarks to his parliamentary colleagues. "The U.S. will lose its status as the superpower of the world financial system," Peer Steinbruck declared. "This world will become multipolar. The world will never be the same again." Governments long criticized by the United States for intervening in their own economies were reveling in the spectacle of U.S. policymakers wading into their own financial markets in ways that even some socialist leaders would never have dreamt of.

There is some wishful thinking or gloating in these remarks, but they represent a massive wave of reassessments from every corner of the world and every point on the political spectrum. The strongly pro-market *Financial Times* declared that we are at the end of the era of American "laissez-faire capitalism." From Beijing, Pan Wei of the Center for Chinese and Global Affairs speculated that the crisis would have a leveling effect: "My belief is that, in 20 years, we will look the Americans straight in the eye—as equals. But maybe it will come sooner than that." (Indeed, it seemed he might already be right. Reports from within the Treasury suggested that the U.S. government intervened in the financial sector, at least in part in response to Chinese threats to reconsider their policy of buying U.S. debts unless Washington moved to stabilize the markets.)

Buried in all of this, one can hear a refrain with eerie echoes of 9/11: that the United States "had it coming." Indeed, one of the factors that links 9/11, the war in Iraq, and this financial crisis is a sense that all of them are tied to the world's changing view of America—a view that is growing darker. While the "blame America" justification for terror is as odious as it is indefensible, we deserve our full share of the blame for the market disaster. An important dimension of this new anti-Americanism relates to Washington's role as the architect, champion, and primary beneficiary of a global system that was widely seen to benefit the few at the expense of the many.

Much of the anger at the bailout and financial melt-down stems from the basic fact that we live in a world in which 1,000 people at the top control assets worth double those held by the bottom 2.5 billion, a world in which the top 10 percent own 85 percent of everything. And it's a world in which that anger is likely to keep growing, because there are other dimensions of this crisis that suggest it could have a broader impact than any financial crisis in three-quarters of a century. This is only the tip of the iceberg.

The collapse in mortgage-backed securities represents a problem with just one subset of a global financial system that has dramatically changed its character in just a generation. Not only is a deep U.S. downturn upon us, but it will hit countries such as China, which trade with the United States; their slumps will depress commodity prices that in turn could cause crises in the emerging world. Complex securities invented by financial "rocket scientists" and understood by very few people have long ago surpassed currency as the primary repositories of "value" on the planet. But they are not issued by governments. Most are not regulated. The risks associated with them are hidden. According to polls of big institutional investors by the National Strategic Investment Dialogue, more than 80 percent do not feel their boards even understand the risks inherent in their portfolios—as they are required to do by law—primarily because of the complexity of these instruments and the new global financial system they dominate.

In other words, this crisis—which, on Sept. 29 alone, wiped away more value from the market than the Congressional Budget Office estimates has been spent to date in Iraq and Afghanistan—is not important merely because it represents the most profound global ideological watershed since the fall of Soviet communism. It is important because it is a harbinger of massive global threats that fester within the system. And our leaders, from [former President George W. Bush] on down, have been caught as flat-footed by these new dangers as they were complacent and complicit in their hatching.

We now know that the costs to the U.S. government associated with this crash will surpass those associated with the wars in Iraq and Afghanistan. We also know that all such government cost estimates tend to be on the low side. We further know that the U.S. economy recovered quickly after 9/11 and that we are in the midst now of a global downturn that may last for many months and perhaps years. We know that there have been vastly more job losses— 600,000 recorded thus far [in 2008] in the United States—than were associated with the 9/11 shocks, and that the global job-loss totals that a recession is likely to bring will be measured in the millions. Among the poorest, the likely shocks to emerging markets caused by the United States' inability to spend freely will take a devastating human toll.

By all the metrics available to us, then, the current financial crisis easily exceeds the post-9/11 war on terror in economic terms. Human costs are harder to measure, of course, and the tolls of both events have been devastating. But the financial crisis will certainly touch many more people in many more countries than did 9/11. And even greater crises may loom ahead, thanks to our unwitting creation of a financial Frankenstein's monster of unregulated, risk-laden, global derivatives markets.

As the dithering U.S. governing class is grappling with the disposal of "toxic assets" in the U.S. economy, the world is moving on to debate what is widely seen as a toxic ideology: a form of market fundamentalism that promotes inequality.

While 9/11 and this financial crisis are very different, the stakes of an economic battle can be as high as those in any military conflict. The era of "leave it to the markets" is clearly over. The question during [the 2008] election is whether the United States will choose leaders who support a new American capitalism—a long-overdue successor to the bankrupt dogma of Reagan–Thatcherism—that includes effective regulation, recognizes the need for new international institutions to manage our complex global economy, and accepts that government must act with markets to promote growth and to protect those whom markets leave behind.

This is the central challenge that will fall to the president [elected in 2008]: not just to navigate the shoals of a treacherous global economy but to reinvent what has been one of our proudest exports—our distinctly American vision of how to create a just society.

14

A Real Cure for the Global Economic Crackup

JOSEPH E. STIGLITZ

Nobel Laureate Stiglitz writes about the financial failure of U.S. economic institutions in light of the global recession of 2008, and offers solutions for stabilizing both local economies and the global economy.

This is not only the worst global economic downturn of the post–World War II era; it is the first serious global downturn of the modern era of globalization. America's financial markets failed to do what they should have done—manage risk and allocate capital well—and these failures have had a major impact all over the world. Globalization, too, did not work the way it was supposed to. It helped spread the consequences of the failures of U.S. financial markets around the world. September 11, 2001, taught us that with globalization not only do good things travel more easily across borders; bad things do too. September 15, 2008, has reinforced that lesson.

A global downturn requires a global response. But so far our responses—to stimulate and regulate the global economy—have largely been framed at the national level and often take insufficient account of the effect on others. The result is that there is less coordination than there should be, as well as a smaller and less well-designed stimulus than is optimal. A poorly designed and insufficient stimulus means that the downturn will last longer, the recovery will be slower and there will be more innocent victims. Among these victims are the many developing countries—including those that have had far better regulatory and macroeconomic policies than the United States and some European countries. In the United States a financial crisis transformed itself into an economic crisis; in many developing countries the economic downturn is creating a financial crisis.

The world has two choices: either we move to a better global regulatory system, or we lose some of the important benefits that have resulted from globalization. But continuing the status quo management of globalization is no longer tenable; too many countries have had to pay too high a price. The G-20's response to the global economic crisis, crafted at meetings in November [2008] in Washington and in April [2009] in London, was a beginning—but just a

SOURCE: From Joseph E. Stiglitz, "A Real Cure for the Global Economic Crackup,"
The Nation (July 13, 2009), pp. 12–14. Used with permission.

beginning. It did not do enough to address the short-term problems nor did it put in place the long-term restructuring necessary to prevent another crisis.

A United Nations meeting in late June [2009] hopes to continue the global discussion begun at earlier G-20 meetings and to extend this discussion to what went wrong in the first place so that we can do a better job of preventing another crisis. The global politics of this meeting are complex. Many of the 173 countries that are not members of the G-20 argue that decisions affecting the lives of their citizens should not be made by a self-selected club that lacks political legitimacy. Some members of the G-20—including new members brought into the discussion for the first time as the G-8 expanded to the G-20—like things the way they are; they like being in the inner circle and argue that enlarging it will only complicate matters. Many from the advanced industrial countries would like to avoid overly harsh criticism of their banks, which played a pivotal role in the crisis, or of the international economic institutions that not only failed to prevent the crisis but pushed the deregulatory policies that contributed so much to it and its rapid spread around the world. Indeed, the G-20's response to the crisis in developing countries relied centrally on the IMF.

I chair the UN Commission of Experts, which was given the task by the General Assembly of preparing an interim report before the June [2009] meeting. This report will, I hope, have some influence on the discussions. It is too soon to tell whether it will, or if anything concrete will come from the meeting. The international community should realize, however, that much more needs to be done than has so far been undertaken by the G-20.

Our preliminary report lists ten policies that need to be implemented immediately. These include strong stimulus efforts from developed countries, providing additional funding for developing countries, creating more policy space for developing countries, avoiding protectionism, opening advanced countries' markets to the least developed countries' exports, and improving coordination of global economic policies. In addition, the commission recommends ten deeper reforms to the global financial system on which work needs to begin.

The United States may have the resources to bail out its banks and stimulate its economy, but developing countries do not. Developing countries have been important engines for economic growth in recent years, and it is hard to see a robust global recovery in which they do not play an important role. There is a consensus that all countries should provide strong stimulus packages, but many of the poorer developing countries don't have the resources to do so. Many in the developed world are worried about the debt burdens resulting from stimulus packages, but for those still scarred by debt crises, taking on additional debt may involve an unacceptable burden. Assistance has to be provided in grants, not just loans.

In the past, the IMF provided assistance accompanied by "conditions." In many cases it demanded that countries raise interest rates to high (sometimes very, very high) levels and reduce deficits by cutting expenditures and/or raising taxes—just the opposite of U.S. and European policies. This led to a weakening of national economies, when the point of IMF assistance was to strengthen them. Although those providing assistance want to be sure their money is used well, these kinds of conditions are counterproductive and make many developing countries reluctant to accept help. A condition imposed on international institutions that provide assistance to developing countries should be that they not engage in such "conditionality."

To help fund the large amount of assistance required, developed countries should set aside 1 percent of their stimulus package to help developing countries. The funds have to be distributed through a variety of channels, including regional institutions and possibly a newly created credit facility whose governance better reflects new potential donors (Asian and Middle East countries) and recipients.

The G-20 did make significant efforts to expand the IMF's lending capacity— partly, some suspect, because of the role the IMF may play in rescuing Eastern Europe rather than because of its desire to help the least developed countries. One clever way of doing so was a new issue of IMF money (to the tune of $250 billion) called "special drawing rights," a positive move, but too little of it will wind up in the hands of the poorest countries.

Although the G-20 made grand statements at its November meeting about avoiding protectionism, the World Bank notes that since then seventeen members have undertaken protectionist measures. Developing countries have to be protected from protectionism and its consequences, especially when it discriminates against them. The United States, for example, included a "buy American" provision in its stimulus bill, but many advanced industrial countries are exempt from this provision due to a WTO government procurement agreement. This means that America, in effect, discriminates against poor countries.

We know that subsidies distort free and fair trade as much as tariffs, but subsidies are even worse than tariffs, because developing countries can ill afford them. The massive bailouts and guarantees provided by the United States and other wealthy countries give their firms an unfair competitive advantage. It is one thing for firms from poor countries to compete against well-capitalized U.S. firms; it is another to compete against Washington. Such subsidies, bailouts, and guarantees are understandable, but the adverse impacts on developing countries must be recognized, and we must find some way of compensating them to offset this unfair advantage.

International cooperation is also required if we are to devise an effective regulatory regime. There is international agreement on ten issues. First, the crisis was caused by excesses of deregulation and deficiencies in the enforcement of existing regulations. Second, self-regulation will not suffice. Third, regulation is required because failures in a large financial institution or the financial system more generally can have "externalities," adverse effects on workers, homeowners, taxpayers, and others worldwide. Fourth, more than transparency is required—even full disclosure of the complex derivatives and other financial products might not have allowed for an adequate risk assessment. Fifth, perverse incentives that encouraged excessive risk-taking and shortsighted behavior contributed to bad banking practices. Sixth, deficiencies in corporate governance contributed to flawed incentive structures. Seventh, so too did the fact that many banks had grown "too big to fail"—which meant that if they gambled and won, they walked away with the gains, but if they lost, taxpayers picked up the losses.

Eighth, unless regulation is comprehensive there can be a "race to the bottom," with countries with lax regulation competing to attract financial services. Ninth, if that race happens, countries will have to take action to protect their economies—they cannot allow bad practices elsewhere to harm their citizens. And tenth, regulation has to be comprehensive across financial institutions. As we have seen, if we regulate the banking system but not the shadow banking system, business will migrate to where it is less well regulated and less transparent.

Despite this broad consensus, the G-20 said little or nothing about some key issues: what to do with banks that have grown not only too big to fail but (according to the Obama administration) too big to be financially restructured? The G-20 failed to ask the hard questions: If these big banks' shareholders and bondholders are insulated from the risk of default, how can there be market discipline? What will replace that discipline? The G-20 has talked about the rapid return of "private capital," but what does this bode if private capital returns without market discipline? There was also talk of continuing to allow over-the-counter derivatives-trading with no transparency. But without transparency of each trade—to assess the nature of the counter-party risk—how can there be market discipline?

The G-20 did take long-overdue action on nontransparent offshore banking centers. The large amount of banking in these centers is not a result of these countries' comparative advantage in providing banking services. It is because they avoid and evade taxes and regulations. But these problems, while important, played little if any role in the current crisis. Why was so much effort spent on these extraneous issues rather than on those more directly related to the crisis?

From the perspective of the developing countries, though, not enough was done about bank secrecy in offshore as well as onshore centers. Developing nations are often criticized for corruption, but secret bank accounts wherever they may be facilitate corruption, providing safe haven for stolen funds. Developing countries want this money returned and want access to information that will allow them to detect secret accounts.

Financial and capital market liberalization—as well as banking deregulation—contributed to the crisis and to the spread of the crisis from the United States to developing countries. Advanced industrial nations are reluctant to admit that these policies, which they pushed so hard on developing countries, are part of the problem. No wonder, then, that the G-20 did not argue for a reconsideration of these longstanding policies.

The global economic crisis highlights the deficiencies of existing international institutions. As I noted, the IMF and the Financial Stability Forum—created in the aftermath of the last global financial crisis, in 1997–98—did not prevent the crisis. In some cases they pushed policies that are now recognized as root causes. Although some of the proposals are moves in the right direction, others (such as changing the name of the Financial Stability Forum to the

Financial Stability Board) are unlikely to have much effect, and as a package they are unlikely to suffice.

If we are to make our global economic system work better, we have to have better systems of global economic governance. It is important to move from ad hoc arrangements to more inclusive and representative institutional frameworks. We need a global economic coordinating council within the UN, not only to coordinate economic policies (e.g., the size of the stimulus and regulatory structures) but also to identify and rectify gaps in the global economic institutional structure. For instance, this crisis will almost surely be marked by some sovereign debt defaults. Despite extensive discussions at the time of Argentina's 2001 default, there was no progress in creating a sovereign debt-restructuring mechanism. The IMF—dominated by the creditor countries—cannot play a central role in designing such a mechanism (any more than we in the United States should turn to our banks to design a good bankruptcy law).

One of the alleged reasons for not "playing by the rules" and forcing troubled international banks to go through financial restructuring (instead, bailing them out) was that it would give rise to huge cross-border complications. Citibank, for example, operates worldwide, and depositors in many countries are not insured. What responsibility do U.S. taxpayers have to depositors abroad if Citibank fails? They didn't pay deposit insurance; there is no contract committing us to pick up the pieces. Yet some claim it would do irreparable harm to America's image if we took no responsibility. Iceland's banking problems illustrate the potential seriousness of these cross-border problems. Its citizens' standard of living may be impaired for decades because of the bankruptcies of its banks and the Icelandic government's decision to assume some responsibility for these failures. And yet, again, nothing is being done to address these problems.

Most important, the UN commission calls attention to the need for reform in the dollar-based global reserve system; it advocates the creation of a global reserve system. Not only is the current system fraying; it contributes to an insufficiency of global aggregate demand and to global instability. Every year developing countries set aside hundreds of billions of dollars to protect themselves against the costs of such instability, made so evident by the East Asia crisis. The commission has argued persuasively that this problem must be addressed if we are to have a robust global recovery. Recent statements from the BRIC nations (Brazil, Russia, India and China), expressing their concerns about the dollar reserve system, have added immediacy to the commission's recommendation. This is an old idea—Keynes argued strongly for the creation of a global reserve currency more than sixty years ago—whose time has come.

Those who would like us to go back to the world as it was before the crisis will find some of the questions being asked at the UN summit uncomfortable. They would be happier with a few harsh words for the offshore islands, a few cosmetic reforms to banking regulation, a few lectures about hedge funds (which, like offshore banking centers, were not at the center of this crisis), a new name and a couple of new members for the Financial Stability Forum—and then for us to move on. Many developing countries will be less content to accept these "reforms" as going to the heart of the matter.

As developed countries struggle to ensure a quick recovery, they need to think of the effects of their actions on developing countries. It is time to begin the restructuring of our global economic and financial system in ways that ensure that the fruits of prosperity are more widely shared and that the system is more stable. This task will not be accomplished overnight. But it is a task that must be begun, now.

REFLECTION QUESTIONS FOR CHAPTER 4

1. What is the evidence of the global economy penetrating your community? Who benefits from it? Who does not?

2. How does Wal-Mart affect your community? How does it affect workers everywhere?

3. Joseph Stiglitz argues that the world's economic system requires international cooperation and effective means to regulate economic activities. Both of these suggestions are opposed vigorously by certain interest groups in the United States. Who is opposed and why? Whose side are you on?

Chapter 5

Political Globalization

The world is divided into nation-states. Each of the states has a sovereign government, that is, it is the ultimate authority within its territory. Each state assumes responsibility for education, the economy, environmental protection, foreign policy, and military defense for its territory and people (Lechner and Boli, 2000:196-197)[1]. The latest wave of globalization, however, has altered the political patterns of nation-states. For example:

- The war on terrorism is not a war against a nation or nations but rather against religious and ethnic groups that have organized both within nations and across national boundaries.
- Nation-states alone are often incapable of combating terrorism, pollution, narcotics trafficking, and transnational crime networks.
- Global warming and environmental degradation, the products of activities within states, have transnational consequences.
- The world economy is beyond the control of nations as a result of the power of transnational corporations, the massive number and magnitude of transnational financial transactions, and interconnected markets. Hence, an economic downturn or natural disaster, or the devaluing of a currency or other event in one nation, has important ramifications for markets, jobs, and stability in many other countries.
- Transnational political entities deal with global problems. Among these entities are the World Court, the United Nations, and agencies associated with the UN such as the World Health Organization and the International Labor Organization. The World Trade Organization and the International Monetary Fund exist primarily to manage the world economy.
- Recently the nations of Europe have consolidated in a new political unit—the European Union—with a common currency and a parliament.

The primacy of the nation-state is not over, but globalization is changing its significance. In the first essay, Joseph E. Stiglitz argues that the integration of nation-states with the global economy works well when sovereign countries define the terms of that integration. It does not work when globalization is managed for nation-states by the International Monetary Fund and other international economic institutions.

Tina Rosenberg emphasizes the government's role in forcing small Mexican farmers off the land through free trade treaties, lowered import barriers, and subsidized agriculture. Her brief essay underscores the devastating effects of U.S. and Mexican policies for the 18 million Mexicans who live on small farms, some of who are eventually forced to move to already overcrowded cities.

The next two articles bring a political perspective to today's economic shifts. Harold James addresses the policy reversals in the United States and elsewhere that have resulted from the economic meltdown that began in 2007. To illustrate how financial globalization and politics intersect, he draws parallels with the political dynamics of the Great Depression, when big public sector action was needed to repair economic collapse. James argues that in today's globalized world, only strong states can effectively stabilize massive economic disruptions.

In the final reading, Nazli Choucri and Dinsha Mistree make important connections between international politics and migration. Although today's sovereign states are porous entities, these authors argue that the movement of people across territorial boundaries fundamentally challenges governmental policies. As these two readings show, twenty-first century globalization requires an active role for the nation-state.

ENDNOTE

1.   Lechner, Frank J., and John Boli, eds. *The Globalization Reader* (Malden, MA: Blackwell Publishers, 2000).

15

Globalism's Discontents

JOSEPH E. STIGLITZ

Stiglitz discusses the uneven governance of globalization and the disadvantages it has wrought for developing countries. He identifies both the benefits of globalization and its darker side to suggest ways of creating a more just global society.

Few subjects have polarized people throughout the world as much as globalization. Some see it as the way of the future, bringing unprecedented prosperity to everyone, everywhere. Others, symbolized by the Seattle protestors of December 1999, fault globalization as the source of untold problems, from the destruction of native cultures to increasing poverty and immiseration. In this article, I want to sort out the different meanings of globalization. In many countries, globalization has brought huge benefits to a few with few benefits to the many. But in the case of a few countries, it has brought enormous benefit to the many. Why have there been these huge differences in experiences? The answer is that globalization has meant different things in different places.

The countries that have managed globalization on their own, such as those in East Asia, have, by and large, ensured that they reaped huge benefits and that those benefits were equitably shared; they were able substantially to control the terms on which they engaged with the global economy. By contrast, the countries that have, by and large, had globalization managed for them by the International Monetary Fund and other international economic institutions have not done so well. The problem is thus not with globalization but with how it has been managed.

The international financial institutions have pushed a particular ideology—market fundamentalism—that is both bad economics and bad politics; it is based on premises concerning how markets work that do not hold even for developed countries, much less for developing countries. The IMF has pushed these economic policies without a broader vision of society or the role of economics within society. And it has pushed these policies in ways that have undermined emerging democracies.

More generally, globalization itself has been governed in ways that are undemocratic and have been disadvantageous to developing countries, especially the poor within those countries. The Seattle protestors pointed to the absence of

SOURCE: Reprinted with permission from Joseph E. Stiglitz, "Globalism's Discontents." *The American Prospect*: January 2001. Volume 13, Issue 1. http://www.prospect. org. *The American Prospect*, 1710 Rhode Island Avenue, NW, 12th Floor, Washington, DC 20036. All rights reserved.

democracy and of transparency, the governance of the international economic institutions by and for special corporate and financial interests, and the absence of countervailing democratic checks to ensure that these informal and *public* institutions serve a general interest. In these complaints, there is more than a grain of truth.

BENEFICIAL GLOBALIZATION

Of the countries of the world, those in East Asia have grown the fastest and done most to reduce poverty. And they have done so, emphatically, via "globalization." Their growth has been based on exports—by taking advantage of the global market for exports and by closing the technology gap. It was not just gaps in capital and other resources that separated the developed from the less-developed countries but differences in knowledge. East Asian countries took advantage of the "globalization of knowledge" to reduce these disparities. But while some of the countries in the region grew by opening themselves up to multinational companies, others, such as Korea and Taiwan, grew by creating their own enterprises. Here is the key distinction: each of the most successful globalizing countries determined its own pace of change; each made sure as it grew that the benefits were shared equitably; each rejected the basic tenets of the "Washington Consensus," which argued for a minimalist role for government and rapid privatization and liberalization.

In East Asia, government took an active role in managing the economy. The steel industry that the Korean government created was among the most efficient in the world—performing far better than its private-sector rivals in the United States (which, though private, are constantly turning to the government for protection and for subsidies). Financial markets were highly regulated. My research shows that those regulations promoted growth. It was only when these countries stripped away the regulations, under pressure from the U.S. Treasury and the IMF, that they encountered problems.

During the 1960s, 1970s, and 1980s, the East Asian economies not only grew rapidly but were remarkably stable. Two of the countries most touched by the 1997–1998 economic crisis had had in the preceding three decades not a single year of negative growth; two had only one year—a better performance than the United States or the other wealthy nations that make up the Organization for Economic Cooperation and Development (OECD). The single most important factor leading to the troubles that several of the East Asian countries encountered in the late 1990s—the East Asian crisis—was the rapid liberalization of financial and capital markets. In short, the countries of East Asia benefited from globalization because they made globalization work for them; it was when they succumbed to the pressures from the outside that they ran into problems that were beyond their own capacity to manage well.

Globalization can yield immense benefits. Elsewhere in the developing world, globalization of knowledge has brought improved health, with life spans increasing at a rapid pace. How can one put a price on these benefits of globalization? Globalization has brought still other benefits: today there is the beginning of a globalized civil society that has begun to succeed with such reforms as the Mine Ban Treaty and debt forgiveness for the poorest highly indebted countries (the Jubilee movement). The globalization protest movement itself would not have been possible without globalization.

THE DARKER SIDE OF GLOBALIZATION

How then could a trend with the power to have so many benefits have produced such opposition? Simply because it has not only failed to live up to its potential but frequently has had very adverse effects. But this forces us to ask, why has it had such adverse effects? The answer can be seen by looking at each of the economic elements of globalization as pursued by the international financial institutions and especially by the IMF.

The most adverse effects have arisen from the liberalization of financial and capital markets—which has posed risks to developing countries without commensurate rewards. The liberalization has left them prey to hot money pouring into the country, an influx that has fueled speculative real-estate booms; just as suddenly, as investor sentiment changes, the money is pulled out, leaving in its wake economic devastation. Early on, the IMF said that these countries were being rightly punished for pursuing bad economic policies. But as the crisis spread from country to country, even those that the IMF had given high marks found themselves ravaged.

The IMF often speaks about the importance of the discipline provided by capital markets. In doing so, it exhibits a certain paternalism, a new form of the old colonial mentality: "We in the establishment, we in the North who run our capital markets, know best. Do what we tell you to do, and you will prosper." The arrogance is offensive, but the objection is more than just to style. The position is highly undemocratic: there is an implied assumption that democracy by itself does not provide sufficient discipline. But if one is to have an external disciplinarian, one should choose a good disciplinarian who knows what is good for growth, who shares one's values. One doesn't want an arbitrary and capricious taskmaster who one moment praises you for your virtues and the next screams at you for being rotten to the core. But capital markets are just such a fickle taskmaster; even ardent advocates talk about their bouts of irrational exuberance followed by equally irrational pessimism.

LESSONS OF CRISIS

Nowhere was the fickleness more evident than in the last global financial crisis. Historically, most of the disturbances in capital flows into and out of a country are not the result of factors inside the country. Major disturbances arise, rather, from influences outside the country. When Argentina suddenly faced high interest rates in 1998, it wasn't because of what Argentina did but because of what happened in Russia. Argentina cannot be blamed for Russia's crisis.

Small developing countries find it virtually impossible to withstand this volatility. I have described capital-market liberalization with a simple metaphor: small countries are like small boats. Liberalizing capital markets is like setting them loose on a rough sea. Even if the boats are well captained, even if the boats are sound, they are likely to be hit broadside by a big wave and capsize. But the IMF pushed for the boats to set forth into the roughest parts of the sea before they were seaworthy, with untrained captains and crews, and without life vests. No wonder matters turned out so badly!

To see why it is important to choose a disciplinarian who shares one's values, consider a world in which there were free mobility of skilled labor. Skilled labor would then provide discipline. Today, a country that does not treat capital well will find capital quickly withdrawing; in a world of free labor mobility if a country did not treat skilled labor well, it too would withdraw. Workers would worry about the quality of their children's education and their family's health care, the quality of their environment and of their own wages and working conditions. They would say to the government: If you fail to provide these essentials, we will move elsewhere. That is a far cry from the kind of discipline that free-flowing capital provides.

The liberalization of capital markets has not brought growth. How can one build factories or create jobs with money that can come in and out of a country overnight? And it gets worse: prudential behavior requires countries to set aside reserves equal to the amount of short-term lending; so if a firm in a poor country borrows $100 million at, say, 20 percent interest rates short-term from a bank in the United States, the government must set aside a corresponding amount. The reserves are typically held in U.S. Treasury bills—a safe, liquid asset. In effect, the country is borrowing $100 million from the United States and lending $100 million to the United States. But when it borrows, it pays a high interest rate, around 4 percent. This may be great for the United States, but it can hardly help the growth of the poor country. There is also a high *opportunity* cost of the reserves; the money could have been much better spent on building rural roads or constructing schools or health clinics. But instead, the country is, in effect, forced to lend money to the United States.

Thailand illustrates the true ironies of such policies: there, the free market led to investments in empty office buildings, starving other sectors—such as education and transportation—of badly needed resources. Until the IMF and the U.S. Treasury came along, Thailand had restricted bank lending for speculative real estate. The Thais had seen the record: such lending is an essential part of the boom-bust cycle that has characterized capitalism for 200 years. It wanted to be sure that the scarce capital went to create jobs. But the IMF nixed this intervention in the free market. If the free market said, "Build empty office buildings," so be it! The market knew better than any government bureaucrat who mistakenly might have thought it wiser to build schools or factories.

THE COSTS OF VOLATILITY

Capital-market liberalization is inevitably accompanied by huge volatility, and this volatility impedes growth and increases poverty. It increases the risks of investing in the country, and thus investors demand a risk premium in the form of higher-than-normal profits. Not only is growth not enhanced but poverty is increased through several channels. The high volatility increases the likelihood of recessions—and the poor always bear the brunt of such downturns. Even in developed countries, safety nets are weak or nonexistent among the self-employed and in the rural sector. But these are the dominant sectors in developing countries. Without adequate safety nets, the recessions that follow from capital-market liberalization lead to impoverishment. In the name of imposing budget discipline and reassuring investors, the IMF invariably demands expenditure reductions, which almost inevitably result in cuts in outlays for safety nets that are already threadbare.

But matters are even worse—for under the doctrines of the "discipline of the capital markets," if countries try to tax capital, capital flees. Thus, the IMF doctrines inevitably lead to an increase in tax burdens on the poor and the middle classes. Thus, while IMF bailouts enable the rich to take their money out of the country at more favorable terms (at the overvalued exchange rates), the burden of repaying the loans lies with the workers who remain behind.

The reason that I emphasize capital-market liberalization is that the case against it—and against the IMF's stance in pushing it—is so compelling. It illustrates what can go wrong with globalization. Even economists like Jagdish Bhagwati, strong advocates of free trade, see the folly in liberalizing capital markets. Belatedly, so too has the IMF—at least in its official rhetoric, though less so in its policy stances—but too late for all those countries that have suffered so much from following the IMF's prescriptions.

But while the case for trade liberalization—when properly done—is quite compelling, the way it has been pushed by the IMF has been far more problematic. The basic logic is simple: trade liberalization is supposed to result in resources moving from inefficient protected sectors to more efficient export sectors. The problem is not only that job destruction comes before the job creation—so that unemployment and poverty result—but that the IMF's "structural adjustment programs" (designed in ways that allegedly would reassure global investors) make job creation almost impossible. For these programs are often accompanied by high interest rates that are often justified by a single-minded focus on inflation. Sometimes that concern is deserved; often, though, it is carried to an extreme. In the United States, we worry that small increases in the interest rate will discourage investment. The IMF has pushed for far higher interest rates in countries with a far less hospitable investment environment. The high interest rates mean that new jobs and enterprises are not created. What happens is that trade liberalization, rather than moving workers from low-productivity jobs to high-productivity ones, moves them from low-productivity jobs to unemployment. Rather than enhanced growth, the effect is increased poverty. To make matters even worse, the unfair trade-liberalization agenda forces poor countries to compete with highly subsidized American and European agriculture.

THE GOVERNANCE OF GLOBALIZATION

As the market economy has matured within countries, there has been increasing recognition of the importance of having rules to govern it. One hundred fifty years ago, in many parts of the world, there was a domestic process that was in some ways analogous to globalization. In the United States, government promoted the formation of the national economy, the building of the railroads, and the development of the telegraph—all of which reduced transportation and communication costs within the United States. As that process occurred, the democratically elected national government provided oversight: supervising and regulating, balancing interests, tempering crisis, and limiting adverse consequences of this very large change in economic structure. So, for instance, in 1863 the U.S. government established the first financial-banking regulatory authority—the Officer of the Comptroller of Currency—because it was important to have strong national banks, and that requires strong regulation.

The United States, among the least statist of the industrial democracies, adopted other policies. Agriculture, the central industry of the United States in the mid-nineteenth century, was supported by the 1862 Morrill Act, which established research, extension, and teaching programs. That system worked extremely well and is widely credited with playing a central role in the enormous increases in agricultural productivity over the last century and a half. We established an industrial policy for other fledgling industries, including radio and civil aviation. The beginning of the telecommunications industry, with the first telegraph line between Baltimore and Washington, D.C., was funded by the federal government. And it is a tradition that has continued, with the U.S. government's founding of the Internet.

By contrast, in the current process of globalization we have a system of what I call global governance without global government. International institutions like the World Trade Organization, the IMF, the World Bank, and others provide an ad hoc system of global governance, but it is a far cry from global government and lacks democratic accountability. Although it is perhaps better than not having any system of global governance, the system is structured not to serve general interests or assure equitable results. This not only raises issues of whether broader values are given short shrift; it does not even promote growth as much as an alternative might.

GOVERNANCE THROUGH IDEOLOGY

Consider the contrast between how economic decisions are made inside the United States and how they are made in the international economic institutions. In this country, economic decisions within the administration are undertaken largely by the National Economic Council, which includes the secretary of labor, the secretary of commerce, the chair of the Council of Economic Advisers, the treasury secretary, the assistant attorney general for antitrust, and the

U.S. trade representative. The Treasury is only one vote and often gets voted down. All of these officials, of course, are part of an administration that must face Congress and the democratic electorate. But in the international arena, only the voices of the financial community are heard. The IMF reports to the ministers of finance and the governors of the central banks, and one of the important items on its agenda is to make these central banks more independent— and less democratically accountable. It might make little difference if the IMF dealt only with matters of concern to the financial community, such as the clearance of checks; but in fact, its policies affect every aspect of life. It forces countries to have tight monetary and fiscal policies: it evaluates the trade-off between inflation and unemployment, and in that trade-off it always puts far more weight on inflation than on jobs.

The problem with having the rules of the game dictated by the IMF—and thus by the financial community—is not just a question of values (though that is important) but also a question of ideology. The financial community's view of the world predominates—even when there is little evidence in its support. Indeed, beliefs on key issues are held so strongly that theoretical and empirical support of the positions is viewed as hardly necessary.

Recall again the IMF's position on liberalizing capital markets. As noted, the IMF pushed a set of policies that exposed countries to serious risk. One might have thought, given the evidence of the costs, that the IMF could offer plenty of evidence that the policies also did some good. In fact, there was no such evidence; the evidence that was available suggested that there was little if any positive effect on growth. Ideology enabled IMF officials not only to ignore the absence of benefits but also to overlook the evidence of the huge costs imposed on countries.

AN UNFAIR TRADE AGENDA

The trade-liberalization agenda has been set by the North, or more accurately, by special interests in the North. Consequently, a disproportionate part of the gains has accrued to the advanced industrial countries, and in some cases the less-developed countries have actually been worse off. After the last round of trade negotiations, the Uruguay Round that ended in 1994, the World Bank calculated the gains and losses to each of the regions of the world. The United States and Europe gained enormously. But sub-Saharan Africa, the poorest region of the world, lost by about 2 percent because of terms-of-trade effects: the trade negotiations opened their markets to manufactured goods produced by the industrialized countries but did not open up the markets of Europe and the United States to the agricultural goods in which poor countries often have a comparative advantage. Nor did the trade agreements eliminate the subsidies to agriculture that make it so hard for the developing countries to compete.

The U.S. negotiations with China over its membership in the WTO displayed a double standard bordering on the surreal. The U.S. trade representative, the chief negotiator for the United States, began by insisting that China was a developed country. Under WTO rules, developing countries are allowed longer transition periods in which state subsidies and other departures from the WTO strictures are permitted. China certainly wishes it were a developed country, with Western-style per capita incomes. And since China has a lot of "capitals," it's possible to multiply a huge number of people by very small average incomes and conclude that the People's Republic is a big economy.... But China is not only a developing economy; it is a low-income developing country. Yet the United States insisted that China be treated like a developed country! China went along with the fiction; the negotiations dragged on so long that China got some extra time to adjust. But the true hypocrisy was shown when U.S. negotiators asked, in effect, for developing-country status for the United States to get extra time to shelter the American textile industry.

Trade negotiations in the service industries also illustrate the unlevel nature of the playing field. Which service industries did the United States say were *very* important? Financial services—industries in which Wall Street has a comparative advantage. Construction industries and maritime services were not on the agenda, because the developing countries would have a comparative advantage in these sectors.

Consider also intellectual-property rights, which are important if innovators are to have incentives to innovate (though many of the corporate advocates of intellectual property exaggerate its importance and fail to note that much of the most important research, as in basic science and mathematics, is not patentable). Intellectual-property rights, such as patents and trademarks, need to balance the interests of producers with those of users—not only users in developing countries, but researchers in developed countries. If we underprice the profitability of innovation to the inventor, we deter invention. If we overprice its cost to the research community and the end user, we retard its diffusion and beneficial effects on living standards.

In the final stages of the Uruguay negotiations, both the White House Office of Science and Technology Policy and the Council of Economic Advisers worried that we had not got the balance right—that the agreement put producers' interests over users'. We worried that, with this imbalance, the rate of progress and innovation might actually be impeded. After all, knowledge is the most important input into research, and overly strong intellectual-property rights can, in effect, increase the price of this input. We were also concerned about the consequences of denying lifesaving medicines to the poor. This issue subsequently gained international attention in the context of the provision of AIDS medicines in South Africa.... The international outrage forced the drug companies to back down—and it appears that, going forward, the most adverse consequences will be circumscribed. But it is worth noting that initially, even the Democratic U.S. administration supported the pharmaceutical companies.

What we were not fully aware of was another danger—what has come to be called "biopiracy," which involves international drug companies patenting traditional medicines. Not only do they seek to make money from "resources" and knowledge that rightfully belong to the developing countries, but in doing so they squelch domestic firms who long provided these traditional medicines. While it is not clear whether these patents would hold up in court if they were effectively challenged, it is clear that the less-developed countries may not have the legal and financial resources required to mount such a challenge. The issue has become the source of enormous emotional, and potentially economic, concern throughout the developing world. This fall, while I was in Ecuador visiting a village in the high Andes, the Indian mayor railed against how globalization has led to biopiracy.

GLOBALIZATION AND SEPTEMBER 11

September 11 brought home a still darker side of globalization—it provided a global arena for terrorists. But the ensuing events and discussions highlighted broader aspects of the globalization debate. They made clear how untenable American unilateralist positions were. President Bush, who had unilaterally rejected the international agreement to address one of the long-term global risks perceived by countries around the world—global warming, in which the United States is the largest culprit—called for a global alliance against terrorism. The administration realized that success would require concerted action by all.

One of the ways to fight terrorists, Washington soon discovered, was to cut off their sources of funding. Ever since the East Asian crisis, global attention had focused on the secretive offshore banking centers. Discussions following that crisis focused on the importance of good information—transparency, or openness—but this was intended for the developing countries. As international discussions turned to the lack of transparency shown by the IMF and the offshore banking centers, the U.S. Treasury changed its tune. It is not because these secretive banking havens provide better services than those provided by banks in New York or London that billions have been put there; the secrecy serves a variety of nefarious purposes—including avoiding taxation and money laundering. These institutions could be shut down overnight—or forced to comply with international norms—if the United States and the other leading countries wanted. They continue to exist because they serve the interests of the financial community and the wealthy. Their continuing existence is no accident. Indeed, the OECD drafted an agreement to limit their scope—and before September 11, the Bush administration unilaterally walked away from this agreement too. How foolish this looks now in retrospect! Had it been embraced, we would have been further along the road to controlling the flow of money into the hands of the terrorists.

There is one more aspect to the aftermath of September 11 worth noting here. The United States was already in recession, but the attack made matters worse. It used to be said that when the United States sneezed, Mexico caught a cold. With globalization, when the United States sneezes, much of the rest of the world risks catching pneumonia. And the United States now has a bad case of the flu. With globalization, mismanaged macroeconomic policy in the United

States—the failure to design an effective stimulus package—has global consequences. But around the world, anger at the traditional IMF policies is growing.
The developing countries are saying to the industrialized nations: "When you
face a slowdown, you follow the precepts that we are all taught in our economic
courses: you adopt expansionary monetary and fiscal policies. But when we face
a slowdown, you insist on contractionary policies. For you, deficits are okay; for
us, they are impermissible—even if we can raise the funds through 'selling forward,' say, some natural resources." A heightened sense of inequity prevails,
partly because the consequences of maintaining contractionary policies are so
great.

GLOBAL SOCIAL JUSTICE

Today, in much of the developing world, globalization is being questioned. For
instance, in Latin America, after a short burst of growth in the early 1990s, stagnation and recession have set in. The growth was not sustained—some might
say, was not sustainable. Indeed, at this juncture, the growth record of the so-
called post-reform era looks no better, and in some countries much worse, than
in the widely criticized import-substitution period of the 1950s and 1960s when
Latin countries tried to industrialize by discouraging imports. Indeed, reform
critics point out that the burst of growth in the early 1990s was little more
than a "catch-up" that did not even make up for the lost decade of the 1980s.

Throughout the region, people are asking: "Has reform failed or has globalization failed?" The distinction is perhaps artificial, for globalization was at the
center of the reforms. Even in those countries that have managed to grow,
such as Mexico, the benefits have accrued largely to the upper 30 percent and
have been even concentrated in the top 10 percent. Those at the bottom have
gained little; many are even worse off. The reforms have exposed countries to
greater risk, and the risks have been borne disproportionately by those least able
to cope with them. Just as in many countries where the pacing and sequencing
of reforms has resulted in job destruction outmatching job creation, so too has
the exposure to risk outmatched the ability to create institutions for coping with
risk, including effective safety nets.

In this bleak landscape, there are some positive signs. Those in the North
have become more aware of the inequalities of the global economic architecture.
The agreement at Doha to hold a new round of trade negotiations—the "Development Round"—promises to rectify some of the imbalances of the past. There
has been a marked change in the rhetoric of the international economic institutions—at least they talk about poverty. At the World Bank, there have been
some real reforms; there has been some progress in translating the rhetoric into
reality—in ensuring that the voices of the poor are heard and the concerns of the
developing countries are listened to. But elsewhere, there is often a gap between
the rhetoric and the reality. Serious reforms in governance, in who makes decisions and how they are made, are not on the table. If one of the problems at the

IMF has been that the ideology, interests, and perspectives of the financial community in the advanced industrialized countries have been given disproportionate weight (in matters whose effects go well beyond finance), then the prospects for success in the current discussions of reform, in which the same parties continue to predominate, are bleak. They are more likely to result in slight changes in the shape of the table, not changes in who is *at* the table or what is on the agenda.

September 11 has resulted in a global alliance against terrorism. What we now need is not just an alliance *against* evil, but an alliance *for* something positive—a global alliance for reducing poverty and for creating a better environment, an alliance for creating a global society with more social justice.

16

Why Mexico's Small Corn Farmers
Go Hungry

TINA ROSENBERG

Rosenberg shows how U.S. and Mexican policies on trade, imports, and agriculture are crushing small farming in Mexico.

Macario Hernandez's grandfather grew corn in the hills of Puebla, Mexico. His father does the same. Mr. Hernandez grows corn, too, but not for much longer. Around his village of Guadalupe Victoria, people farm the way they have for centuries, on tiny plots of land watered only by rain, their plows pulled by burros. Mr. Hernandez, a thoughtful man of 30, is battling to bring his family and neighbors out of the Middle Ages. But these days modernity is less his goal than his enemy.

This is because he, like other small farmers in Mexico, competes with American products raised on megafarms that use satellite imagery to mete out fertilizer. These products are so heavily subsidized by the government that many are exported for less than it costs to grow them. According to the Institute for Agriculture and Trade Policy in Minneapolis, American corn sells in Mexico for 25 percent less than its cost. The prices Mr. Hernandez and others receive are so low that they lose money with each acre they plant.

In January [2003], campesinos from all over the country marched into Mexico City's central plaza to protest. Thousands of men in jeans and straw hats jammed the Zocalo, alongside horses and tractors. Farmers have staged smaller protests around Mexico for months. The protests have won campesino organizations a series of talks with the government. But they are unlikely to get what they want: a renegotiation of the North American Free Trade Agreement, or NAFTA, protective temporary tariffs and a new policy that seeks to help small farmers instead of trying to force them off the land.

The problems of rural Mexicans are echoed around the world as countries lower their import barriers, required by free trade treaties and the rules of the World Trade Organization. When markets are open, agricultural products flood in from wealthy nations, which subsidize agriculture and allow agribusiness to export crops cheaply. European farmers get 35 percent of their income in

SOURCE: From Tina Rosenberg, "Why Mexico's Small Corn Farmers Go Hungry," *New York Times* (March 3, 2003). Used with permission.

government subsidies, American farmers 20 percent. American subsidies are at record levels, and last year, Washington passed a farm bill that included a $40 billion increase in subsidies to large grain and cotton farmers.

It seems paradoxical to argue that cheap food hurts poor people. But three-quarters of the world's poor are rural. When subsidized imports undercut their products, they starve. Agricultural subsidies, which rob developing countries of the ability to export crops, have become the most important dispute at the W.T.O. Wealthy countries do far more harm to poor nations with these subsidies than they do good with foreign aid.

While such subsidies have been deadly for the 18 million Mexicans who live on small farms—nearly a fifth of the country—Mexico's near-complete neglect of the countryside is at fault, too. Mexican officials say openly that they long ago concluded that small agriculture was inefficient, and that the solution for farmers was to find other work. "The government's solution for the problems of the countryside is to get campesinos to stop being campesinos," says Victor Suarez, a leader of a coalition of small farmers.

But the government's determination not to invest in losers is a self-fulfilling prophecy. The small farmers I met in their fields in Puebla want to stop growing corn and move into fruit or organic vegetables. Two years ago Mr. Hernandez, who works with a farming cooperative, brought in thousands of peach plants. But only a few farmers could buy them. Farm credit essentially does not exist in Mexico, as the government closed the rural bank, and other bankers do not want to lend to small farmers. "We are trying to get people to rethink and understand that the traditional doesn't work," says Mr. Hernandez. "But the lack of capital is deadly."

The government does subsidize producers, at absurdly small levels compared with subsidies in the United States. Corn growers get about $30 an acre. Small programs exist to provide technical help and fertilizer to small producers, but most farmers I met hadn't even heard of them.

Mexico should be helping its corn farmers increase their productivity or move into new crops—especially since few new jobs have been created that could absorb these farmers. Mexicans fleeing the countryside are flocking to Houston and swelling Mexico's cities, already congested with the poor and unemployed. If Washington wants to reduce Mexico's immigration to the United States, ending subsidies for agribusiness would be far more effective than beefing up the border patrol.

17

The Late, Great Globalization

HAROLD JAMES

James tackles the implications of the current economic downturn for the world's political order. He illustrates parallels between this economic crisis and the political dynamics associated with the Great Depression, and sketches a future in which state capitalism will be a powerful force.

At the moment we are suffering the effects of a financial meltdown unparalleled since the Great Depression. The collapse will have long-lasting impacts on markets, corporations, and states. Indeed, as a consequence of the credit collapse and the shape of the responses taken, we are looking at the end of the era of financially driven globalization. This will have profound implications not only for the international economic system but also for the world's political order.

Economic historians often like to note that some knowledge of history would have been helpful in avoiding or mitigating a financial upheaval. That also goes for the discussion of solutions. In working out where the problems lie, and what solutions may be envisaged, a longer-term perspective is helpful. Crises, history suggests, are frequently an opportunity for making dramatic policy shifts. The urgency of the situation means that debates and discussions get cut short, and we often settle for stopgap measures. In practice, however, such measures create a momentum that leads to permanent policy shifts. If the temporary measure is flawed, the long-term policy aftermath will be even more harmful.

The extent of the current crisis, together with the fear of further losses, has already prompted dramatic policy reversals in the United States and elsewhere. In September 2008, the U.S. Treasury made clear its opposition to bailouts of financial institutions. Its willingness to let the investment bank Lehman Brothers fail was intended as a signal that the financial system was fundamentally robust and could stand a quite severe shock. Immediately after the failure of Lehman, the defenders of classical liberal positions in newspapers such as the *Financial Times* and the *Neue Zürcher Zeitung* argued that the stance in favor of liquidation was compelling. But then it became clear that the collapse of a major financial institution set off a train of unanticipated events.

A rescue of a giant insurer, AIG, was needed; then the establishment of a massive $700 billion program to take bad assets off the balance sheets of banks.

SOURCE: From Harold James, "The Late, Great Globalization." *Current History* 108 (January, 2009). Reprinted with permission, Current History, Inc.

When this proved also to be inadequate, the crisis reverberated around the world, first in Europe and then, with even more drama and brutality, in the emerging markets. Then the U.S. Treasury abandoned as too complex the strategy of taking bad assets off balance sheets, and instead simply used public funds to recapitalize banks and directly to support badly affected groups of the population. More recently, the Treasury reversed course again and decided to buy up some bad assets after all—to bail out Citigroup.

It is both the institutional collapses and the widening responses to them that have brought an end to the era of financially driven globalization. In this era, which lasted from the 1970s to the 2000s, expectations of earnings and profitability largely drove corporate behavior. Financial markets rested on short-term considerations, to the exclusion of long-term perspectives and consideration of systemic risks. Maybe the markets were simply shaped by the experiences of the market makers and market players. The working lives of financial service providers had not included experiences of profound breakdown and systemic reordering.

In any case, the politics as well as the finance of globalization has now been profoundly altered. U.S. policy makers are treating a new state capitalism as the emerging global reality. In continental Europe, politicians find it attractive to suggest that capitalism (which they usually identify with a so-called "Anglo-Saxon" model of financially driven globalization) is discredited. French President Nicolas Sarkozy has suggested "the return of politics" and a much more active role for the state in nationalizing strategic industries to prevent them from falling under foreign control. He recently stood leafing through the pages of *Das Kapital* for the benefit of a photographer. The German finance minister, Peer Steinbrück, has suggested that Karl Marx was "not all that incorrect" in his analysis of the periodically crisis-prone character of capitalism. A German filmmaker, Alexander Kluge, is even proposing to make a movie of *Das Kapital*.

DÉJÀ VU

Over the past decades, many have recited a reassuring mantra that globalization—the mutually dependent movements around the world of goods, people, and capital—is irreversible. This view is historically unsustainable. There is nothing inexorable about globalization, just as there is nothing new about globalization. One of the comfort blankets that modern people sometimes clutch is the idea that only one big simultaneous world depression ever occurred, and it was produced by such an odd confluence of causes as to be quite unique. The legacy of the First World War and of the financial settlement of reparations and war debt; the chaotic banking system of the largest economy in the world, that of the United States; and inexperience in handling monetary policy in a world that was still pining for metallic money all conspired to create the Great Depression, and since these circumstances were unique, they cannot occur again. Historians should point out that this reasoning may be quite wrong.

Describing the very dynamic global trade of the second half of the nineteenth century and the early twentieth century has become a standard part of the repertoire of economic historians. But that era was not the only episode of globalization before the latest one. Archaeological evidence indicates the global reach of trade during the Roman Empire, with Roman coins being traded in coastal regions of Sri Lanka and Vietnam. And there were subsequent expansions of global trade and finance. In many of them, ideas from classical antiquity and from the Roman age of globalization (and global rule) were revived, as in the economic rebound of the late fifteenth and early sixteenth centuries (the economic backdrop to the Renaissance), or the eighteenth century, when improved technology and increased ease of communications opened the way to global empires (for Britain and France).

All of these previous globalization episodes came to an end, almost always with wars that were accompanied by highly disruptive and contagious financial crises. Globalization is often supposed to have produced a universalization of peace, since only in a peaceful world can trade and an interchange of ideas really flourish. But in practice, globalization of goods, capital, and people often leads to a globalization of violence.

Moments of backlash also have generally produced moral indignation—frequently inspired by religion—about the corruption of long-distance trade. The widening world of the Renaissance inspired such a response from the Dominican friar Girolamo Savonarola, whose followers staged "bonfires of the vanities" in which they consigned to flame the excrescences of the Florentine luxury trade. Martin Luther replicated Savonarola's outrage north of the Alps, and wrote tracts against long-distance trade in which he started with Saint Paul's comment that "greed is a root of all evils." Luther complained that Germans were sending their gold and silver to foreign countries, enriching these but creating poverty at home. The American Revolution reproduced the same condemnations, with Puritan divines inveighing against the ills of foreign commerce and the predations of the East India Company.

VULNERABILITY AND CRISIS

Global integration, global violence, and a crisis of values are—from a historical perspective—closely enmeshed with each other. We should not think them strange when we encounter them in our own age of globalization. Even before the financial meltdown, advocates for globalization in the business and political world of the advanced industrial countries had become deeply worried, as globalization has become a concept with overwhelmingly negative connotations. In the industrial countries, it seems to be responsible for job losses and pay reductions. It also is held responsible for apparently illegitimate rewards for the owners of scarce resources, in particular superstars with a reputation, whether sports or entertainment stars, or CEOs who market themselves like superstars.

Whereas until recently the most dramatic effects of globalization were seen in the market for unskilled labor, and thus most policy thinkers simply saw better training as an answer, it has become clear that skilled service jobs (most conspicuously in computer software but also in medical and legal analysis) can also be "outsourced." Consequently, the gigantic Western middle class—the great winner of the twentieth century—is now extremely alarmed by the prospect that it might be overtaken by an even larger (and harder working) middle class in emerging market countries at the same time that it is separated by a great gulf of wealth from a new global elite.

The "middle classes" and their fears were at the center of the political campaigns of both U.S. presidential candidates in 2008. Sarkozy too has spoken of a "revolt of the popular and middle classes who reject a globalization that they consider not a promise but a threat." The political backlash drives an intense populist concern in the rich industrial countries about corporate governance, corporate abuses, and the excesses of executive pay.

The new backlash naturally terrifies business leaders, who want to devise some appropriate response that will not hurt them too much. Events such as the annual World Economic Forum in Davos, Switzerland, formerly parodied as a fiesta for pro-globalization fanatics, are now packed with presentations by globalization critics and choruses about corporate social responsibility. Even the language of religion has made a comeback. The founder and patron of the World Economic Forum, Klaus Schwab, began the 2008 meeting by saying that it was "time to pay for our sins."

Indeed, it is hard to find defenders of classical rule-bound liberalism at events such as Davos. The readiness with which global captains of business embrace their opponents reminds me rather of the way in which the Florentine ruling and banking house of Medici sponsored Savonarola, the most vociferous and radical critic of commercial culture; or of the decision by Henry Ford, at the depth of the Great Depression and the height of labor discontent, to commission the Mexican artist Diego Rivera to paint Marxist-inspired murals at the Detroit Institute of Arts.

Two phenomena today are overlapping: one is a crisis in the financial system that drove global integration over the past four decades; the other is a worry about the character of globalization.

THE WORLD TRANSFORMED, AGAIN

The recent episode of globalization was driven by a dramatic expansion of financial flows. Although the pre-1914 era witnessed very substantial capital flows, and the developing countries of the time were able to sustain larger levels of borrowing than they do today, the modern era is characterized by a much larger volume of transactions. The daily turnover in foreign exchange markets rose from $200 billion in the mid-1980s to $800 billion in the mid-1990s and to almost $4 trillion today. Financial assets now amount to four times the annual value of world output.

Like previous globalization eras, the latest one was a time of accelerated innovation. The results of globalization produced a great deal of relatively widely dispersed wealth and well-being, with the conquest of global poverty looking more feasible at the beginning of the twenty-first century than at any previous moment. But there may have been too much innovation to digest.

Previous globalizations were also characterized by the special dynamism of one sector of the economy. Eighteenth-century expansion was driven by big productivity increases in agriculture, which led to increased purchasing power and made possible a consumer revolution during the so-called "first Industrial Revolution." The second half of the nineteenth century saw a transformation of the world because the expansion of manufacturing, especially in iron and steel, made possible a transportation revolution with the steam engine and the iron-hulled steam ship.

Economic historians consequently identify two big shifts in activity and employment over the past two hundred years: the shift from agriculture to manufacturing in the nineteenth century, and a move away from the "old" manufacturing into services in the twentieth century. But in each case, a dramatic crisis created the conditions for change. The 1840s witnessed the last big hunger crisis in traditional rural Europe, with crop failures over most of the continent leading to famine and starvation. The impoverishment of agriculture was sufficient to produce a major downturn for everyone else. When the crisis was over, many farmers moved to other activities, and a major boom in the core businesses of the first Industrial Revolution began.

The Great Depression of the 1930s had a similar transformative effect on economic structures. In the short run the Depression looked devastating. But widespread industrial unemployment pushed workers into new occupations and underlined the importance of the skills and education that proved vital for the development of the service economy of the late twentieth century.

In each instance of economic and employment transition, older types of business did not stop. They were just carried out more efficiently. Though it has a tiny farm population, the United States remains a major agricultural producer and exporter. Despite all the discussion of deindustrialization and America's loss of relative competitiveness since the 1970s and 1980s, the United States is also still (just) the world's largest manufacturer.

The twentieth century movement into services can be broken into two phases. First, and in large part as a response to the Great Depression, as well as to the Second World War, all industrial countries saw a massive growth in government employment. The enthusiasm for state-run economies reached a peak in the 1970s, but soon there was a widespread realization that the public sector simply could not go on growing. Electoral pressures produced a new emphasis on efficiency in the provision of public services. Privatization became the obvious source of efficiency, and required a new approach to the financing of public goods. Secondly, then, a new sort of economic growth set in. From the 1970s, the breaking down of national frontiers helped promote financial globalization and with it an explosive growth in the demand for financial services.

The credit crunch of 2007–2008 is a turning point because it has revealed some serious weaknesses in the growth sector of the past decades. It was precisely in the world's most financially sophisticated countries—the United States, the United Kingdom, France, and Switzerland—that the flaws in bank supervision were most apparent. The flaws were not just of external supervision. Internal risk management in financial institutions was frequently inadequate. Italy and Spain, on the other hand, which had dynamic banking systems but were still widely regarded as fundamentally sleepy, had much better and tighter supervision.

How will adaptation to the new environment occur? A conventional response to financial disorder is a demand for more regulation. There are analogies in the previous waves of economic development. Faced by agricultural distress, well-intentioned people called for a greater regulation of agricultural prices. In the industrial chaos of the 1930s, forced amalgamations, cartels, and state supervision seemed like a good answer. But the better answer has always been technical change. Raising productivity in agriculture or industry through the use of new techniques and new equipment meant that these activities could become much more productive, while employing fewer people and generating less macroeconomic vulnerability.

THE INNOVATION OPTION

Innovation is likely to be the long-run answer to banking problems as well. Regulatory and supervisory functions will be handled—as they have been in the wake of the credit crunch—by an intensified testing of responses to hypothetical situations (the failure of a big counter-party, a geopolitical upheaval, and so forth). Conflict-of-interest situations can be handled by algorithms that authorize or prohibit transactions. In the same way as trading floors are now mostly obsolete, most banking functions can and will be handled simply by machines. Financial intermediation will become simply an interaction of software systems.

As in previous transition phases, individuals who work in the industry will try to produce convincing arguments that their business depends on the human touch. Harvesting machines were supposed to lower the quality of the grain collected, since it was no longer subject to immediate inspection by the human eye. British train drivers at one time insisted that two people were needed to ensure transportation safety. It took a long time to convince the public that one-man trains were safe. Today some mass transit systems work with driverless trains. Advanced automobile plants are now largely collections of robots. A personal banker is a status symbol, but nothing more, and certainly not a necessity.

One of the achievements of the recent application of psychology to economics has been the demonstration of how irrational many human decisions are. The story of the contemporary credit crunch has also, surprisingly often, been a tale familiar from much older financial crises, that of the flawed individual or the rogue trader. The application of electronics would eliminate much of the potential for the errors and flaws of humans.

But which of these outcomes is most likely in the short and medium term—the state-centered story of increased regulation, or a benign story of continued technical change? Past globalization episodes indicate that change perceived as too fast or too incomprehensible often provokes revulsion. In such cases, very rapid innovations are followed by regulatory clampdown—as during the economic crisis of the 1930s. In this sense, we are likely to see more and more parallels with the political dynamics of the world of the Great Depression.

BACK TO POWER POLITICS

The rapid reversal of financial flows has profoundly destabilizing consequences for the political systems that try to produce a response. Two surprising conclusions are emerging in today's policy discussions, but only one has been fully digested. First, big public sector action is needed. Second, such action is complicated because in a globalized world the need for assistance spans borders.

Private sector solutions have been tried but have failed in a breathtakingly short space of time. The most frequent consolation for such failures is that a really bad crisis is purgative. Insolvent businesses close, bad loans are written off, and lenders can lend with new confidence. Henry Paulson, who came to America's Treasury from the strongest U.S. investment bank, Goldman Sachs, made the purgation gamble in allowing Lehman to go under. He argued that the United States could not tolerate a bailout culture. A firm denial by the government should be seen as a sign that most of the American economy is fundamentally sound, and that U.S. financial markets are sophisticated enough to be able to identify sound business practices.

The Treasury secretary in the Great Depression was also a titan of finance, Andrew Mellon. Mellon's immediate conclusion in the face of the 1929 stock market panic has subsequently become notorious: "liquidate labor, liquidate stocks, liquidate the farmers, liquidate real estate … purge the rottenness out of the system." It very quickly became clear that the high-risk bet of 2008 did not pay off, any more than it did in 1929.

In a financial system that suffers a radical loss of confidence, only institutions with more or less infinite resources can stem the tide. Such institutions can conceivably be self-help organizations, such as pools of powerful banks. That was the case in 1907. The U.S. Treasury indeed tried to put together such a pool on Sunday, September 14, 2008. However, in a climate of profound uncertainty, self-help is not enough. Governments or central banks are needed, because only they are both big and quick enough. Only they could quickly come to the assistance of the giant U.S. housing finance institutions Fannie Mae and Freddie Mac, and then deal with AIG.

Now the call for a stronger state is a global phenomenon. A recent study by the U.S. National Intelligence Council on what the world will look like in 2025 sketches a future in which state capitalism of the kind practiced in China or particularly Russia will be a powerful force to be reckoned with.

The question of state involvement having been settled, a second question is what kind of government can do the job of putting the pieces back together. Not just any government will do. In the heyday of modern globalization in the 1990s, it looked as if small open states would be the winners of globalization: New Zealand, Chile, Ireland, the Baltic Republics, Slovakia, or Slovenia. But already by the 2000s, a new vision had emerged in which the main beneficiaries would be very large emerging markets with quite powerful states, the group that Goldman Sachs dubbed the BRICs (Brazil, Russia, India, and China).

In a world in which there is a new preference for power, even states of moderate size—such as the traditional big European states or Japan—are not big enough to act effectively on their own. Their helplessness was already visible in recent debates over European energy policy. Britain, France, or Germany (let alone the much smaller central European countries) cannot tackle issues such as the politics of pipelines crossing Russian territory without a collective negotiating stance.

The absence of an effective Europe-wide solution has been even more apparent in the responses to the 2008 financial crisis. Mid-sized European governments possibly can rescue mid-sized European institutions. But in the case of really major financial conglomerates at the heart of the world's financial system, there are probably only two governments that have the firepower: the United States and China.

In the similar circumstances of a financial meltdown in 1931, there were also only a limited number of governments that could be effective. The old economic superpower, Britain, was too exhausted and strained to help anyone else. The world's reserves were massively accumulated in the United States. Thus, the only plausible case for a way out of the worldwide depression in 1931 lay, as the great economic historian Charles Kindleberger emphasized, with some step taken by the United States. At the time, there were all kinds of convincing reasons that Americans should not want to take on the burden of a worldwide rescue: sending more money to Europe might be seen as pouring money down a drain; had not the Europeans fought a World War that had been the fount of the financial mess? Economically, such action would have made a great deal of sense from a long-term perspective. Politically, with no short-term payoff at all, it was a non-starter.

CHINA'S CHOICE

China is the America of this century. Indeed, the initial stages of the credit crunch in 2007 were managed with such apparent painlessness because sovereign wealth funds from the Middle East, but above all from China, were willing to step in and recapitalize the debt of American and European institutions. The pivotal moment in today's events came when a Chinese sovereign wealth fund, China Investment Corporation (CIC), was unwilling to go further in its exploration of buying Lehman Brothers. CIC's turning back will be seen in the future as a moment when history could have turned in a different direction.

Today there will be plenty of reasons why the Chinese might want to pull back from a global commitment. Some of the arguments reverberating around

Beijing are very reasonable: there is a great deal of uncertainty, and the sovereign wealth funds might lose a lot of money. CIC would have initially lost some money with Lehman. Some lines of thought are more emotional: might not 2008 be a payback for the American bungling of the 1997–1998 East Asian financial crisis?

We are about to see what stake China really has in the survival of the globalized world economy. As in 1931, the political arguments are all against a rescue operation. Only the far-sighted will see that the economic case for such an operation is compelling. The arguments that convince, however, will be fundamentally political rather than economic. Just before the Asia-Europe Meeting of October 24, 2008, Chinese President Hu Jintao stated that China would behave "with a sense of responsibility." But that responsibility has a distinct price, at least in the short term.

There were already before the summer of 2007, when the financial crisis began, many signs that we were at the beginning of a new era, in which the "globalization thesis" was being rolled back once more. In this new world, differences between countries and between different national styles of capitalism (and differing approaches to state control) become more important. Business leaders focus on the way they have "traditionally" done business. Individuals see risks rather than opportunities coming from the outside. Citizens detect corruption and distrust elites. Countries are willing to fight trade and currency wars, and to resist foreign interventions in corporate affairs.

Political leaders in this world focus on redesigning the trading and monetary systems to alter the balance of political and economic power. Office holders attempt to use their new control of the financial system to shape the national economic destiny. State capitalism becomes more widespread. Politicians treat foreigners both in and outside their countries as scapegoats. Power and hegemony matter again as political decision makers see a zero-sum game rather than the attractions of collaboration. In this world, conflict tends to escalate, and conflicts destroy the basis of prosperity and international order.

18

Globalization, Migration, and New Challenges to Governance

NAZLI CHOUCRI AND DINSHA MISTREE

This article shows how international migration is imposing increasing demands on political institutions and policy makers. The authors explain why migration is now a powerful shaper of international politics.

The movement of people across national borders—along with the cross-border flow of ideas, goods, services, and pollutants—has reached unprecedented levels in recent decades. As a result, sovereign states find themselves under increasing pressure to manage these flows and respond to the challenges that the flows create, while balancing the interests of various constituencies, both national and international.

Few countries can escape these pressures on governance; even fewer have been able to manage cross-border flows as if they were routine matters of "low politics"—that is, the kind of politics that contends with issues not critical to a state's survival. The reason for this is that we live in an increasingly globalized world, one in which territorial boundaries are ever more porous. And perhaps in no other arena is countries' lack of effective control over borders and national access so striking as in the realm of international migration.

Given that the theoretical sanctity of national borders—borders that delineate the limits of sovereign jurisdiction—has long been a defining factor in the international system, we tend to assume that borders are known, fixed, and permanent. Under such conditions, migration means crossing boundaries. But such conditions can in fact be a "variable" in international politics: borders are not always fully known, they are not necessarily fixed, and they are not permanent.

At this point we can identify two modes of initiatives that nations often take in an effort to govern flows of people, goods, services, and so forth. One mode is coordinated international action through various international institutions. The World Trade Organization, governing goods and services, is one example. In this regard the International Organization for Migration, an intergovernmental body, is among a number of organizations committed to facilitating the orderly and humane

SOURCE: From Nazli Choucri and Dinsha Mistree, "Globalization, Migration, and New Challenges to Governance." *Current History* 108 (April, 2009). Reprinted with permission, Current History, Inc.

movement of people. In the second mode, states take matters into their own hands by seeking to buttress their control over entry and access across boundaries.

It is too early to tell which of these two trends will eventually dominate, or if either will be superseded by some other model of governance, but we can almost certainly expect borders in the future to remain at least as porous as they have been in the past. Therefore, to take stock of twenty-first century international migration in all its various manifestations, it is important to consider the movements of people in light of their interconnections with other contemporary forms of globalization.

Globalization per se is not new. What is new, however, is the nature of twenty-first century globalization—characterized as it is by considerably greater scale, scope, and intensity than earlier episodes in history, and by its pervasive penetration into societies and across national jurisdictions worldwide.

Likewise, the movement of people is not new. Since time immemorial humans have been on the move—in search of better opportunities or in response to threats, perceived or actual. The movement of people within national boundaries is in theory the prerogative of the nation itself and its government, but flows of people across territorial boundaries are clearly an issue of international concern. And when outward migration results from internal violence, the consequences become both national and international. Importantly, in all of these cases today, migrations are imposing increasing demands on institutions and policy makers.

As with globalization, the scale and scope of contemporary migrations are unprecedented. But in addition, the various forms, causes, and consequences of migration are all in a state of flux. Consequently, the movement of people that had for so long been relegated to the realm of low politics is now lodged clearly within the bounds of "high politics"—where national security is threatened— with commensurate degrees of social salience and intensity.

The most obvious example is the case of the United States in relation to Mexican migrants. Proponents of a border "wall" suggest it is needed not just to prevent people from crossing, but also to address national security interests raised by the accompanying flow of illicit goods and activities.

SHIFTING BOUNDARIES

A cursory look at any historical world atlas will show a wide range of globalization patterns spanning almost all of recorded time. Ancient empires were formed through globalization, and their expansion extended cross-border flows. The demise of old empires led to the creation of new states, and with these came (by definition) the establishment of territorial boundaries. However, while territorial borders formally delimit sovereignty in a legal sense, they seldom provide fully effective control of entry and exit. This is as true of authoritarian states as it is of more open societies. No one has devised a foolproof method for regulating population movements (just as no one is able to imagine a way in which pollution can be contained within national boundaries).

Several distinct periods of globalization were observable in the twentieth century, from the outbreak of the world wars to the large-scale process of decolonization to the enormous expansion of international transactions in recent decades. In such cases, the flows of goods and services and even of people across borders were shaped by competition and coercion, but at the same time they fostered conditions for cooperation and collaboration. This has always been so. European nations, for example, competed in their territorial expansion during the initial stages of imperialism, but they also negotiated, in the Treaty of Utrecht, to divide their spheres of influence and colonies.

In the same way, countries have gone to war with each other but at the same time have developed common understandings and legal precepts regarding the treatment of combatants. Over time, we have seen the evolution of a large body of international law and the establishment of a large number of international institutions—covering an ever-widening domain of cross-border activities—that display various degrees of organization and effectiveness. This historical experience and the potential similarities between the past and the present are instructive in helping us appreciate the nature of twenty-first century globalization and the close relationship between globalization and migrations.

It would be nearly impossible to list all existing definitions of globalization. At best we can highlight a few illustrative descriptions. For much of this decade, many have framed globalization as a process of ever-deepening international economic integration. While McGill University's Mark Brawley, for example, recognizes that this integration is driven by various factors—including technological and policy changes, falling costs of transport and communications, and greater reliance on markets—he argues that other cross-border flows follow primarily from greater economic integration.

Other scholars go so far as to declare that framing globalization as anything other than driven by economic factors would render the term useless. And still others have attempted to concentrate largely on the domestic impacts of cross-border movements of goods and services. On balance, the consensus is that any social, political, or cultural transformations that have resulted from globalization are secondary to economic transactions.

IT'S NOT JUST THE ECONOMY

This commonly held view is parsimonious and to the point. The utility of parsimony is appreciated—simplification is often a good way to get at the heart of a matter. We believe, however, that such parsimony misses the fundamental nature of globalization and the truly complex interconnections created by the overall dynamics of the process.

Our view is that globalization is not simply a process of economic activity that leads to other kinds of cross-border flows. If that were the case, globalization would have ended with the global economic downturn that began in 2007. Yet there are many reasons to expect trends identified with globalization to continue in the coming years.

International migration is a case in point. It is true that many people will migrate if opportunities presented in another nation are greater than the opportunities available in their native countries. Much of the migration that took place during the past several years of global economic growth resulted from people choosing to leave their homes in pursuit of better opportunities elsewhere. And evidence today suggests that some migrants are finding it necessary to return home as jobs dry up. There is also every indication, however, that many people are continuing to migrate away from their home countries because conditions there have grown progressively worse, making other nations seem more appealing.

In other words, during good times people migrate to find better opportunities; during bad times people migrate to escape more difficult circumstances. Each of these situations holds implications for globalization, but perhaps in different ways and with different consequences. To understand better the movements of people across national boundaries, it is necessary to see them as taking place within a broader system of cross-border movements. This in turn requires a broader conception of globalization.

GLOBALIZATION DYNAMICS

We consider globalization to be *transformations in socioeconomic and political structures and processes* shaped by the movement of tangible and intangible factors, ranging from goods and labor to ideas and services, across territorial borders. Increasing economic interconnections are critically important, to be sure. But they do not tell the entire story, nor are they the only source for fostering other consequences. Indeed, many kinds of international flows occur simultaneously, and to try to figure out which begets which can be a fool's errand. When we examine current trends in cross-border movements, a more complex picture of globalization emerges, one that includes economic integration but is not limited to it.

Twenty-first century globalization, in our view, is a process generated by uneven growth and development within and across states. This uneven development leads to (a) the movement of people, goods, services, ideas, pollutants, and other factors across national borders. The cross-border movements contribute to (b) transformations of socioeconomic and political structures within and across states, and also (c) create pressures on prevailing modes of governance. The international flows thus (d) generate demands for changes in governance, which in turn (e) contribute to further changes in patterns of growth and development— leading us right back to (a). Put differently, the fundamental logic of globalization is one of feedback dynamics rather than linear sequencing. A simplified representation is shown below.

The Globalization Process

```
              ┌──────────────────────┐
              │   Uneven Growth      │
              │   and Development    │
              └──────────────────────┘

┌──────────────────────────┐      ┌──────────────────────┐
│ Transformations of        │      │ Cross-Border Movements│
│ Governance                │      │ and Impacts           │
│ Structures and Processes  │      │                       │
└──────────────────────────┘      └──────────────────────┘

              ┌──────────────────────┐
              │   New Pressures on    │
              │   Modes of Governance │
              └──────────────────────┘
```

THE GLOBALIZATION PROCESS

The post–World War II period in Europe provides a good example of the dynamics presented in the figure above. The end of the war created a need for large-scale reconstruction, and this reconstruction effort was complicated by manpower shortages. Labor shortages were addressed by importing foreign workers who, in some countries, were called "guest workers." The understanding was that foreign laborers would eventually return to their homeland. But two or three generations later, these workers had become part of the receiving countries' social fabric.

As a result, governments—in France and Germany, for example—faced conflicting pressures. On one hand, the workers wanted to be treated like all other members of society. On the other hand, the "native" population did not fully accept the foreigners, and some pressured their governments to resist demands for full citizenship. Over time, the democratic ethos in France and Germany created an institutional demand to manage these pressures in ways oriented toward greater equality. The governments developed new policies to facilitate fuller integration. In practice, however, effective integration is contingent on the overall well-being of the economy.

The current situation in some ways parallels that of post–World War II Europe. Today almost everyone is affected by the global downturn. New unevenness in growth and development will almost surely create added pressures on the movements of people and the policies of governments.

Already we see a new phase in the familiar scenario being worked out in the Gulf region of the Middle East. Oil-rich countries, to facilitate their extraordinary economic development, have imported nearly all of their labor requirements. But with the recent downturn in the global economy, these countries no longer want many of the workers. And the workers' own countries do not wish to pay for their return home. As a result they are unemployed, without discretionary income, and unable to leave the region. This situation has given rise to a new label, signaling a new migration experience: the "stranded migrants."

DRIVERS OF CHANGE

The differential rates of growth and development, which propel both globalization and migration, are themselves driven by changes in technology, resources, and population. These factors thus dictate the parameters of globalization politics: technology, resources, and population vary according to conditions within states, and they also affect power distributions and relations among states.

Consider some of the implications of this way of looking at the drivers of globalization. Imagine a China with its population as we know it, but with the resources of Saudi Arabia and the technology of the United States. Imagine a Saudi Arabia with the resources of Chad and the population of India. Imagine a Japan with the petroleum reserves of Iraq. Most important of all, remember that it was foreign technology that made it possible for Saudi Arabia to identify its oil reserves, exploit those reserves, and then produce and export this highly valuable resource. And doing all of this required Saudi Arabia (and other oil-rich but population-poor countries) to import foreign labor in order to "redress" their "imbalances" and embark on large, complex projects.

More to the point, imagine an Iran with the technology to develop nuclear weapons, or a North Korea that could actually build a nuclear device. The key point here is that each of the "master" variables—technology, resources, and population—can be manipulated, and at times has been effectively manipulated (with different degrees of effort and variable time constants, of course, not to mention real costs as well as opportunity costs).

An important corollary with respect to migration is this: depending on scale and scope, the movement of people across territorial boundaries can fundamentally alter the nature of a nation-state's politics, economy, and policies. A related effect is that the cross-border movements prompted by uneven growth and development generate increasing pressures for new management, regulation, and governance.

A CROWDED WORLD

The various aspects of globalization, is depicted in the figure, operate at different time intervals depending on the historical context and influence of the critical drivers. Our world today is still defined by the principles of the Treaty of Westphalia, with the sovereign state formally enfranchised to speak on behalf of its citizens. But a number of distinctive features characterize the unique contours of international relations in the twenty-first century—and each of these is reinforced by accelerated migration.

First, the international community consists of more member states than ever before. There are also greater numbers and more types of "voices" in various international arenas. And there is greater representation of individual views on almost all issues, within and across states. This crowding adds to the complexity of governing international migration.

Second, a large and growing number, as well as a greater variety, of recognized intergovernmental entities have emerged, supporting diverse missions and mandates. While there are international institutions designed to help manage migration, they are far from adequate.

Third, a global civil society has emerged, with a flourishing network of nongovernmental interests, agents, and institutions. Many nongovernmental organizations now recognize the salience of migration at all levels, from local to global.

Fourth, much of the foregoing has resulted in increasing numbers of decision-making entities worldwide, with greater global reach, driven by increasingly diverse objectives and strategies. This decision-making density is reinforced by growth in the number of actors and agents involved in border and international exchanges.

Fifth, growth in innovation and increasing social transformation allow more flexibility in production of goods—in the composition of production, the physical distribution of assembly locations, and the like—thus significantly changing the production process. Such changes notwithstanding, the demand for migrant workers continues to grow in labor-importing countries.

Sixth, increasingly differentiated products and increasingly powerful consumers have resulted in an expansion of consumerism, further driving consumption. It goes without saying that consumption levels respond to the flow of migrants.

Seventh, for the first time in history, an urgent threat to the environmental conditions and life-supporting properties of the entire planet is recognized as the cumulative effect of human actions, insofar as they are altering the global climate. Environmental and resource pressures appear certain to increase migration further.

Eighth, the globalization process has contributed to the forging of a new virtual arena, cyberspace, in which individuals can interact and, in so doing, engage in political discourse in ways that were earlier restricted to states. Earlier phases of globalization that created new spaces of interaction, control, or conquest (for example, colonies, the polar regions, and outer space) provided opportunities mainly for the few and the powerful. By contrast, in principle at least, cyber venues create opportunities for everyone. The evidence suggests that many migrants avail themselves of the conveniences, and incur the costs, of cyber venues.

Ninth, unrelated to any single factor above, globalization is accompanied more and more by recognition of localization. Indeed, the term "glocalization" is now part of international institutional discourse. This term refers to the sensitivity of local as well as global contexts to the dynamics of globalization. Thus migration, for example, has an impact on the sending community as well as the receiving community. In each case, effects are felt at a very local level and they may also reflect and even generate broader, more global, influences.

And finally, as already noted, all of these factors are further reinforced by the accelerated movement of people across territorial boundaries—voluntary and nonvoluntary, organized and non-organized—from developing to industrial nations, as well as among developing countries.

Accelerated migration in combination with the other factors listed above creates new pressures for governance of one sort or another, in the effort to manage some of the unforeseen or complex correlates and consequences of globalization.

VARIETIES OF MOBILITY

The movement of people across territorial borders, as a manifestation of globalization, can also be traced to the earliest forms of organized societies—earlier than the state system of the modern era. Migration, in contrast to trade, is seldom encouraged as a form of sustained international transaction or interaction. It is also highly variable in its manifestations, as it is in its sources and consequences. The usual distinction between voluntary and nonvoluntary mobility represents only the tip of a highly complex iceberg of social relations, economic correlates, and political impacts—at both ends of the migration stream.

The impacts of migrations are enormous and long-lasting. Indeed, when coupled with settlement intents and practices, the large-scale movements of people across long distances often constitute the foundations for a new social order and a source of state building, a base on which subsequent generations have pursued and reinforced the visions of the early settlers.

On the basis of this statement, given that such processes take place largely without conflict, one might get the impression that migration tends to flow to "empty" lands, and that the net effect is to provide new residents with new opportunities in areas that otherwise would be devoid of human settlement. Nothing could be more misleading. For the most part, permanent settlements resulting from migration are usually preceded by violent interaction between the migrants and the "natives." The proclamation of a new state is then the last step in a highly complex dynamic of conflict, competition, and domination by the newcomers.

And permanent settlement, of course, is only one of many forms of voluntary mobility. There are many others, as we shall note further along, and there are also many forms of forced, induced, coerced, or otherwise nonvoluntary movements of people. To simplify, different forms of migration contribute differently to pressures for governance, as do different modes of globalization at different times and in different regions of the world.

Yet only a very few and distinctive forms of population movements have been formally regulated by states or made routine by international institutions. Even in the few instances in which mobility is taken into account in various official statistics, record keeping usually focuses on formal immigration, tourism, organized travel, pilgrimage, and other routine movements.

On the coercion side of the ledger, migration is the subject of international management under very distinctive and even restrictive definitions. For example, the status of "refugee" is a formal one defined by international law, but it does not cover all the people who cross boundaries because they are persecuted or pushed out, or because they are caught in the crossfire of hostilities. International agencies that manage refugees are not usually expected to extend their services to other displaced persons who do not qualify for this (relatively fortunate) status.

MIGRATION MOTIVES

A basic assumption of our view of the world is that each statistic is an indicator—and consequence—of discrete decisions by individual humans, as governed by their preferences. Population growth, for example, is in fact the outcome of a large number of discrete private decisions (whether free or coerced) over which policy makers or national governments are not likely to have direct control.

In this connection, if there is any determinism in the logic that results in uneven growth and development, it is a determinism driven by individual decision making. Whole societies may drift in a certain way, as characterized by certain reproductive patterns, and regardless of leaders' influence, because millions of private citizens are behaving in normal and often most legitimate ways. The connection to migration is this: to the extent that population growth exceeds a society's employment potential, the probability is very high that people will move to other countries in search of jobs.

Indicators of technology, like those of population, are also the observed outcomes of a number of widely dispersed decisions by individual actors (such as developers, inventors, scientists, investors, manufacturers, and so forth). This may be less true in authoritarian systems than in open societies, but the difference may be more of degree than of kind. While the development of technology is influenced more by organizations and bureaucratic decisions than are trends in population, the fact remains that individual actions are critical in shaping the success or failure of technological ventures. The connection to migration is this: countries that are resource-rich, but technology-poor and small in population, seek to draw migrants. They compensate for population constraints by importing foreign labor; they counterbalance limitations in technology by relying on highly skilled workers from other countries.

To simplify a complex situation, we find it useful to differentiate among people crossing borders, first, in terms of voluntary versus nonvoluntary migration, and then in terms of motivation for mobility and duration of stay. Taking these criteria into account, the most obvious patterns of international migration today include the following: migration for employment; seasonal mobility for employment; permanent settlements; refugees who are forced to migrate; resettlement; state-sponsored movements; tourism and ecotourism; brain drains and "reversals" of brain drain; smuggled and trafficked people; people returning to their countries of origin; environmental migration and refugees from natural shortages or crises; nonlegal migration; and religious pilgrimage.

These types are not necessarily mutually exclusive, a fact that complicates any simple accounting of migration. The dynamic status of migrants also adds to the complexity of cross-border movements. For example, migrants may begin their experience with an official employment status, but if their contracts expire they become nonlegal migrants. If a war erupts and they are forced to move again, the same migrants then become refugees. On the other hand, mobility that is nonvoluntary—and regarded simply as refugeeism—may result from a variety of circumstances, including conflict and violence, droughts and environmental degradation, or shifts in territorial boundaries. The situation for migrants is seldom clear-cut and it is almost always changing.

This consideration is relevant for two reasons. First, it serves as a reminder of the relational basis of migration in a globalized world. Contextual factors enter into the definition of migration types and are formalized, sometimes arbitrarily, with reference to both the legal status of migrants and the policies of state institutions. The second reason is that it shows how difficult it is to obtain an internally consistent accounting of international migration by source, cause, category, duration, and so on. This further complicates issues of jurisdiction over shifting forms of cross-border mobility.

The sheer scale and scope of people on the move, in combination with the blurring of borders, creates new challenges for the conduct of international relations and strategies for sustainable development. By extension, assigning responsibility for the remediation of problems generated by past migration-related policies becomes especially difficult. When sovereignties are diluted (or when new ones are created), lines of authority and responsibility also become blurred. Determining who is responsible for what, when, and how becomes particularly vexing.

The leakage of territorial boundaries—the evident inability to regulate and control access across national borders—is one legacy of the current phase of globalization, which began in the twentieth century, and it has become a powerful shaper of international politics. The sovereign state today is a porous entity. Not only do resources extracted in one location usually find their way to another place for utilization and consumption, as mobile capital scours the world for investments. And not only does technology (embodied in human skills or in machinery) move from one place to another (often referred to as the "transfer" of technology), as information moves as rapidly as we wish around the globe. People also are moving across borders in unprecedented numbers.

A second legacy of globalization from the twentieth century is the remarkable expansion of sovereign entities, an expansion that greatly complicates the challenges of international coordination and global management. We must also appreciate that migration can take place, in effect, even when people themselves do not actually move, but when territorial boundaries are changed. For example, the demise of the Soviet Union—leading to a proliferation of sovereign entities—left people in place but shifted jurisdictions.

As cross-border linkages become more prevalent—and as political institutions find it more difficult to regulate cross-boundary flows—states become more interconnected, and governments find it more difficult to exercise control over domestic politics and economies. They become more vulnerable to decisions made elsewhere, as well as to unanticipated shocks or crises—natural or social—that originate outside their formal boundaries. Invariably, states are forced to acknowledge external pressures, yet they may experience greater difficulties in coordinating policies with other states even when they wish to do so.

The challenge to governance created by the conjunction of migration and globalization is difficult to overstate. As people move across borders, they encounter the rules and regulations of the recipient community. Given the diversity of causes and consequences of such mobility, as well as the diversity of the recipient community's rules and regulations, it should come as no surprise that friction may rise and that the social contract in the community of destination may become stressed.

MANAGING CHANGE

In this context, the task of governance is to manage the changes and ease the transitions wrought by globalized migrations. Over time the changes penetrate deep into the social fabric of nations, potentially transforming them in important ways, even altering the landscape of the state system as a whole. But trying to oversee this process is complicated because, as a result of globalization, the structures and functions of governance are affected by processes that operate across levels of analysis and between social systems and political jurisdictions.

Obviously, the pressures on governance are not always met by requisite and appropriate responses. Still, we can recognize that pervasive dynamics of uneven growth and development affect the institutional arrangements that evolve in the course of seeking to manage these dynamics. Further examining these evolving relationships can help us understand the structure of the international system and adapt more readily to the transforming effects of globalization—including the accelerated movements of people. We know that the world is changing. We hope that our institutions manage these changes as effectively as can be reasonably expected.

REFLECTION QUESTIONS FOR CHAPTER 5

1. Social scientists say that the elements of globalization—capitalism, trade, and technological revolutions—are breaking down old obstacles and mindsets. With that in mind, is the notion of national sovereignty an outmoded concept?

2. Using the essay by Stiglitz as a guide, identify the benefits of globalization. What countries have benefited the most from globalization and why?

3. Explain Rosenberg's contention that cheap food hurts poor people. How do the U.S. and Mexican governments contribute to this paradox?

4. What does James mean by *state capitalism?* Why does he think it will become widespread in the coming decades?

5. Do you agree with Choucri and Mistree when they say that twenty-first century population changes will be a key factor in shaping the parameters of global politics?

Chapter 6

Cultural Globalization

Tribes have a culture. Societies have a culture. But is there such a thing as a global culture? The concept of culture has traditionally meant the knowledge that a people share. This knowledge includes all manner of shared meanings, such as religion, myths, technology, political ideology, language, music, art, fashion, and consumption patterns. Using this conventional view of culture, there are no global equivalencies—no global myths, legends, or symbols that unite the world's people. On the other hand, English is at least a second language throughout much of the world, Western movies and television shows are shown almost everywhere, and products such as McDonald's, Coca-Cola, Disney, and Nike proliferate worldwide. There are global icons from the worlds of entertainment (like Madonna, Tom Cruise, Beyoncé, U2, and the Rolling Stones) and sports (such as David Beckham, Tiger Woods, and Michael Jordan). In short, while there are real differences across national boundaries, some elements of culture transcend those boundaries.

While local or national culture is tied to place and time, "global culture is free of these constraints: as such it is 'disconnected,' 'disembedded' and 'deterritorialized,' existing outside the usual reference to geographical territory."[1] In global culture there is a continuous and rapid flow of images, ideas, information, tastes, and products across geographical, political, and linguistic borders.

Two major issues arise from this form of global integration. First, the world appears to be becoming culturally homogenized along a westernized model, which raises the question, Is this evidence of progressive cosmopolitanism or oppressive imperialism? Second, will a global culture diminish or even demolish the uniqueness, diversity, and richness of local cultures? In other words, does globalization lead to cultural homogenization or does it result in heterogenization as various groups resist the global onslaught on their way of life?

The first selection in this chapter, by political scientist Manfred B. Steger, addresses the question of whether global culture results in sameness or difference. Next, Ian Condry provides a case study for exploring the globalization of popular culture. His ethnography of the hip-hop music scene in Tokyo reveals how "global" popular culture takes on local and distinctly Japanese characteristics.

The final article in this section is about beauty ideals and skin bleaching. Sociologist Evelyn Nakano Glenn examines the growing preference for lighter skin and the consumption of skin-lightening products around the world. She shows how transnational pharmaceutical and cosmetics companies fuel the desire for lighter skin through print, internet, and television ads.

ENDNOTES

1. John Beynon and David Dunkerley (eds.), *Globalization: The Reader* (New York: Routledge, 2000), p. 13.
2. Jan Aart Scholte, *Globalization: A Critical Introduction* (New York: Palgrave, 2000), p. 23.

19

Global Culture: Sameness or Difference?

MANFRED B. STEGER

Steger explains why the various perspectives on the question of whether global culture results in sameness of difference are not incompatible. Sameness or difference coexist, he argues, because cultural globalization is always transformed by local contexts.

Does globalization make people around the world more alike or more different? This is the question most frequently raised in discussions on the subject of cultural globalization. A group of commentators we might call "pessimistic hyperglobalizers" argue in favour of the former. They suggest that we are not moving towards a cultural rainbow that reflects the diversity of the world's existing cultures. Rather, we are witnessing the rise of an increasingly homogenized popular culture underwritten by a Western "culture industry" based in New York, Hollywood, London, and Milan. As evidence for their interpretation, these commentators point to Amazonian Indians wearing Nike training shoes, denizens of the Southern Sahara purchasing Texaco baseball caps, and Palestinian youths proudly displaying their Chicago Bulls sweatshirts in downtown Ramallah. Referring to the diffusion of Anglo-American values and consumer goods as the "Americanization of the world," the proponents of this cultural homogenization thesis argue that Western norms and lifestyles are overwhelming more vulnerable cultures. Although there have been serious attempts by some countries to resist these forces of "cultural imperialism"—for example, a ban on satellite dishes in Iran, and the French imposition of tariffs and quotas on imported film and television—the spread of American popular culture seems to be unstoppable.

But these manifestations of sameness are also evident inside the dominant countries of the global North. American sociologist George Ritzer coined the term "McDonaldization" to describe the wide-ranging sociocultural processes by which the principles of the fast-food restaurant are coming to dominate more and more sectors of American society as well as the rest of the world. On

SOURCE: From "Global Culture: Sameness or Difference?" by Manfred B. Steger from *Globalization–A Very Short Introduction*. Oxford University Press, 2009. Used with permission.

the surface, these principles appear to be rational in their attempts to offer efficient and predictable ways of serving people's needs. However, looking behind the facade of repetitive TV commercials that claim to "love to see you smile," we can identify a number of serious problems. For one, the generally low nutritional value of fast-food meals—and particularly their high fat content—has been implicated in the rise of serious health problems such as heart disease, diabetes, cancer, and juvenile obesity. Moreover, the impersonal, routine operations of "rational" fast-service establishments actually undermine expressions of forms of cultural diversity. In the long run, the McDonaldization of the world amounts to the imposition of uniform standards that eclipse human creativity and dehumanize social relations.

Perhaps the most thoughtful analyst in this group of pessimistic hyperglobalizers is American political theorist Benjamin Barber. In his popular book on the subject, he warns his readers against the cultural imperialism of what he calls "McWorld"—a soulless consumer capitalism that is rapidly transforming the world's diverse populations into a blandly uniform market. For Barber, McWorld is a product of a superficial American popular culture assembled in the 1950s and 1960s, driven by expansionist commercial interests. Music, video, theatre, books, and theme parks are all constructed as American image exports that create common tastes around common logos, advertising slogans, stars, songs, brand names, jingles, and trademarks.

Barber's insightful account of cultural globalization also contains the important recognition that the colonizing tendencies of McWorld provoke cultural and political resistance in the form of "Jihad"—the parochial impulse to reject and repel the homogenizing forces of the West wherever they can be found.... Jihad draws on the furies of religious fundamentalism and ethnonationalism which constitute the dark side of cultural particularism. Fueled by opposing universal aspirations, Jihad and McWorld are locked in a bitter cultural struggle for popular allegiance. Barber asserts that both forces ultimately work against a participatory form of democracy, for they are equally prone to undermine civil liberties and thus thwart the possibility of a global democratic future.

Optimistic hyperglobalizers agree with their pessimistic colleagues that cultural globalization generates more sameness, but they consider this outcome to be a good thing. For example, American social theorist Francis Fukuyama explicitly welcomes the global spread of Anglo-American values and lifestyles, equating the Americanization of the world with the expansion of democracy and free markets. But optimistic hyperglobalizers do not just come in the form of American chauvinists who apply the old theme of manifest destiny to the global arena. Some representatives of this camp consider themselves staunch cosmopolitans who celebrate the Internet as the harbinger of a homogenized "techno-culture." Others are free-market enthusiasts who embrace the values of global consumer capitalism.

It is one thing to acknowledge the existence of powerful homogenizing tendencies in the world, but it is quite another to assert that the cultural diversity existing on our planet is destined to vanish. In fact, several influential commentators offer a contrary assessment that links globalization to new forms of cultural expression. Sociologist Roland Robertson, for example, contends that global cultural

flows often reinvigorate local cultural niches. Hence, rather than being totally obliterated by the Western consumerist forces of sameness, local difference and particularity still play an important role in creating unique cultural constellations. Arguing that cultural globalization always takes place in local contexts, Robertson rejects the cultural homogenization thesis and speaks instead of "glocalization"—a complex interaction of the global and local characterized by cultural borrowing. The resulting expressions of cultural "hybridity" cannot be reduced to clear-cut manifestations of "sameness" or "difference." [S]uch processes of hybridization have become most visible in fashion, music, dance, film, food, and language.

In my view, the respective arguments of hyperglobalizers and sceptics are not necessarily incompatible. The contemporary experience of living and acting across cultural borders means both the loss of traditional meanings and the creation of new symbolic expressions. Reconstructed feelings of belonging coexist in uneasy tension with a sense of placelessness. Cultural globalization has contributed to a remarkable shift in people's consciousness. In fact, it appears that the old structures of modernity are slowly giving way to a new "postmodern" framework characterized by a less stable sense of identity and knowledge.

The American Way of Life

Number of types of packaged bread available at a Safeway in Lake Ridge, Virginia	104
Number of those breads containing no hydrogenated fat or diglycerides	0
Amount of money spent by the fast-food industry on television advertising per year	$3 billion
Amount of money spent promoting the National Cancer Institute's "Five A Day" programme, which encourages the consumption of fruits and vegetables to prevent cancer and other diseases	$1 million
Number of "coffee drinks" available at Starbucks, whose stores accommodate a stream of over 5 million customers per week, most of whom hurry in and out	26
Number of "coffee drinks" in the 1950s coffee houses of Greenwich Village, New York City	2
Number of new models of cars available to suburban residents in 2001	197
Number of convenient alternatives to the car available to most such residents	0
Number of U.S. daily newspapers in 2000	1,483
Number of companies that control the majority of those newspapers	6
Number of leisure hours the average American has per week	35
Number of hours the average American spends watching television per week	28

SOURCE: Eric Schlosser, *Fast Food Nation* (Houghton Mifflin, 2001), p. 47; www.naa.org/info/facts00/11htm; *Consumer Reports Buying Guide, 2001* (Consumers Union, 2001), pp. 147–163; Laurie Garrett, *Betrayal of Trust* (Hyperion, 2000), p. 353; www.roper.com/news/content/news169.htm; *The World Almanac and Book of Facts 2001* (World Almanac Books, 2001), p. 315; www.starbucks.com.

Given the complexity of global cultural flows, one would actually expect to see uneven and contradictory effects. In certain contexts, these flows might change traditional manifestations of national identity in the direction of a popular culture characterized by sameness; in others they might foster new expressions of cultural particularism; in still others they might encourage forms of cultural hybridity. Those commentators who summarily denounce the homogenizing effects of Americanization must not forget that hardly any society in the world today possesses an "authentic," self-contained culture. Those who despair at the flourishing of cultural hybridity ought to listen to exciting Indian rock songs, admire the intricacy of Hawaiian pidgin, or enjoy the culinary delights of Cuban-Chinese cuisine. Finally, those who applaud the spread of consumerist capitalism need to pay attention to its negative consequences, such as the dramatic decline of communal sentiments as well as the commodification of society and nature.

20

Japanese Hip-Hop and the Globalization of Popular Culture

IAN CONDRY

This selection is an anthropological study of hip-hop music in Japan. Condry shows how hip-hop, a transnational form of entertainment, is adapted to local Japanese culture.

INTRODUCTION

Japanese hip-hop, which began in the 1980s and continues to develop today, is an intriguing case study for exploring the globalization of popular culture. Hip-hop is but one example among many of the transnational cultural styles pushed by entertainment and fashion industries, pulled by youth eager for the latest happening thing, and circulated by a wide range of media outlets eager to draw readers and to sell advertising. In Tokyo, a particular combination of local performance sites, artists, and fans points to ways that urban areas are crucibles of new, hybrid cultural forms. Hip-hop was born in the late 1970s in New York City as a form of street art: rapping on sidewalk stoops, outdoor block parties with enormous sound systems, graffiti on public trains, and breakdancing in public parks. In its voyage to Japan, the street ethic of hip-hop remains, but it is performed most intensely in all-night clubs peppered around Tokyo. This paper examines these nightclubs as an urban setting that helps us grasp the cultural dynamics of Japanese hip-hop. In particular, the interaction between artist-entrepreneurs and fans in live shows demonstrates how "global" popular culture is still subject to important processes of localization.

Anthropologists have a special role in analyzing such transnational forms because of their commitment to extended fieldwork in local settings. Ethnography aims to capture the cultural practices and social organization of a people. This offers a useful way of seeing how popular culture is interwoven with everyday life. Yet there is a tension between ethnography and globalization, because in many ways they seem antithetical to each other. While ethnography attempts to evoke the distinctive texture of local experience, globalization is often seen as erasing local differences. An important analytical challenge for today's media-saturated world is finding a way to understand how local culture interacts with such global media flows.

On one hand, it seems as if locales far removed from each other are becoming increasingly the same. It is more than a little eerie to fly from New York to Tokyo and see teenagers in both places wearing the same kinds of fashion characteristic of rap fans: baggy pants with boxers on display, floppy hats or baseball caps, and immaculate space-age Nike sneakers. In Tokyo stores for youth, rap music is the background sound of choice. Graffiti styled after the New York City aerosol artists dons numerous walls, and breakdancers can be found in public parks practicing in the afternoon or late at night. In all-night dance clubs throughout Tokyo, Japanese rappers and DJs take to the stage and declare that they have some "extremely bad shit" (geki yaba shitto)—meaning "good music"—to share with the audience. For many urban youth, hip-hop is the defining style of their era. In 1970s Japan, the paradigm of high school cool was long hair and a blistering solo on lead guitar. Today, trendsetters are more likely to sport "dread" hair and show off their scratch techniques with two turntables and a mixer. In the last few years, rap music has become one of the best-selling genres of music in the United States and around the world as diverse youth are adapting the style to their own messages and contexts.

But at the same time, there are reasons to think that such surface appearances of sameness disguise differences at some deeper level. Clearly, cultural setting and social organization have an impact on how movies and television shows are viewed. Yet if we are to understand the shape of cultural forms in a world that is increasingly connected by global media and commodity flows, we must situate Japanese rappers in the context of contemporary Japan. When thinking about how hip-hop is appropriated, we must consider, for example, that most Japanese rappers and fans speak only Japanese. Many of them live at home with their parents, and they all went through the Japanese education system. Moreover, even if the origin of their beloved music genre is overseas, they are caught up in social relations that are ultimately quite local, animated primarily by face-to-face interactions and telephone calls. So while these youth see themselves as "hip-hoppers" and "B-Boys" and "B-Girls," and associate themselves with what they call a "global hip-hop culture," they also live in a day-to-day world that is distinctly Japanese.

For those interested in studying the power of popular culture, there is also a more practical question of research methods. How does a lone researcher go about studying something as broad and unwieldy as the globalization of mass culture? One of the tenets of anthropological fieldwork is that you cannot understand a people without being there, but in the case of a music genre, where is "there"? In the fall of 1995, I began a year and a half of fieldwork in Tokyo, and the number of potential sites was daunting. There were places where the music was produced: record companies, recording studios, home studios, and even on commuter trains with handheld synthesizers. There were places where the music was promoted: music magazines, fashion magazines, TV and radio shows, night-clubs, and record stores. There was the interaction between musicians and fans at live shows, or in mediated form on cassettes, CDs, and 12-inch LPs. To make matters worse, rap music is part of the larger category of "hip-hop." Hip-hop encompasses not only rap, but also breakdance, DJ, graffiti, and fashion. The challenge was to understand the current fascination among Japanese youth with hip-hop music and style, while also considering the role of money-making organizations. How does one situate the experiential pleasures within the broader structures of profit that produce mass culture? As I began interviewing rappers, magazine writers, and record company people, I found a recurring theme that provided a partial answer. Everyone agreed that you cannot understand Japanese rap music without going regularly to the clubs. Clubs were called the "actual site" (genba) of the Japanese rap scene.[1] It was there that rappers performed, DJs learned which songs elicit excitement in the crowd, and breakdancers practiced and competed for attention. In what follows, I would like to suggest that an effective tool for understanding the globalization of popular culture is to consider places like Japanese hip-hop nightclubs in terms of what might be called "genba globalism." By using participant-observation methods to explore key sites that are a kind of media crossroads, we can observe how globalized images and sounds are performed, consumed, and then transformed in an ongoing process. I use the Japanese term "genba" to emphasize that the processing of such global forms happens through the local language and in places where local hip-hop culture is produced. In Japanese hip-hop, these clubs are important not only as places where fans can see live shows and hear the latest releases from American and Japanese groups, but also as places for networking among artists, writers, and record company people. In this essay, I would like to point out some of the advantages of considering key sites as places to understand the cross-cutting effects of globalization. To get a sense of what clubs are about, let's visit one.

GOING TO HARLEM ON THE YAMANOTE LINE

A visit to Tokyo's Harlem is the best place to begin a discussion of Japanese hip-hop. Opened in the summer of 1997, Harlem is one of many all night dance clubs, but as the largest club solely devoted to hip-hop and R&B, it has become the flagship for the Japanese scene (at least, at the time of this writing in February 2001). Nestled in the love hotel area of the Shibuya section of Tokyo, Harlem is representative of the otherworldliness of clubs as well as their location within the rhythms and spaces of mainstream Japan.

If we were visiting the club, we would most likely meet at Shibuya train station around midnight because the main action seldom gets started before 1 a.m. Most all-night revelers commute by train, a practice that links Tokyo residents in a highly punctual dance. The night is divided between the last train (all lines stop by 1 a.m. at the latest) and the first train of the morning (between 4:30 and 5 a.m.). The intervening period is when clubs (*kurabu*) are most active.[2] Shortly after midnight, Shibuya station is the scene for the changing of the guard: those heading home, running to make their connections for the last train, and those like us heading out, dressed up, and walking leisurely because we will be spending all night on the town. The magazine stands are closing. Homeless men are spread out on cardboard boxes on the steps leading down to the subways. The police observe the masses moving past each other in the station square towards their respective worlds. Three billboard-size TVs looming overhead, normally spouting pop music videos and snowboard ads during the day, are now dark and silent. The throngs of teenagers, many in their school uniforms, that mob Shibuya during afternoons and all weekend have been replaced by a more balanced mix of college students, "salarymen" and "career women," and of course more than a few B-Boys and B-Girls—the hip-hop enthusiasts in baggy pants and headphones. The sidewalks are splashed with light from vending machines—cigarettes, soda, CDs, beer (off for the night), and "valentine call" phone cards. A few drunken men are being carried by friends or lie in their suits unconscious on the sidewalk.

To get to Harlem, we walk uphill among Dôgenzaka Avenue toward a corner with a large neon sign advertising a capsule hotel, where businessmen who have missed their last train can sleep in coffin-like rooms. We pass disposable lamp-post signs and phone booth stickers advertising various sex services. An elderly man in the back of a parked van is cooking and selling *takoyaki* (octopus dumplings) to those with the late-night munchies. The karaoke box establishments advertise cheaper rates at this hour. Turning right at a Chinese restaurant, we move along a narrow street packed with love hotels, which advertise different prices for "rest" or "stay." In contrast to the garish yellow sign advertising the live music hall, On-Air East, about fifty meters ahead, a nondescript door with a spiffy, long-haired bouncer out front is all that signals us that Harlem is inside. It seems there are always a couple of clubbers out front talking on their tiny cell phones. Up the stairs, past a table filled with flyers advertising upcoming hip-hop events, we pay our ¥3000 each (around $25, which may seem expensive, but is only about half again as much as a movie ticket). We move into the circulating and sweaty mass inside.

Traveling to a club instills a sense of moving against the mainstream in time and space. Others are going home to bed as the clubber heads out. When the clubber returns home in the morning, reeking of smoke and alcohol, the train cars hold early-bird workers as well. So the movement to and from the club, often from the distant suburbs, gives clubbers a sense of themselves as separate, flaunting their leisure, their costumes, and their consumer habits. During the course of my year and a half of fieldwork, between the fall of 1995 and the spring of 1997, I went to over a hundred club events around Tokyo and I began to see that clubs help one understand not only the pleasures of rap in Japan, but also the social organization of the scene and the different styles that have emerged. This becomes clear as you spend time inside the clubs.

INSIDE THE CLUB

Inside the club, the air is warm and thick, humid with the breath and sweat of dancing bodies. Bone-thudding bass lines thump out of enormous speakers. There is the scratch-scratch of a DJ doing his turntable tricks, and the hum of friends talking, yelling really, over the sound of the music. The lighting is subdued, much of it coming from a mirrored ball slowly rotating on the ceiling. The fraternity house smell of stale beer is mostly covered up by the choking cigarette haze, but it is best not to look too closely at what is making the floor alternately slippery and sticky. The darkness, low ceiling, black walls, and smoky murk create a space both intimate and claustrophobic. Almost everyone heads for the bar as soon as they come in. An important aspect of clubbing is the physical experience of the music and crowded setting.

Harlem is a larger space than most of the Tokyo clubs, and can hold upwards of one thousand people on a crowded weekend night. On the wall behind the DJ stage, abstract videos, *anime* clips, or edited Kung Fu movies present a background of violence and mayhem, albeit with an Asian flavor. Strobe lights, steam, and moving spotlights give a strong sense of the space, and compound the crowded, frenetic feeling imposed by the loud music. The drunken revelry gives clubs an atmosphere of excitement that culminates with the live show and the following freestyle session. But an important part of clubbing is also the lull before and after the show, when one circulates among the crowd, flirting, networking, gossiping, or simply checking out the scene. Clubs are a space where the diffuse network of hip-hop fans comes together in an elusive effort to have fun. To the extent that a "community" emerges in the hip-hop scene, it revolves around specific club events and the rap groups that draw the crowd.

Much of the time is spent milling around, talking, drinking, and dancing. The live show often produces a welcome rise in the excitement level of the clubbers. Some events feature several live acts, often followed by a freestyle session. The rap show will usually begin between 1:00 and 1:30 a.m. Formats vary depending on the club and the event. "B-Boy Night" at R-Hall (organized by Crazy-A) was held one Sunday a month and would start with a long breakdancing show, with many groups each doing a five-minute routine. Then a series of rap groups would come on, each doing two or three songs. At other shows, like FG Night, sometimes a series of groups would perform, while on other nights only one group would do a show followed by a more open-ended freestyle. Nevertheless, there were many similarities, and a characteristic live show would proceed as follows. Two rappers take the stage (or step up into the DJ booth), as the DJ prepares the music. For people enamored of live bands, the live show of a rap concert may strike one as a bit lifeless. The music is either pre-recorded on a digital audio tape (DAT) or taken from the breakbeats section of an album.[3] The flourish of a lead guitar, bass, or drum solo is replaced in the hip-hop show by the manic scratching of a record by a DJ who deftly slides a record back and forth across the slip mat laid on the turntable and works the mixer to produce the rhythmic flurries of sampled sound.

The rappers begin with a song introducing themselves as a group. Every group seems to have its own introductory song of self-promotion:

rainutsutaa ga rainut shi ni yatte kita	Rhymester has come to rhyme
doko ni kita? Shibuya!	Where are we? Shibuya!
hai faa za dopesuto da	We are By Phar the Dopest
oretachi kyo cho gesuto da	We are tonight's super guests
makka na me o shita fuktuô	The red-eyed owl [You the Rock]
ore tojo	I've arrived on stage

These songs tend to be brief, only a couple of minutes long. Between the first and second song, the rappers ask the audience how they feel. A common catch-phrase was "How do you feel/My crazy brothers."[4] The group will introduce by name the rappers and DJ, and also make sure everyone remembers the name of the group. The rappers will comment about how noisy the crowd is. Crowds are more often criticized for not being worked up enough rather than praised for their excitement.

The second song tends to be the one the group is most famous for. On stage, each rapper holds a cordless microphone right up to his mouth, and a rapper might steady the mic by holding his index finger under his nose. The other arm is gesticulating, palm out in a waving motion at the audience. A bobbing motion in the head and shoulders can be more or less pronounced.

Between the second and third song, the group will usually demand some call-and-response from the audience, almost always as follows:

Call	Response	Call	Response
ie yo ho	*ho*	Say, ho	ho
ie yo ho ho	*ho ho*	Say, ho ho	ho ho
ie yo ho ho ho	*ho ho ho*	Say, ho ho ho	ho ho ho
sawage!	[screams]	Make noise!	[screams]

The third and usually final song tends to be a new song, often introduced in English as "brand new shit," a revolting image for English-speakers perhaps, but apparently heard by the audience as a cooler way of saying "new song" than the Japanese *shinkyoku*. If the song is about to be released as a record or CD, this information is also announced before the song's performance. If there are other rap groups in the audience, this is also the time for "shout outs" (praise for fellow hip-hoppers) as in "Shakkazombie in da house" or "Props to King Giddra" and once even "Ian Condry in da house." After the third song, there is seldom talk besides a brief goodbye in English: "Peace" or "We out." Encores are rare, but freestyle sessions, discussed below, are ubiquitous. After the show, rappers retreat backstage or to the bar area, but never linger around the stage after performing. The year 1996 was also a time of a "freestyle boom," when most shows were closed with an open-ended passing of the microphone. Anyone could step on stage and try his or her hand at rapping for a few minutes. This has been an important way for younger performers to get the attention of more established acts. There is a back-and-forth aspect of performance in the clubs that shows how styles are developed, honed, and reworked in a context where the audience is knowledgeable, discriminating, and at times participates in the show itself.

It is important to understand that over the years, this kind of feedback loop has helped determine the shape of current Japanese rap styles. One of my main sites was a weekly Thursday night event that featured another collection of rap groups called Kitchens. Hip-hop collectives such as Kitchens, Little Bird Nation, Funky Grammar Unit, and Rock Steady Crew Japan are called "families" (*famirii*, in Japanese). The different groups often met at clubs or parties, at times getting acquainted after particularly noteworthy freestyle sessions. Over time some would become friends, as well as artistic collaborators, who performed together live or in the studio for each others' albums. Such families define the social organization of the "scene." What is interesting is how they also characterize different aesthetic takes on what Japanese hip-hop should be. Kitchens, for example, aim to combine a pop music sensibility with their love for rap music, and, like many such "party rap" groups, they appeal to a largely female audience. The Funky Grammar Unit aims for a more underground sound that is nonetheless accessible, and they tend to have a more even mix of men and women in the audience. Other families like Urbarian Gym (UBG) are less concerned with being accessible to audiences than with conveying a confrontational, hard-core stance. The lion's share of their audiences are young males, though as UBG's leader, Zeebra, breaks into the pop spotlight, their audiences are becoming more diverse.

The lull that precedes and follows the onstage performance is a key time for networking to build these families. In all, the live show is at most an hour long, at times closer to twenty minutes, and yet there is nowhere for the clubbers to go until the trains start running again around 5 a.m. It is not unusual for music magazine writers to do interviews during club events, and record company representatives often come to shows as well, not only as talent scouts but also to discuss upcoming projects. I found that 3:00 to 4 a.m. was the most productive time for fieldwork because by then the clubbers had mostly exhausted their supply of stories and gossip to tell friends, and were then open to finding out what this strange foreigner jotting things in his notebook was doing in their midst.

Japanese cultural practices do not disappear just because everyone is wearing their hip-hop outfits and listening to the latest rap tunes. To give one example, at the first Kitchens event after the New Year, I was surprised to see all the clubbers who knew each other going around and saying the traditional New Year's greeting in very formal Japanese: "Congratulations on the dawn of the New Year. I humbly request your benevolence this year as well." There was no irony, no joking atmosphere in these statements. This is a good example of the way that globalization may appear to overshadow Japanese culture, but one needs to spend time in clubs with the people to see how surface appearances can be deceiving.

In many ways, then, it is not surprising that rappers, DJs, breakdancers, record company people, and magazine editors all agree you cannot understand the music unless you go to the clubs. There is an intensity of experience in hearing the music at loud volume, surrounded by a crush of dancing people, while drinking alcohol and staying out all night, that gives the music an immediacy and power it lacks when heard, say, on headphones in the quiet of one's room. Indeed, it is difficult to convey in words the feeling of communal excitement during a particularly good show, when one gets wrapped up in a surge of energy that is palpable yet intangible. It is this emotional experience that in many ways counteracts any fears that it is all "merely imitation," which is the most common criticism of the music.

At the same time, going to a club involves a strange mix of the extraordinary and the routine. On one hand, you visit a place with bizarre interior design, listen to music at exceedingly high volume, stay out all night and, often, get drunk. It is a sharp contrast to an ordinary day of school or work. We must also recognize, however, that while a club may strive to be a fantastic microcosm, it is still embedded in Japan's political-economic structures, characteristic social relations, and the contemporary range of cultural forms. It is not by chance that clubs tend to attract people of specific class, age, sexuality, and to some extent locale. Moreover, if you go regularly to clubs, after a while it becomes just another routine. It is largely predictable what kind of pleasures can be expected, and also the generally unpleasant consequences for work or school after a night without sleep.[5] Clubbing offers freedom and constraints. This tension is the key to understanding how clubs socialize the club-goers by structuring pleasure in characteristic ways.

I have only suggested some of the ways that clubs offer insight into the ways that global hip-hop becomes transformed into a local form of Japanese hip-hop, but we can see how an idea of "genba globalism" can help us understand the process of localization. Globalism is refracted and transformed in important ways through the actual site of urban hip-hop clubs. Japanese rappers perform for local audiences in the Japanese language and use Japanese subjects to build their base of fans. In contrast to club events with techno or house music, hip-hop events emphasize lyrics in the shows and the freestyle sessions. There is a wide range of topics addressed in Japanese hip-hop, but they all speak in some way to the local audience. Dassen 3 uses joking lyrics ridiculing school and television. Scha Dara Parr is also playful, emphasizing things like their love of video games and the kind of verbal repartee characteristic of close buddies. When Zeebra acts out his hard-core stance, he tells of drug use in California, expensive dates with girlfriends, and abstract lyrics about hip-hop as a revolutionary war. Rhymester's lyrics are often set in a club or just after a show, for example, describing an imagined, fleeting love affair with a girl on a passing train. Some songs refer to cultural motifs going back centuries, such as a song performed by Rima and Umedy about a double-suicide pact between lovers, remade as a contemporary R&B and hip-hop jam.

UNDERSTANDING GLOBALIZATION

Rap music in Japan offers an interesting case study of the way popular culture is becoming increasingly global in scope, while at the same time becoming domesticated to fit with local ideas and desires. At the dawn of the twenty-first century, entertainment industries are reaching wider markets and larger audiences. The film *Titanic*, for example, grossed over $1.5 billion, the largest amount ever for a film, and two-thirds of this income came from overseas. In music, there are global pop stars too, like Britney Spears and Celine Dion. In rap music, the Fugees could be considered global stars. Their 1996 album "The Score" sold over 17 million copies worldwide. More recently, Lauryn Hill's 1998 album revealed that the transnational market for hip-hop is still growing, and most major rap stars do promotional tours in Japan. An important feature of pop culture commodities is that they tend to be expensive to produce initially, but then relatively cheap to reproduce and distribute. Compact disks are one of the most striking examples. Although studio time is expensive (between $25,000 for a practically homemade album to upwards of $250,000 for state-of-the-art productions), the CDs themselves cost about eighty cents to produce, including the packaging. Obviously, the more one can sell, the higher the return, and this helps explain the eagerness of entertainment businesses to develop new markets around the world.

Less clear are the kinds of effects such globalized pop culture forms might have. The fluidity of culture in the contemporary world raises new questions about how we are linked together, what we share and what divides us. The spread of popular culture seems in some ways linked with a spread in values, but we must be cautious in our assessment of how and to what extent this transfer takes place. It is safe to say that the conventional understanding of globalization is that it is producing a homogenization of cultural forms. From this perspective, we are witnessing the McDonaldization and the Coca-Cola colonization of the periphery by powerful economic centers of the world system. The term "cultural imperialism" captures this idea, that political and economic power is being used "to exalt and spread the values and habits of a foreign culture at the expense of the native culture."[6] In some ways, anthropology as a discipline emerged at a time when there was a similar concern that the forces of modernity (especially missionaries and colonial officials) were wiping out "traditional cultures," and thus one role for ethnographers was to salvage, at least in the form of written documents, the cultures of so-called "primitive peoples." Many people view globalization, and particularly the spread of American pop culture, as a similar kind of invasion, but the idea that watching a Disney movie automatically instills certain values must be examined and not simply assumed. In some ways the goals of anthropology—combatting simplistic and potentially dangerous forms of ethnocentrism—remain as important today as when the discipline was born.

The example of Japanese hip-hop gives us a chance to examine some recent theorizing on globalization. The sociologist Malcolm Waters offers a useful overview of globalization, which he defines as follows:

A social process in which the constraints of geography on social and cultural arrangements recede, and in which people become increasingly aware that they are receding.[7]

A key aspect of this definition is not only that the world is increasingly becoming one place, but that people are becoming increasingly aware of that. This awareness may lead to a heightened sense of risk, such as global warming or the "love bug" virus, or to a rosy view of increased opportunities, for example, to get the most recent hip-hop news in real time or to download the latest music instantly via the Internet.

It is important to recognize, however, that globalization involves much more than Hollywood movies and pop music. Waters does a good job of analyzing three aspects of globalization, namely, economic, political, and cultural. He contends that globalization processes go back five hundred years, and that the relative importance of economic, political, and cultural exchanges has varied over that time.[8] From the sixteenth to nineteenth centuries, economics was key. In particular, the growth of the capitalist world system was the driving force in linking diverse regions. During the nineteenth and twentieth centuries, politics moved to the fore. Nation-states produced a system of international relations that characterized global linkages with multinational corporations and integrated national traditions. Now, at the dawn of the twenty-first century, cultural forms are leading global changes in both politics and economics. Waters argues that a "global idealization" is producing politics based on worldwide values (e.g., human rights, the environment, anti-sweatshop movement) and economic exchanges centered on lifestyle consumerism. The key point is that while economics and then politics were the driving forces in globalization of previous centuries, it is cultural flows that are increasingly important today. If he is right, and I would argue he is, this points to the importance of studying the kinds of ideals that are spread around the globe.

What ideals are spread by hip-hop in Japan? Clubbing certainly promotes an attitude that stresses leisure, fashion, and consumer knowledge of music over other kinds of status in work and school. Although it is important to recognize that the effects of lyrics are somewhat complicated, it is worth considering, to some extent, the messages carried by the music. Although rappers deal with a wide variety of subjects, one theme appears again and again, namely, that youth need to speak out for themselves. As rapper MC Shiro of Rhymester puts it, "If I were to say what *hip-hop* is, it would be a 'culture of the first person singular.' In hip-hop, … rappers are always yelling, 'I'm this.'" Such a message may seem rather innocuous compared to some of the hard-edged lyrics one is likely to hear in the United States, but it is also a reflection of the kind of lives these Japanese youth are leading. In Japan, the education system tends to emphasize rote memorization and to track students according to exams. Sharply age-graded hierarchies are the norm, and may be especially irksome in a situation where the youth are likely to live with their parents until they get married. Moreover, the dominant ideology that harmony of the group should come before individual expression ("the nail that sticks up gets hammered down") makes for a social context in which the hip-hop idea that one should be speaking for oneself is, in some limited sense, revolutionary. At the very least, it shows how global pop culture forms are leading not to some simple homogenization, but rather adding to a complex mix that in many ways can only be studied ethnographically through extended research in local sites.

Another important theorist of globalization is Arjun Appadurai, who proposes that we consider contemporary cultural flows in terms of movement in five categories: people or ethnicity, ideology, finance, technology, and media. He adds the suffix "-scape" to each to highlight that the deterritorialization of cultural forms is accompanied by new landscapes of cultural exchange, thus we have "ethnoscapes, idcoscapes, financescapes, technoscapes, and mediascapes" (others have added "sacriscapes" to describe the spread of religion, and one might add "leisurescapes" for the spread of popular culture). The key point about these landscapes is that they are "non-isomorphic," that is, they don't map evenly onto each other. Appadurai notes, for example, that the "Japanese are notoriously hospitable to ideas and are stereotyped as inclined to export (all) and import (some) goods, but they are notoriously closed to immigration."[9] Migration and electronic mass media are the main driving forces to Appadurai's theorization. One of the problems with Appadurai's theory, however, is that the notion of "-scapes" draws us away from considering how flows of technology, media, finance, and people are connected. An alternative is to consider key sites, *genba* if you will, of various sorts depending on one's interests as a way to see how new, hybrid forms of culture are produced. "Genba globalism" aims to show how artists, fans, producers, and media people are actively consuming and creating these new forms.

One thing that anthropologists offer to the advancement of human knowledge is a clear sense of the ways people interact in specific places. At one time, anthropologists would choose a village or island to map in elaborate detail. Now in a media-filled world, we face different analytical challenges, but the techniques of fieldwork—learning the language, participating in daily life, observing rituals, and so on—can still be used. One of Appadurai's conclusions is that exchanges along these different "-scapes" are leading not only to a deterritorialization of cultural forms, but also to an increased importance of the "work of the imagination."[10] In other words, as identities can be picked up from a variety of media sources, the construction of "who we are" arises increasingly from how we imagine ourselves, rather than from where we live. Life in urban areas seems to make this aspect of identity—as imagination and as performance—all the more salient. What I hope I have drawn attention to is the way the hip-hop nightclubs give us a chance to bring some of this work of the imagination down to the level of daily life.

CONCLUSION: GLOBAL POP AND CULTURAL CHANGE

In the end, the globalization of popular culture needs to be understood as two related yet opposing trends of greater massification and deeper compartmentalization.

On one hand, the recording industry is reaching larger and larger markets, both within Japan and around the world, as mega-hits continue to set sales records. On the other hand, there is an equally profound if less visible process by which niche scenes are becoming deeper and more widely connected than before, and in the process, new forms of heterogeneity are born. Although I have only been able to touch on a few of its aspects here, Japanese rap music is a revealing case study of the social location, cultural role, and capitalist logic of such micro-mass cultures. It is important to recognize, however, that these micro-mass cultures also have the potential to move into the mainstream.

The distinction between "scene" and "market" highlights what is at stake when we try to analyze the cultural and capitalist transformations associated with globalization. Information-based and service industries are growing rapidly, promising to reorganize the bounds of culture and commodities, yet we need close readings of how such emerging economies influence everyday lives. Although B-Boys and B-Girls go to great lengths to distinguish the "cultural" from the "commercial" in their favored genre, it is rather the linkage of the two in the circuits of popular culture that offers the deepest insight. In the end, the winds of global capitalism that carried the seeds of rap music to Japan can only be grasped historically with a close attention to social spaces, media forms, and the rhythms of everyday life.

Walking to a hip-hop club in Tokyo, one is confronted with a tremendous range of consumer options, and it is this heightened sense of "you are what you buy" that has in many ways become the defining feature of identity in advanced capitalist nations, at least among those people with the money to consume their preferred lifestyle. At the same time, it is important to be sensitive to the ways that, outward appearances notwithstanding, the consumers of things like hip-hop are embedded in a quite different range of social relations and cultural meanings. It makes a difference that B-Boys and B-Girls, listening to American hip-hop records, still feel it is important to go around to their friends and associates with the traditional New Year's greeting of deference and obligation. This is an example of the ways social relations within the Japanese rap scene continue to carry the weight of uniquely Japanese practices and understandings.

It is likely, too, that "global pop" will become more heterogeneous as the entertainment industries in other countries develop. There are reasons to think that, in music at least, the domination of American popular music as the leading "global" style seems likely to be a temporary situation. In the immediate postwar period in Japan, Western music initially dominated sales. But sales of Japanese music steadily grew and in 1967 outpaced Western sales. Today, three-fourths of Japan's music market is Japanese music to one-fourth Western. Moreover, although American music currently constitutes about half of global sales, this is down from 80 percent a decade ago. It is quite possible that as local record companies mature in other countries, they will, as in Japan, come to dominate local sales. Certainly, multinational record companies are moving in this direction of developing local talent and relying less on Western pop stars. Moreover, although Japan is a ravenous importer of American popular culture, it has some notable exports as well. Some are more familiar than others, but they include the Mighty Morphin Power Rangers, karaoke, "Japanimation," *manga* (Japanese comic books), mechanical pets, Nintendo or the Sony PlayStation video games, and of course, Pokémon.

Just as it would seem strange to Americans if someone claimed Pokémon is making U.S. kids "more Japanese," it is dangerous to assume that mass culture goods by themselves threaten to overwhelm other cultures. The anthropologist Daniel Miller has been a proponent for taking a closer look at the ways such goods are woven into everyday lives. He argues that mass commodities are better analyzed in terms of an "unprecedented diversity created by the differential consumption of what had once been thought to be global and homogenizing institutions." Miller's emphasis on the active and creative aspects of consumption is characteristic of a broad trend within the social sciences to view global commodities in terms of their local appropriations, and to represent local consumers with a greater degree of agency than found in other works that emphasize "cultural imperialism." It is this perspective that seems to me the best characterization of what is going on in Japan.

It is easy to see how the "sameness" aspect of globalization is promoted. Music magazines, TV video shows, and record stores promote similar artists whether in Japan or the United States. A new album by Nas is met with a flurry of publicity in the Japanese and English-language hip-hop magazines available in Tokyo. This relates in part to the structure of record companies and their marketing practices. In this sense, the widening and increasingly globalized market for popular culture does appear to be leading to greater homogenization. But it is primarily a process of homogenizing what is available, regardless of where you are. I would argue that the global marketing blitz of megahit productions like the film *Titanic* and the music of Celine Dion and Lauryn Hill reflect a homogenization of *what* is available, but not *how* it is interpreted. Although it is more difficult to see, in part because it is hidden beneath similar clothes, hairstyles, and consumption habits, different interpretations are generated in different social contexts. By attending to "actual sites" of cultural production and consumption, we can more clearly gauge the ways local contexts alter the meanings of globalization. "Keeping it real" for hip-hoppers in Japan means paying attention to local realities.

ENDNOTES

1. The word "*genba*" is made up of the characters "to appear" and "place," and it is used to describe a place where something actually happens, like the scene of an accident or of a crime, or a construction site. In the hip-hop world the term is used to contrast the intense energy of the club scene with the more sterile and suspect marketplace.

2. Hip-hop is not the only style for club music. Techno, House, Reggae, Jungle/Drum 'n' Bass, and so on, are some of the other popular club music styles. Live

music tends to be performed earlier in the evening, usually starting around 7 p.m., and finishing in time for the audience to catch an evening train home. In contrast to "clubs," "discos" must by law close by 1 a.m.

3. The term "breakbeats" refers to the section of a song where only the drums, or drums and bass, play. It is the break between the singing and the melodies of the other instruments, hence "breakbeats." This section can be looped by a DJ using two turntables and a mixer with cross-fader, and produces a backing track suitable for rapping.

4. In Japanese, *chôoshi wa dô dailikarera kyôdai*. The masculinity of Japanese rap is here indexed by the calling out to "brothers" and also by the use of the masculine slang *dai* instead of *da*.

5. Youthful Japanese clubbers use the mixed English-Japanese construction "all *suru*" (do all) to mean "stay out all night in a club." For example, the following exchange occurred between two members of the female group Now. Here, the sense of routine outweighs the excitement. A: *konban mo ooru suru ka na*? (Are we staying out all night again?) B: *Tabun*. (Probably) A: *Yabai*. (That sucks.)

6. John Tomlinson, *Cultural Imperialism: A Critical Introduction* (London: Printer Publishers, 1991), p. 3.

7. Malcolm Waters, *Globalization* (London: Routledge, 1995), p. 3.

8. Waters, pp. 157–164.

9. Arjun Appadurai, *Modernity at Large: Cultural Dimensions of Globalization* (Minneapolis: University of Minnesota Press, 1996), p. 37.

10. Appadurai, p. 3.

21

Yearning for Lightness: Transnational Circuits in the Marketing and Consumption of Skin Lighteners

EVELYN NAKANO GLENN

Colorism—social ranking based on gradations of skin tone—is found throughout the world. In this article, Glenn examines the transnational "yearning for lightness" and the ways in which global pharmaceutical and cosmetic companies work to link light skin with modernity, social mobility, and youth.

With the breakdown of traditional racial categories in many areas of the world, colorism, by which I mean the preference for and privileging of lighter skin and discrimination against those with darker skin, remains a persisting frontier of intergroup and intragroup relations in the twenty-first century. Sociologists and anthropologists have documented discrimination against darker-skinned persons and correlations between skin tone and socioeconomic status and achievement in Brazil and the United States (Hunter 2005; Sheriff 2001; Telles 2004). Other researchers have revealed that people's judgments about other people are literally colored by skin tone, so that darker-skinned individuals are viewed as less intelligent, trustworthy, and attractive than their lighter-skinned counter-parts (Herring, Keith, and Horton 2003; Hunter 2005; Maddox 2004).

One way of conceptualizing skin color, then, is as a form of symbolic capital that affects, if not determines, one's life chances. The relation between skin color and judgments about attractiveness affect women most acutely, since women's worth is judged heavily on the basis of appearance. For example, men who have wealth, education, and other forms of human capital are considered "good catches," while women who are physically attractive may be considered desirable despite the lack of other capital. Although skin tone is usually seen as a form of fixed or unchangeable capital, in fact, men and women may attempt to acquire light-skinned privilege. Sometimes this search takes the form of seeking light-skinned marital partners to raise one's status and to achieve

SOURCE: Evelyn Nakano Glenn, "Yearning for Lightness: Transnational Circuits in the Marketing and Consumption of Skin Lighteners," *Gender & Society* 22(3), June 2008, pp. 281-302. Used with permission.

intergenerational mobility by increasing the likelihood of having light-skinned children. Often, especially for women, this search takes the form of using cosmetics or other treatments to change the appearance of one's skin to make it look lighter.

This article focuses on the practice of skin lightening, the marketing of skin lighteners in various societies around the world, and the multinational corporations that are involved in the global skin-lightening trade. An analysis of this complex topic calls for a multilevel approach. First, we need to place the production, marketing, and consumption of skin lighteners into a global political-economic context. I ask, How is skin lightening interwoven into the world economic system and its transnational circuits of products, capital, culture, and people? Second, we need to examine the mediating entities and processes by which skin lighteners reach specific national/ethnic/racial/class consumers. I ask, What are the media and messages, cultural themes and symbols, used to create the desire for skin-lightening products among particular groups? Finally, we need to examine the meaning and significance of skin color for consumers of skin lighteners. I ask, How do consumers learn about, test, and compare skin-lightening products, and what do they seek to achieve through their use?

The issue of skin lightening may seem trivial at first glance. However, it is my contention that a close examination of the global circuits of skin lightening provides a unique lens through which to view the workings of the Western-dominated global system as it simultaneously promulgates a "white is right" ideology while also promoting the desire for and consumption of Western culture and products.

SKIN LIGHTENING AND GLOBAL CAPITAL

Skin lightening has long been practiced in many parts of the world. Women concocted their own treatments or purchased products from self-styled beauty experts offering special creams, soaps, or lotions, which were either ineffective sham products or else effective but containing highly toxic materials such as mercury or lead. From the perspective of the supposedly enlightened present, skin lightening might be viewed as a form of vanity or a misguided and dangerous relic of the past.

However, at the beginning of the twenty-first century, the search for light skin, free of imperfections such as freckles and age spots, has actually accelerated, and the market for skin-lightening products has mushroomed in all parts of the world. The production and marketing of products that offer the prospect of lighter, brighter, whiter skin has become a multi-billion-dollar global industry. Skin lightening has been incorporated into transnational flows of capital, goods, people, and culture. It is implicated in both the formal global economy and various informal economies. It is integrated into both legal and extralegal transnational circuits of goods. Certain large multinational corporations have become major players, spending vast sums on research and development and on advertising and marketing to reach both mass and specialized markets. Simultaneously, actors in informal or underground economies, including smugglers, transnational migrants, and petty traders, are finding unprecedented opportunities in producing, transporting, and selling unregulated lightening products.

One reason for this complex multifaceted structure is that the market for skin lighteners, although global in scope, is also highly decentralized and segmented along socioeconomic, age, national, ethnic, racial, and cultural lines. Whether the manufacturers are multi-billion-dollar corporations or small entrepreneurs, they make separate product lines and use distinct marketing strategies to reach specific segments of consumers. Ethnic companies and entrepreneurs may be best positioned to draw on local cultural themes, but large multinationals can draw on local experts to tailor advertising images and messages to appeal to particular audiences.

The Internet has become a major tool/highway/engine for the globalized, segmented lightening market. It is the site where all of the players in the global lightening market meet. Large multinationals, small local firms, individual entrepreneurs, skin doctors, direct sales merchants, and even eBay sellers use the Internet to disseminate the ideal of light skin and to advertise and sell their products. Consumers go on the Internet to do research on products and shop. Some also participate in Internet message boards and forums to seek advice and to discuss, debate, and rate skin lighteners. There are many such forums, often as part of transnational ethnic Web sites. For example, IndiaParenting.com and sukh-dukh.com, designed for South Asians in India and other parts of the world, have chat rooms on skin care and lightening, and Rexinteractive.com, a Filipino site, and Candymag.com, a site sponsored by a magazine for Filipina teens, have extensive forums on skin lightening. The discussions on these forums provide a window through which to view the meaning of skin color to consumers, their desires and anxieties, doubts and aspirations. The Internet is thus an important site from which one can gain a multilevel perspective on skin lightening.

CONSUMER GROUPS AND MARKET NICHES

Africa and African Diaspora

In Southern Africa, colorism is just one of the negative inheritances of European colonialism. The ideology of white supremacy that European colonists brought included the association of Blackness with primitiveness, lack of civilization, unrestrained sexuality, pollution, and dirt. The association of Blackness with dirt can be seen in a 1930 French advertising poster for Dirtoff. The poster shows a drawing of a dark African man washing his hands, which have become white, as he declares, "Le Savon Dirtoff me blanchit!" The soap was designed not for use by Africans but, as the poster notes, *pour mécaniciens automobilistes et ménagères*—French auto mechanics and housewives. Such images showing Black people "dramatically losing their pigmentation as a result of the cleansing process," were common in late nineteenth- and early twentieth-century soap advertisements, according to art historian Jean Michel Massing (1995, 180).

Some historians and anthropologists have argued that precolonial African conceptions of female beauty favored women with light brown, yellow, or reddish tints. If so, the racial hierarchies established in areas colonized by Europeans cemented and generalized the privilege attached to light skin (Burke 1996; Ribane 2006, 12). In both South Africa and Rhodesia/Zimbabwe, an intermediate category of those considered to be racially mixed was classified as "coloured" and subjected to fewer legislative restrictions than those classified as "native." Assignment to the coloured category was based on ill-defined criteria, and on arrival in urban areas, people found themselves classified as native or coloured on the basis of skin tone and other phenotypic characteristics. Indians arriving in Rhodesia from Goa, for example, were variously classified as "Portuguese Mulatto" or coloured. The multiplication of discriminatory laws targeting natives led to a growing number of Blacks claiming to be coloured in both societies (Muzondidya 2005, 23–24).

The use of skin lighteners has a long history in Southern Africa, which is described by Lynn Thomas and which I will not recount here (in press). Rather, I will discuss the current picture, which shows both a rise in the consumption of skin-lightening products and concerted efforts to curtail the trade of such products. Despite bans on the importation of skin lighteners, the widespread use of these products currently constitutes a serious health issue in Southern Africa because the products often contain mercury, corticosteroids, or high doses of hydroquinone. Mercury of course is highly toxic, and sustained exposure can lead to neurological damage and kidney disease. Hydroquinone (originally an industrial chemical) is effective in suppressing melanin production, but exposure to the sun—hard to avoid in Africa—damages skin that has been treated. Furthermore, in dark-skinned people, long-term hydroquinone use can lead to ochronosis, a disfiguring condition involving gray and blue-black discoloration of the skin (Mahe, Ly, and Dangou 2003). The overuse of topical steroids can lead to contact eczema, bacterial and fungal infection, Cushing's syndrome, and skin atrophy (Margulies n.d.; Ntambwe 2004).

Perhaps the most disturbing fact is that mercury soaps used by Africans are manufactured in the European Union (EU), with Ireland and Italy leading in the production of mercury soap. One company that has been the target of activists is Killarney Enterprises, Ltd., in County Wicklow, Ireland. Formerly known as W&E Products and located in Lancashire, England, the company was forced to close following out-of-court settlements of suits filed by two former employees who had given birth to stillborn or severely malformed infants due to exposure to mercury. However, W&E Products then secured a 750,000-pound grant from the Irish Industrial Development Authority to relocate to Ireland, where it changed its name to Killarney Enterprises, Ltd. The company remained in business until April 17, 2007, producing soaps under the popular names Tura, Arut, Swan, Sukisa Bango, Meriko, and Jeraboo (which contained up to 3 percent mercuric iodide). Distribution of mercury soap has been illegal in the EU since 1989, but its manufacture has remained legal as long as the product is exported

(Chadwick 2001; Earth Summit 2002, 13–14). These soaps are labeled for use as antiseptics and to prevent body odor; however, they are understood to be and are used as skin bleaches. To complete the circuit, EU–manufactured mercury soaps are smuggled back into the EU to sell in shops catering to African immigrant communities. An Irish journalist noted that the very same brands made by Killarney Enterprises, including Meriko and Tura (banned in both the EU and South Africa), could easily be found in African shops in Dublin (De Faoite 2001; O'Farrell 2002).

As a result of the serious health effects, medical researchers have conducted interview studies to determine how prevalent the practice of skin lightening is among African women. They estimate that 25 percent of women in Bamaki, Mali; 35 percent in Pretoria, South Africa; and 51 percent in Dakar, Senegal, use skin lighteners, as do an astonishing 77 percent of women traders in Lagos, Nigeria (Adebajo 2002; del Guidice and Yves 2002; Mahe, Ly, and Dangou 2003; Malangu and Ogubanjo 2006).

There have been local and transnational campaigns to stop the manufacture of products containing mercury in the EU and efforts to inform African consumers of the dangers of their use and to foster the idea of Black pride. Governments in South Africa, Zimbabwe, Nigeria, and Kenya have banned the import and sale of mercury and hydroquinone products, but they continue to be smuggled in from other African nations (Dooley 2001; Thomas 2004).

Despite these efforts, the use of skin lighteners has been increasing among modernized and cosmopolitan African women. A South African newspaper reported that whereas in the 1970s, typical skin lightener users in South Africa were rural and poor, currently, it is upwardly mobile Black women, those with technical diplomas or university degrees and well-paid jobs, who are driving the market in skin lighteners. A recent study by Mictert Marketing Research found that 1 in 13 upwardly mobile Black women aged 25 to 35 used skin lighteners. It is possible that this is an underestimation, since there is some shame attached to admitting to using skin lighteners (Ntshingila 2005).

These upwardly mobile women turn to expensive imported products from India and Europe rather than cheaper, locally made products. They also go to doctors to get prescriptions for imported lighteners containing corticosteroids, which are intended for short-term use to treat blemishes. They continue using them for long periods beyond the prescribed duration, thus risking damage (Ntshingila 2005). This recent rise in the use of skin lighteners cannot be seen as simply a legacy of colonialism but rather is a consequence of the penetration of multinational capital and Western consumer culture. The practice therefore is likely to continue to increase as the influence of these forces grows.

African America

Color consciousness in the African American community has generally been viewed as a legacy of slavery, under which mulattos, the offspring of white men and slave women, were accorded better treatment than "pure" Africans. While slave owners considered dark-skinned Africans suited to fieldwork, lighter-skinned mulattos were thought to be more intelligent and better suited for indoor work as servants and artisans. Mulattos were also more likely to receive at least rudimentary education and to be manumitted. They went on to form the nucleus of many nineteenth-century free Black communities. After the Civil War, light-skinned mulattos tried to distance themselves from their darker-skinned brothers and sisters, forming exclusive civic and cultural organizations, fraternities, sororities, schools, and universities (Russell, Wilson, and Hall 1992, 24–40). According to Audrey Elisa Kerr, common folklore in the African American community holds that elite African Americans used a "paper bag" test to screen guests at social events and to determine eligibility for membership in their organizations: anyone whose skin was darker than the color of the bag was excluded. Although perhaps apocryphal, the widespread acceptance of the story as historical fact is significant. It has been credible to African Americans because it was consonant with their observations of the skin tone of elite African American society (Kerr 2005).

The preference and desire for light skin can also be detected in the longtime practice of skin lightening. References to African American women using powders and skin bleaches appeared in the Black press as early as the 1850s, according to historian Kathy Peiss. She notes that *American Magazine* criticized African Americans who tried to emulate white beauty standards: "Beautiful black and brown faces by application of rouge and lily white are made to assume unnatural tints, like the vivid hue of painted corpses" (Peiss 1998, 41). How common such practices were is unknown. However, by the 1880s and 1890s, dealers in skin bleaches were widely advertising their wares in the African American press. A Crane and Company ad in the *Colored American Magazine* (1903) promised that use of the company's "wonderful Face Bleach" would result in a "peach-like complexion" and "turn the skin of a black or brown person five or six shades lighter and of a mulatto person perfectly white" (Peiss 1998, 41, 42).

Throughout the twentieth century, many African American leaders spoke out against skin bleaching, as well as hair straightening, and the African American press published articles decrying these practices. However, such articles were far outnumbered by advertisements for skin bleaches in prominent outlets such as the *Crusader, Negro World*, and the *Chicago Defender*. An estimated 30 to 40 percent of advertisements in these outlets were for cosmetics and toiletries including skin bleaches. Many of the advertised lighteners were produced by white manufacturers; for example, Black and White Cream was made by Plough Chemicals (which later became Plough-Shearing), and Nadolina was made by the National Toilet Company. A chemical analysis of Nadolina Bleach conducted in 1930 found it contained 10 percent ammoniated mercury, a concentration high enough to pose a serious health risk. Both brands are still marketed in African American outlets, although with changed ingredients (Peiss 1998, 210, 212).[1]

The manufacture and marketing of Black beauty products, including skin lighteners, provided opportunities for Black entrepreneurs. Annie Turnbo Malone, who founded the Poro brand, and Sara Breedlove, later known as Madam C. J. Walker, who formulated and marketed the Wonder Hair Grower, were two of the most successful Black entrepreneurs of the late nineteenth and early twentieth centuries. Malone and Walker championed African American causes and were benefactors of various institutions (Peiss 1998, 67–70; see also Bundles 2001). Significantly, both refused to sell skin bleaches or to describe their hair care products as hair straighteners. After Walker died in 1919, her successor, F. B. Ransom, introduced Tan-Off, which became one of the company's best sellers in the 1920s and 1930s. Other Black-owned companies, such as Kashmir (which produced Nile Queen), Poro, Overton, and Dr. Palmer, advertised and sold skin lighteners. Unlike some white-produced products, they did not contain mercury but relied on such ingredients as borax and hydrogen peroxide (Peiss 1998, 205, 212, 213).

Currently, a plethora of brands is marketed especially to African Americans, including Black and White Cream, Nadolina (sans mercury), Ambi, Palmer's, DR Daggett and Ramsdell (fade cream and facial brightening cream), Swiss Whitening Pills, Ultra Glow, Skin Success, Avre (which produces the Pallid Skin Lightening System and B-Lite Fade Cream), and Clear Essence (which targets women of color more generally). Some of these products contain hydroquinone, while others claim to use natural ingredients.

Discussions of skin lightening on African American Internet forums indicate that the participants seek not white skin but "light" skin like that of African American celebrities such as film actress Halle Berry and singer Beyoncé Knowles. Most women say they want to be two or three shades lighter or to get rid of dark spots and freckles to even out their skin tones, something that many skin lighteners claim to do. Some of the writers believe that Halle Berry and other African American celebrities have achieved their luminescent appearance through skin bleaching, skillful use of cosmetics, and artful lighting. Thus, some skin-lightening products, such as the Pallid Skin Lightening System, purport to offer the "secret" of the stars. A Web site for Swiss Lightening Pills claims that "for many years Hollywood has been keeping the secret of whitening pills" and asks, rhetorically, "Have you wondered why early childhood photos of many top celebs show a much darker skin colour than they have now?"[2]

India and Indian Diaspora

As in the case of Africa, the origins of colorism in India are obscure, and the issue of whether there was a privileging of light skin in precolonial Indian societies is far from settled. Colonial-era and postcolonial Indian writings on the issue may themselves have been influenced by European notions of caste, culture, and race. Many of these writings expound on a racial distinction between lighter-skinned Aryans, who migrated into India from the North, and darker-skinned "indigenous" Dravidians of the South. The wide range of skin color from North to South and the variation in skin tone within castes make it hard to correlate light skin with high caste. The most direct connection between skin color and social status could be found in the paler hue of those whose position and wealth enabled them to spend their lives sheltered indoors, compared to the darker hue of those who toiled outdoors in the sun (Khan 2008).

British racial concepts evolved over the course of its colonial history as colonial administrators and settlers attempted to make sense of the variety of cultural and language groups and to justify British rule in India. British observers attributed group differences variously to culture, language, climate, or biological race. However, they viewed the English as representing the highest culture and embodying the optimum physical type: they made invidious comparisons between lighter-skinned groups, whose men they viewed as more intelligent and martial and whose women they considered more attractive, and darker-skinned groups, whose men they viewed as lacking intelligence and masculinity, and whose women they considered to be lacking in beauty (Arnold 2004).

Regardless of the origins of color consciousness in India, the preference for light skin seems almost universal today, and in terms of sheer numbers, India and Indian diasporic communities around the world constitute the largest market for skin lighteners. The major consumers of these products in South Asian communities are women between the ages of 16 and 35. On transnational South Asian blog sites, women describing themselves as "dark" or "wheatish" in color state a desire to be "fair." Somewhat older women seek to reclaim their youthful skin color, describing themselves as having gotten darker over time. Younger women tend to be concerned about looking light to make a good marital match or to appear lighter for large family events, including their own weddings. These women recognize the reality that light skin constitutes valuable symbolic capital in the marriage market (Views on Article n.d.).

Contemporary notions of feminine beauty are shaped by the Indian mass media. Since the 1970s, beauty pageants such as Miss World–India have been exceedingly popular viewer spectacles; they are a source of nationalist pride since India has been highly successful in international pageants such as Miss World. As might be expected, the competitors, although varying in skin tone, tend to be lighter than average. The other main avatars of feminine allure are Bollywood actresses, such as Isha Koppikar and Aishwarya Rai, who also tend to be light skinned or, if slightly darker, green eyed (see http://www.indianindustry.com/herbalcosmetics/10275.htm).

Many Indian women use traditional homemade preparations made of plant and fruit products. On various blog sites for Indians both in South Asia and diasporic communities in North America, the Caribbean, and the United Kingdom, women seek advice about "natural" preparations and trade recipes. Many commercial products are made by Indian companies and marketed to Indians around the globe under such names as "fairness cream," "herbal bleach cream," "whitening cream," and "fairness cold cream." Many of these products claim to be based on ayurvedic medicine and contain herbal and fruit extracts such as saffron, papaya, almonds, and lentils (Runkle 2004).

With economic liberalization in 1991, the number of products available on the Indian market, including cosmetics and skin care products, has mushroomed. Whereas prior to 1991, Indian consumers had the choice of two brands of cold cream and moisturizers, today they have scores of products from which to select. With deregulation of imports, the rise of the Indian economy, and growth of the urban middle class, multinational companies see India as a prime target for expansion, especially in the area of personal care products. The multinationals, through regional subsidiaries, have developed many whitening product lines in various price ranges that target markets ranging from rural villagers to white-collar urban dwellers and affluent professionals and managers (Runkle 2005).

Southeast Asia: the Philippines

Because of its history as a colonial dependency first of Spain and then of the United States, the Philippines has been particularly affected by Western ideology and culture, both of which valorize whiteness. Moreover, frequent intermarriage among indigenous populations, Spanish colonists, and Chinese settlers has resulted in a substantially mestizo population that ranges widely on the skin color spectrum. The business and political elites have tended to be dispropor-tionately light skinned with visible Hispanic and/or Chinese appearance. In the contemporary period, economic integration has led to the collapse of traditional means of livelihood, resulting in large-scale emigration by both working-class and middle-class Filipinos to seek better-paying jobs in the Middle East, Asia, Europe, and North America. An estimated 10 million Filipinos were working abroad as of 2004, with more than a million departing each year. Because of the demand for domestic workers, nannies, and care workers in the global North, women make up more than half of those working abroad (Tabbada 2006). Many, if not most, of these migrants remit money and send Western con-sumer goods to their families in the Philippines. They also maintain transnational ties with their families at home and across the diaspora through print media, phone, and the Internet. All of these factors contribute to an interest in and fas-cination with Western consumer culture, including fashion and cosmetics in the Philippines and in Filipino diasporic communities (Parreñas 2001).

Perhaps not surprisingly, interest in skin lightening seems to be huge and growing in the Philippines, especially among younger urban women. Synovate, a market research firm, reported that in 2004, 50 percent of respondents in the Philippines reported currently using skin lightener (Synovate 2004). Young Filipinas participate in several Internet sites seeking advice on lightening pro-ducts. They seek not only to lighten their skin overall but also to deal with dark underarms, elbows, and knees. Judging by their entries in Internet discus-sion sites, many teens are quite obsessed with finding "the secret" to lighter skin and have purchased and tried scores of different brands of creams and pills. They are disappointed to find that these products may have some temporary effects but do not lead to permanent change. They discuss products made in the Philippines but are most interested in products made by large European and American mul-tinational cosmetic firms and Japanese and Korean companies. Clearly, these young Filipinas associate light skin with modernity and social mobility. Interest-ing to note, the young Filipinas do not refer to Americans or Europeans as hav-ing the most desirable skin color. They are more apt to look to Japanese and Koreans or to Spanish- or Chinese- appearing (and light-skinned) Filipina celeb-rities, such Michelle Reis, Sharon Cuneta, or Claudine Barretto, as their ideals.[3]

The notion that Japanese and Korean women represent ideal Asian beauty has fostered a brisk market in skin lighteners that are formulated by Korean and Japanese companies. Asian White Skin and its sister company Yumei Misei, headquartered in Korea, sell Japanese and Korean skin care products in the Philippines both in retail outlets and online. Products include Asianwhiteskin Underarm Whitening Kit, Japanese Whitening Cream Enzyme Q-10, Japan Whitening Fruit Cream, Kang Tian Sheep Placenta Whitening Capsules, and Kyusoku Bhaku Lightening Pills (see http://yumeimise.com/store/index).

East Asia: Japan, China, and Korea

East Asian societies have historically idealized light or even white skin for women. As Intage (2001), a market research firm in Japan, puts it, "Japan has long idolized ivory-like skin that is 'like a boiled egg'—soft, white and smooth on the surface." Indeed, prior to the Meiji Period (starting in the 1860s), men and women of the higher classes wore white-lead powder makeup (along with blackened teeth and shaved eyebrows). With modernization, according to Mikiko Ashikari, men completely abandoned makeup, but middle- and upper-class women continued to wear traditional white-lead powder when dressed in formal kimonos for ceremonial occasions, such as marriages, and adopted light-colored modern face powder to wear with Western clothes. Ashikari finds through observations of 777 women at several sites in Osaka during 1996–1997 that 97.4 percent of women in public wore what she calls "white face," that is, makeup that "makes their faces look whiter than they really are" (2003, 3).

Intage (2001) reports that skin care products, moisturizers, face masks, and skin lighteners account for 66 percent of the cosmetics market in Japan. A perusal of displays of Japanese cosmetics and skin care products shows that most, even those not explicitly stated to be whitening products, carry names that contain the word "white," for example, facial masks labeled "Clear Turn White" or "Pure White." In addition, numerous products are marketed specifically as whiteners. All of the leading Japanese firms in the cosmetics field, Shiseido, Kosa, Kanebo, and Pola, offer multiproduct skin-whitening lines, with names such as "White Lucent" and "Whitissimo." Fytokem, a Canadian company that produces ingredients used in skin-whitening products, reports that Japan's market in skin lighteners topped $5 billion in 1999 (Saskatchewan Business Unlimited 2005). With deregulation of imports, leading multinational firms, such as L'Oreal, have also made large inroads in the Japanese market. French products have a special cachet (Exhibitor Info 2006).

While the Japanese market has been the largest, its growth rate is much lower than those of Korea and China. Korea's cosmetic market as been growing at a 10 percent rate per year while that of China has been growing by 20 percent. Fytokem estimates that the market for skin whiteners in China was worth $1 billion in 2002 and was projected to grow tremendously. A 2007 Nielsen global survey found that 46 percent of Chinese, 47 percent of people in Hong Kong, 46 percent of Taiwanese, 29 percent of Koreans, and 24 percent of Japanese had used a skin lightener in the past year. As to regular users, 30 percent of Chinese, 20 percent of Taiwanese, 18 percent of Japanese and Hong Kongers, and 8 percent of Koreans used them weekly or daily. However, if money were no object, 52 percent of Koreans said they would spend more on skin lightening, compared to 26 percent of Chinese, 23 percent of Hong Kongers and Taiwanese, and 21 percent of Japanese (Nielsen 2007).

Latin America: Mexico and the Mexican Diaspora

Throughout Latin America, skin tone is a major marker of status and a form of symbolic capital, despite national ideologies of racial democracy. In some countries, such as Brazil, where there was African chattel slavery and extensive miscegenation, there is considerable color consciousness along with an elaborate vocabulary to refer to varying shades of skin. In other countries, such as Mexico, the main intermixture was between Spanish colonists and indigenous peoples, along with an unacknowledged admixture with African slaves. *Mestizaje* is the official national ideal. The Mexican concept of mestizaje meant that through racial and ethnic mixture, Mexico would gradually be peopled by a whiter "cosmic race" that surpassed its initial ingredients. Nonetheless, skin tone, along with other phenotypical traits, is a significant marker of social status, with lightness signifying purity and beauty and darkness signifying contamination and ugliness (Stepan 1991, 135). The elite has remained overwhelmingly light skinned and European appearing while rural poor are predominantly dark skinned and indigenous appearing.

Ethnographic studies of Mexican communities in Mexico City and Michoacan found residents to be highly color conscious, with darker-skinned family members likely to be ridiculed or teased. The first question that a relative often poses about a newborn is about his or her color (Farr 2006, chap. 5; Guttman 1996, 40; Martinez 2001). Thus, it should not be a surprise that individuals pursue various strategies to attain light-skinned identity and privileges. Migration from rural areas to the city or to the United States has been one route to transformation from an Indian to a mestizo identity or from a mestizo to a more cosmopolitan urban identity; another strategy has been lightening one's family line through marriage with a lighter-skinned partner. A third strategy has been to use lighteners to change the appearance of one's skin (Winders, Jones, and Higgins 2005, 77–78).

In one of the few references to skin whitening in Mexico, Alan Knight claims that it was "an ancient practice … reinforced by film, television, and advertising stereotypes" (1990, 100). As in Africa, consumers seeking low-cost lighteners can easily purchase mercury-laden creams that are still manufactured and used in parts of Latin America (e.g., Recetas de la Farmacia–Crema Blanqueadora, manufactured in the Dominican Republic, contains 6000 ppm of mercury) (NYC Health Dept. 2005). The use of these products has come to public attention because of their use by Latino immigrants in the United States. Outbreaks of mercury poisoning have been reported in Texas, New Mexico, Arizona, and California among immigrants who used Mexican-manufactured creams such as Crema de Belleza–Manning. The cream is manufactured in

Mexico by Laboratories Vide Natural SA de CV, Tampico, Tamaulipas, and is distributed primarily in Mexico. However, it has been found for sale in shops and flea markets in the United States in areas located along the U.S.–Mexican border in Arizona, California, New Mexico, and Texas. The label lists the ingredient calomel, which is mercurous chloride (a salt of mercury). Product samples have been found to contain 6 to 10 percent mercury by weight (Centers for Disease Control 1996; U.S. Food and Drug Administration 1996).

For high-end products, hydroquinone is the chemical of choice. White Secret is one of the most visible products since it is advertised in a 30-minute, late-night television infomercial that is broadcast nationally almost nightly.[4] Jamie Winders and colleagues (2005), who analyze the commercial, note that the commercial continually stresses that White Secret is "una formula Americana." According to Winders, Jones, and Higgins, the American pedigree and English-language name endow White Secret with a cosmopolitan cachet and "a first worldliness." The infomercial follows the daily lives of several young urban women, one of whom narrates and explains how White Secret cream forms a barrier against the darkening rays of the sun while a sister product transforms the color of the skin itself. The infomercial conjures the power of science, showing cross sections of skin cells. By showing women applying White Secret in modern, well-lit bathrooms, relaxing in well-appointed apartments, and protected from damaging effects of the sun while walking around the city, the program connects skin lightening with cleanliness, modernity, and mobility (Winders, Jones, and Higgins 2005, 80–84).

Large multinational firms are expanding the marketing of skin care products, including skin lighteners, in Mexico and other parts of Latin America. For example, Stiefel Laboratories, the world's largest privately held pharmaceutical company, which specializes in dermatology products, targets Latin America for skin-lightening products. Six of its 28 wholly owned subsidiaries are located in Latin America. It offers Clariderm, an over-the-counter hydroquinone cream and gel (2 percent), in Brazil, as well as Clasifel, a prescription-strength hydroquinone cream (4 percent), in Mexico, Peru, Bolivia, Venezuela, and other Latin American countries. It also sells Claripel, a 4 percent hydroquinone cream, in the United States.[5]

Middle-Aged and Older White Women in North America and Europe

Historically, at least in the United States, the vast majority of skin lightener users have been so-called white women. Throughout the nineteenth and early twentieth centuries, European American women, especially those of Southern and Eastern European origins, sought to achieve whiter and brighter skin through use of the many whitening powders and bleaches on the market. In 1930, J. Walter Thomson conducted a survey and found 232 brands of skin lighteners and bleaches for sale. Advertisements for these products appealed to the association of white skin with gentility, social mobility, Anglo-Saxon superiority, and youth. In large cities, such as New York and Chicago, some Jewish women used skin lighteners and hair straighteners produced by Black companies and frequented Black beauty parlors (Peiss 1998, 85, 149, 224).

By the mid-1920s, tanning became acceptable for white women, and in the 1930s and 1940s, it became a craze. A year-round tan came to symbolize high social status since it indicated that a person could afford to travel and spend time at tropical resorts and beaches. In addition, there was a fad for "exotic" Mediterranean and Latin types, with cosmetics designed to enhance "olive" complexions and brunette hair (Peiss 1998, 150–51, 148–49).

However, in the 1980s, as the damaging effects of overexposure to sun rays became known, skin lightening among whites reemerged as a major growth market. Part of this growth was fueled by the aging baby boom generation determined to stave off signs of aging. Many sought not only toned bodies and uplifted faces but also youthful skin—that is, smooth, unblemished, glowing skin without telltale age spots. Age spots are a form of hyperpigmentation that results from exposure to the sun over many years. The treatment is the same as that for overall dark skin: hydroquinone, along with skin peeling, exfoliants, and sunscreen.[6]

MULTINATIONAL COSMETIC AND PHARMACEUTICAL FIRMS AND THEIR TARGETING STRATEGIES

Although there are many small local manufacturers and merchants involved in the skin-lightening game, I want to focus on the giant multinationals, which are fueling the desire for light skin through their advertisement and marketing strategies. The accounts of the skin-lightening markets have shown that the desire for lighter skin and the use of skin bleaches is accelerating in places where modernization and the influence of Western capitalism and culture are most prominent. Multinational biotechnology, cosmetic, and pharmaceutical corporations have coalesced through mergers and acquisitions to create and market personal care products that blur the lines between cosmetics and pharmaceuticals. They have jumped into the field of skin lighteners and correctors, developing many product lines to advertise and sell in Europe, North America, South Asia, East and Southeast Asia, and the Middle East (Wong 2004).

Three of the largest corporations involved in developing the skin-lightening market are L'Oreal, Shiseido, and Unilever. The French-based L'Oreal, with €15.8 billion in sales in 2006, is the largest cosmetics company in the world. It consists of 21 major subsidiaries including Lancôme; Vichy Laboratories; LaRoche-Posay Laboratoire Pharmaceutique; Biotherm; Garnier; Giorgio Armani Perfumes; Maybelline, New York; Ralph Lauren Fragrances; Skinceuticals; Shu Uemura; Matrix; Redken; and SoftSheen Carlson. L'Oreal is also a 20 percent shareholder of Sanofi-Synthelabo, a major France-based pharmaceutical firm.

Three L'Oreal subsidiaries produce the best-known skin-lightening lines marketed around the world (which are especially big in Asia): Lancôme Blanc Expert with Melo-No Complex, LaRoche-Posay Mela-D White skin lightening daily lotion with a triple-action formula, and Vichy Biwhite, containing procystein and vitamin C.

A second major player in the skin-lightening market is Shiseido, the largest and best-known Japanese cosmetics firm, with net sales of $5.7 billion. Shiseido cosmetics are marketed in 65 countries and regions, and it operates factories in Europe, the Americas, and other Asian countries. The Shiseido Group, including affiliates, employs approximately 25,200 people around the globe. Its two main luxury lightening lines are White Lucent (for whitening) and White Lucency (for spots/aging). Each product line consists of seven or eight components, which the consumer is supposed to use as part of a complicated regimen involving applications of specific products several times a day.[7]

The third multinational corporation is Unilever, a diversified Anglo-Dutch company with an annual turnover of more than €40 billion and net profits of €5 billion in 2006 (Unilever 2006). It specializes in so-called fast-moving consumer goods in three areas: food (many familiar brands, including Hellman's Mayonnaise and Lipton Tea), home care (laundry detergents, etc.), and personal care, including deodorants, hair care, oral care, and skin care. Its most famous brand in the skin care line is Ponds, which sells cold creams in Europe and North America and whitening creams in Asia, the Middle East, and Latin America.

Through its Indian subsidiary, Hindustan Lever Limited, Unilever patented Fair & Lovely in 1971 following the patenting of niacinamide, a melanin suppressor, which is its main active ingredient. Test-marketed in South India in 1975, it became available throughout India in 1978. Fair & Lovely has become the largest-selling skin cream in India, accounting for 80 percent of the fairness cream market. According to anthropologist Susan Runkle (2005), "Fair & Lovely has an estimated sixty million consumers throughout the Indian subcontinent and exports to thirty-four countries in Southeast and Central Asia as well as the Middle East."

Fair & Lovely ads claim that "with regular daily use, you will be able to unveil your natural radiant fairness in just 6 weeks!" As with other successful brands, Fair & Lovely has periodically added new lines to appeal to special markets. In 2003, it introduced Fair & Lovely, Ayurvedic, which claims to be formulated according to a 4,500-year-old Indian medical system. In 2004, it introduced Fair & Lovely Oil-Control Gel and Fair & Lovely Anti-Marks. In 2004, Fair & Lovely also announced the "unveiling" of a premium line, Perfect Radiance, "a complete range of 12 premium skin care solutions" containing "international formulations from Unilever's Global Skin Technology Center, combined with ingredients best suited for Indian skin types and climates." Its ads say "Experience Perfect Radiance from Fair & Lovely. Unveil Perfect Skin." Intended to compete with expensive European brands, Perfect Radiance is sold only in select stores in major cities, including Delhi, Mumbai, Chennai, and Bangalore.[8]

Unilever is known for promoting its brands by being active and visible in the locales where they are marketed. In India, Ponds sponsors the Femina Miss India pageant, in which aspiring contestants are urged to "be as beautiful as you can be." Judging by photos of past winners, being as beautiful as you can be means being as light as you can be. In 2003, partly in response to criticism by the All India Democratic Women's Association of "racist" advertisement of fairness products, Hindustani Lever launched the Fair & Lovely Foundation, whose mission is to "encourage economic empowerment of women across India" through educational and guidance programs, training courses, and scholarships.[9]

Unilever heavily promotes both Ponds and Fair & Lovely with television and print ads tailored to local cultures. In one commercial shown in India, a young, dark-skinned woman's father laments that he has no son to provide for him and his daughter's salary is not high enough. The suggestion is that she could neither get a better job nor marry because of her dark skin. The young woman then uses Fair & Lovely, becomes fairer, and lands a job as an airline hostess, making her father happy. A Malaysian television spot shows a college student who is dejected because she cannot get the attention of a classmate at the next desk. After using Pond's lightening moisturizer, she appears in class brightly lit and several shades lighter, and the boy says, "Why didn't I notice her before?" (BBC 2003).

Such advertisements can be seen as not simply responding to a preexisting need but actually creating a need by depicting having dark skin as a painful and depressing experience. Before "unveiling" their fairness, dark-skinned women are shown as unhappy, suffering from low self-esteem, ignored by young men, and denigrated by their parents. By using Fair & Lovely or Ponds, a woman undergoes a transformation of not only her complexion but also her personality and her fate. In short, dark skin becomes a burden and handicap that can be overcome only by using the product being advertised.

CONCLUSION

The yearning for lightness evident in the widespread and growing use of skin bleaching around the globe can rightfully be seen as a legacy of colonialism, a manifestation of "false consciousness," and the internalization of "white is right" values by people of color, especially women. Thus, one often-proposed solution to the problem is reeducation that stresses the diversity of types of beauty and desirability and that valorizes darker skin shades, so that lightness/whiteness is dislodged as the dominant standard.

While such efforts are needed, focusing only on individual consciousness and motives distracts attention from the very powerful economic forces that help to create the yearning for lightness and that offer to fulfill the yearning at a steep price. The manufacturing, advertising, and selling of skin lightening is no longer a marginal, underground economic activity. It has become a major growth market for giant multinational corporations with their sophisticated means of creating and manipulating needs.

The multinationals produce separate product lines that appeal to different target audiences. For some lines of products, the corporations harness the prestige of science by showing cross-sectional diagrams of skin cells and by displaying images of doctors in white coats. Dark skin or dark spots become a disease for which skin lighteners offer a cure. For other lines, designed to appeal to those who respond to appeals to naturalness, corporations call up nature by emphasizing the use of plant extracts and by displaying images of light-skinned women against a background of blue skies and fields of flowers. Dark skin becomes a veil that hides one's natural luminescence, which natural skin lighteners will uncover. For all products, dark skin is associated with pain, rejection, and limited options; achieving light skin is seen as necessary to being youthful, attractive, modern, and affluent—in short, to being "all that you can be."

ENDNOTES

1. Under pressure from African American critics, Nadolina reduced the concentration to 6 percent in 1937 and 1.5 percent in 1941.

2. Discussions on Bright Skin Forum, Skin Lightening Board, are at http://excoboard.com/exco/forum.php?forumid=65288. Pallid Skin Lightening system information is at http://www.avreskincare.com/skin/pallid/index.html. Advertisement for Swiss Whitening Pills is at http://www.skinbleaching.net.

3. Skin whitening forums are at http://www.candymag.com/teentalk/index.php/topic,131753.0.html and http://www.rexinteractive.com/forum/topic.asp?TOPIC_ID=41.

4. Discussion of the ingredients in White Secret is found at http://www.vsantivirus.com/hoax-white-secret.htm.

5. I say that Stiefel targets Latin America because it markets other dermatology products, but not skin lighteners, in the competitive Asian, Middle Eastern, African, and European countries. Information about Stiefel products is at its corporate Web site, http://www.stiefel.com/why/about.aspx (accessed May 1, 2007).

6. Many of the products used by older white and Asian women to deal with age spots are physician-prescribed pharmaceuticals, including prescription-strength hydroquinone formulas. See information on one widely used system, Obagi, at http://www.obagi.com/article/homepage.html (accessed December 13, 2006).

7. *Shiseido Annual Report 2006*, 34, was downloaded from http://www.shiseido.co.jp/e/annual/html/index.htm. Data on European, American, and Japanese markets are at http://www.shiseido.co.jp/e/story/html/sto40200.htm. World employment figures are at http://www.shiseido.co.jp/e/story/html/sto40200.htm. White Lucent information is at http://www.shiseido.co.jp/e/whitelucent_us/products/product5.htm. White Lucency information is at http://www.shiseido.co.jp/e/whitelucency/ (all accessed May 6, 2007).

8. "Fair & Lovely Launches Oil-Control Fairness Gel" (Press Release, April 27, 2004) is found at http://www.hll.com/mediacentre/release.asp?fl=2004/PR_HLL_042704.

htm (accessed May 6, 2007). "Fair & Lovely Unveils Premium Range" (Press Release, May 25, 2004) is available at http://www.hll.com/mediacentre/release.asp?fl=2004/PR_HLL_052104_2.htm (accessed May 6, 2007).

9. The Pond's Femina Miss World site is http://feminamissindia.indiatimes.com/articleshow/1375041.cms. The All India Democratic Women's Association objects to skin lightening ad is at http://www.aidwa.org/content/issues_of_concern/women_and_media.php. Reference to Fair & Lovely campaign is at http://www.aidwa.org/content/issues_of_concern/women_and_media.php. "Fair & Lovely Launches Foundation to Promote Economic Empowerment of Women" (Press Release, March 11, 2003) is found at http://www.hll.com/mediacentre/release.asp?fl=2003/PR_HLL_031103.htm (all accessed December 2, 2006).

REFERENCES

Adebajo, S. B. 2002. An epidemiological survey of the use of cosmetic skin lightening cosmetics among traders in Lagos, Nigeria. *West African Journal of Medicine* 21 (1): 51–55.

Arnold, David. 2004. Race, place and bodily difference in early nineteenth century India. *Historical Research* 77: 162.

Ashikari, Makiko. 2003. Urban middle-class Japanese women and their white faces: Gender, ideology, and representation. *Ethos* 31 (1): 3, 3–4, 9–11.

BBC. 2003. India debates "racist" skin cream ads. *BBC News World Edition*, July 24. http://news.bbc.co.uk/1/hi/world/south_asia/3089495.stm (accessed May 8, 2007).

Bundles, A'Lelia. 2001. *On Her Own Ground: The Life and Times of Madam C. J. Walker.* New York: Scribner.

Burke, Timothy. 1996. *Lifebuoy Men, Lux Women: Commodification, Consumption, and Cleanliness in Modern Zimbabwe.* Durham, NC: Duke University Press.

Centers for Disease Control and Prevention. 1996. *FDA warns consumers not to use Crema de Belleza.* FDA statement. Rockville, MD: U.S. Food and Drug Administration.

Chadwick, Julia. 2001. Arklow's toxic soap factory. *Wicklow Today*, June. http://www.wicklowtoday.com/features/mercurysoap.htm (accessed April 18, 2007).

De Faoite, Dara. 2001. Investigation into the sale of dangerous mercury soaps in ethnic shops. *The Observer*, May 27. http://observer.guardian.co.uk/uk_news/story/0,6903,497227,00.html (accessed May 1, 2007).

Del Guidice, P., and P. Yves. 2002. The widespread use of skin lightening creams in Senegal: A persistent public health problem in West Africa. *International Journal of Dermatology* 41:69–72.

Dooley, Erin. 2001. Sickening soap trade. *Environmental Health Perspectives*, October.

Earth Summit. 2002. *Telling it like it is: 10 years of unsustainable development in Ireland.* Dublin, Ireland: Earth Summit.

Exhibitor info. 2006. http://www.beautyworldjapan.com/en/efirst.html (accessed May 8, 2007).

Farr, Marcia. 2006. *Rancheros in Chicagoacán: Language and Identity in a Transnational Community.* Austin: University of Texas Press.

Guttman, Matthew C. 1996. *The Meanings of Macho: Being a Man in Mexico City.* Berkeley: University of California Press.

Herring, Cedric, Verna M. Keith, and Hayward Derrick Horton, eds. 2003. *Skin Deep: How Race and Complexion Matter in the "Color Blind" Era.* Chicago: Institute for Research on Race and Public Policy.

Hunter, Margaret. 2005. *Race, Gender, and the Politics of Skin Tone.* New York: Routledge.

Intage. 2001. Intelligence on the cosmetic market in Japan. http://www.intage.co.jp/expess/01_08/market/index1.html (accessed November 2005).

Kerr, Audrey Elisa. 2005. The Paper Bag Principle: The Myth and the Motion of Colorism. *Journal of American Folklore* 118:271–89.

Khan, Aisha. 2008. "Caucasian," "coolie," "Black," or "white"? Color and race in the Indo-Caribbean Diaspora. Unpublished paper.

Knight, Alan. 1990. Racism, Revolution, and Indigenismo: Mexico, 1910–1940. In *The Idea of Race in Latin America, 1870–1940,* edited by Richard Graham. Austin: University of Texas Press.

Maddox, Keith B. 2004. Perspectives on racial phenotypicality bias. *Personality and Social Psychology Review* 8:383–401.

Mahe, Antoine, Fatimata Ly, and Jean-Marie Dangou. 2003. Skin diseases associated with the cosmetic use of bleaching products in women from Dakar, Senegal. *British Journal of Dermatology* 148 (3): 493–500.

Malangu, N., and G. A. Ogubanjo. 2006. Predictors of tropical steroid misuse among patrons of pharmacies in Pretoria. *South African Family Practices* 48 (1): 14.

Margulies, Paul. n.d. Cushing's syndrome: The facts you need to know. http://www.nadf.us/diseases/cushingsmedhelp.org/www/nadf4.htm (accessed May 1, 2007).

Martinez, Ruben. 2001. *Crossing Over: A Mexican Family on the Migrant Trail.* New York: Henry Holt.

Massing, Jean Michel. 1995. From Greek proverb to soap advert: Washing the Ethiopian. *Journal of the Warburg and Courtauld Institutes* 58:180.

Muzondidya, James. 2005. *Walking a Tightrope: Towards a Social History of the Coloured Community of Zimbabwe.* Trenton, NJ: Africa World Press.

Nielsen. 2007. Prairie plants take root. In *Health, beauty & personal grooming: A global Nielsen consumer report,* http://www.acnielsen.co.in/news/20070402.shtml (accessed May 3, 2007).

Ntambwe, Malangu. 2004. Mirror mirror on the wall, who is the fairest of them all? *Science in Africa, Africa's First On-Line Science Magazine,* March. http://www.scienceinafrica.co.za/2004/march/skinlightening.htm (accessed May 1, 2007).

Ntshingila, Futhi. 2005. Female buppies using harmful skin lighteners. *Sunday Times, South Africa,* November 27. http://www.sundaytimes.co.za (accessed January 25, 2006).

NYC Health Dept. 2005. NYC Health Dept. warns against use of "skin lightening" creams containing mercury or similar products which do not list ingredients. http://www.nyc.gov/html/doh/html/pr/pr008-05.shtml (accessed May 7, 2007).

O'Farrell, Michael. 2002. Pressure mounts to have soap plant shut down. *Irish Examiner,* August 26. http://archives.tcm.ie/irishexaminer/2002/08/26/story510455503.asp (accessed May 1, 2007).

Parreñas, Rhacel. 2001. *Servants of Globalization: Women, Migration, and Domestic Work.* Palo Alto, CA: Stanford University Press.

Peiss, Kathy. 1998. *Hope in a Jar: The Making of America's Beauty Culture.* New York: Metropolitan Books.

Ribane, Nakedi. 2006. *Beauty: A Black perspective.* Durban, South Africa: University of KwaZulu-Natal Press.

Runkle, Susan. 2004. Making "Miss India": Constructing gender, power and nation. *South Asian Popular Culture* 2 (2): 145–59.

———. 2005. The beauty obsession. *Manushi* 145 (February). http://www .indiatogether.org/manushi/issue145/lovely.htm (accessed May 5, 2007).

Russell, Kathy, Midge Wilson, and Ronald Hall. 1992. *The Color Complex: The Politics of Skin Color Among African Americans.* New York: Harcourt Brace Jovanovich.

Saskatchewan Business Unlimited. 2005. Prairie plants take root in cosmetics industry. *Saskatchewan Business Unlimited* 10 (1): 1–2.

Sheriff, Robin E. 2001. *Dreaming Equality: Color, Race and Racism in Urban Brazil.* New Brunswick, NJ: Rutgers University Press.

Stepan, Nancy Ley. 1991. *The Hour of Eugenics: Race, Gender, and Nation in Latin America.* Ithaca, NY: Cornell University Press.

Synovate. 2004. In:fact. http://www.synovate.com/knowledge/infact/issues/200406 (accessed March 21, 2007).

Tabbada, Reyna Mae L. 2006. Trouble in paradise. Press release, September 20. http:// www.bulatlat.com/news/6-33/6-33-trouble.htm (accessed May 5, 2007).

Telles, Edward E. 2004. *Race in Another America: The Significance of Skin Color in Brazil.* Princeton, NJ: Princeton University Press.

Thomas, Iyamide. 2004. "Yellow fever": The disease that is skin bleaching. *Mano Vision* 33 (October): 32–33. http://www.manovision.com/ISSUES/ISSUE33/33skin.pdf (accessed May 7, 2007).

Thomas, Lynn M. (in press.) Skin lighteners in South Africa: Transnational entanglements and technologies of the self. In *Shades of Difference: Why Skin Color Matters,* edited by Evelyn Nakano Glenn. Stanford, CA: Stanford University Press.

Unilever. 2006. Annual report. http://www.unilever.com/ourcompany/investorcentre/ annual_reports/archives.asp (accessed May 6, 2007).

U.S. Food and Drug Administration. 1996. *FDA warns consumers not to use Crema de Belleza.* FDA statement, July 23. Rockville, MD: U.S. Food and Drug Administration.

Views on article—Complexion. n.d. http://www.indiaparenting.com/beauty/beauty041book.shtml (accessed November 2005).

Winders, Jamie, John Paul Jones III, and Michael James Higgins. 2005. Making Güeras: Selling white identities on late-night Mexican television. *Gender, Place and Culture* 12 (1): 71–93.

Wong, Stephanie. 2004. Whitening cream sales soar as Asia's skin-deep beauties shun Western suntans. *Manila Bulletin.* http://www.mb.com.ph/issues/2004/08/24/ SCTY2004082416969.html# (accessed March 24, 2007).

REFLECTION QUESTIONS FOR CHAPTER 6

1. What evidence do you find of other cultures in your community? Do they seem to be exactly as they are found in their countries of origin, or have they adapted to the local culture?

2. Some theorists have argued that rather than a global culture, there is or will be a "clash of civilizations" as religious fundamentalists or nationalists resist (sometimes with violence) any attempt to change their cultures. What evidence do you see from world events to support or reject this argument?

3. Does the direction of the globalization of culture move one way—that is, toward the westernization of the globe? Are there products from other countries that have changed our habits? Are there non–U.S. cultural icons from entertainment and sports who have captivated Americans? What about the influence of other countries on fashion, architecture, work arrangements, and the like?

Chapter 7

The Restructuring of Social Arrangements

Gender, Families, and Relationships

B y now it is a truism that some of the most devastating effects of globalization fall on women. We denounce female exploitation in third-world economies, where women and girls are forced to work in jobs with high turnover, low wages, and hazardous working conditions. Indeed, the standard view of the "global assembly line" comes packaged with gendered images. Other features of globalization are well known for their harmful effects on women, including sex trafficking and female enslavement. Yet these realities tell us little about the impact of gender on the global economy.

Few macro structural discussions of globalization put gender at the center of the analysis. Although they "stir" women into the macro picture, they fail to examine how "producers, consumers, and bystanders of globalization are all caught up in processes that are gendered themselves."[1]

The readings in this chapter challenge and broaden mainstream views of the emerging global order by exploring several dimensions of the relationship between globalization and gender. They uncover hidden social and economic global processes that rest on expectations of masculine and feminine in locations around the world. Yet in the process of moving across national boundaries through global systems of production, immigration, and cultural images, gender inequalities are reconstructed and take new shape. At the same time, global movements of gender transform the social institutions with which they come in contact.[2] Looking at global connections and interdependencies through the lens of gender reveals that inequalities between women and men serve as building blocks of the global order, and are simultaneously restructured by global processes.

In the first article, Barbara Ehrenreich and Arlie Russell Hochschild provide an overview of women's migration from the third world to do "women's work" for families around the world. The authors unmask the transfer of women's traditional work—child care, homemaking, and sex—from poor countries to rich ones. While the globalization of "women's work" supports the professional careers and lifestyles of women in affluent countries, it leaves immigrant women of color disenfranchised and separated from their own families. The global transfer of "care work" throws new light on the entire process of globalization and its contradictory effects on women, men, and children in different parts of the world.

Next, Rhacel Parreñas takes the analysis and critique of the global migration of care work a step further. Her study of Filipina migrant mothers and the children they leave behind draws on the stories of family members who struggle to remain "family" even as they are separated by time and space. Their own words are compelling and they offer new insights about transnational families and the costs extracted from them by the global economy.

In the next article, Arlie Russell Hochschild provides a different view of the global transfer of women's service work. Here, she describes commercial surrogacy in a clinic in Anand, Gujarat, India, where poorer women are paid to carry embryos and bear babies for wealthy women. Hochschild illustrates the coming together of science and capitalism and the new connections forged between the rich and poor in the first and third worlds.

John Ross continues with gender and family themes as he describes the feminization of Mexican agriculture. Men's outmigration from states like Michoacan, Jalisco, Guanajuato, and Zacatecas have left women behind to work "triple days" on the farm, in factories, and in their families.

While the first four readings point to women's paid and unpaid work as a key component of globalization, the final reading makes gender in the global arena visible for *men*. R. W. Connell asks the question, "What does masculinity have to do with globalization?" He answers it by untangling key strands in "the world gender order" and unearthing multiple masculinities, all of which are intertwined with global forces.

ENDNOTES

1. Freeman, Carla. "Is Local: Global as Feminine: Masculine? Rethinking the Gender of Globalization," *SIGNS: Journal of Women in Culture and Society*, Vol. 26, No. 4 (Summer 2001)1007–1038.

2. Baca Zinn, Maxine, Pierrette Hondagneu-Sotelo, and Michael A. Messner, eds. *Gender Through the Prism of Difference,* 4th Edition (New York: Oxford University Press, 2010).

22

Global Woman: Nannies, Maids, and Sex Workers in the New Economy

BARBARA EHRENREICH ARLIE RUSSELL HOCHSCHILD

The authors discuss the global transfer of women's labor, in which third world women migrate to affluent countries to do "women's work." Understanding this global division of female services requires that we rethink the notion of dependency, as affluent and middle class families in the first world depend on migrants from poorer countries for childcare, homemaking, and other domestic services.

"Whose baby are you?" Josephine Perera, a nanny from Sri Lanka, asks Isadora, her pudgy two-year-old charge in Athens, Greece.

Thoughtful for a moment, the child glances toward the closed door of the next room, in which her mother is working, as if to say, "That's my mother in there."

"No, you're *my* baby," Josephine teases, tickling Isadora lightly. Then, to settle the issue, Isadora answers, "Together!" She has two mommies—her mother and Josephine. And surely a child loved by many adults is richly blessed.

In some ways, Josephine's story—which unfolds in an extraordinary documentary film, *When Mother Comes Home for Christmas*, directed by Nilita Vachani—describes an unparalleled success. Josephine has ventured around the world, achieving a degree of independence her mother could not have imagined, and amply supporting her three children with no help from her ex-husband, their father. Each month she mails a remittance check from Athens to Hatton, Sri Lanka, to pay the children's living expenses and school fees. On her Christmas visit home, she bears gifts of pots, pans, and dishes. While she makes payments on a new bus that Suresh, her oldest son, now drives for a living, she is also saving for a modest dowry for her daughter, Norma. She dreams of buying a new house in which the whole family can live. In the meantime, her work as a nanny enables Isadora's parents to devote themselves to their careers and avocations.

But Josephine's story is also one of wrenching global inequality. While Isadora enjoys the attention of three adults, Josephine's three children in Sri Lanka have been far less lucky. According to Vachani, Josephine's youngest child, Suminda, was two—Isadora's age—when his mother first left home to work in Saudi Arabia. Her middle child, Norma, was nine; her oldest son, Suresh, thirteen. From Saudi Arabia, Josephine found her way first to Kuwait, then to Greece. Except for one two-month trip home, she has lived apart from her children for ten years. She writes them weekly letters, seeking news of relatives, asking about school, and complaining that Norma doesn't write back.

Although Josephine left the children under her sister's supervision, the two youngest have shown signs of real distress. Norma has attempted suicide three times. Suminda, who was twelve when the film was made, boards in a grim, Dickensian orphanage that forbids talk during meals and showers. He visits his aunt on holidays. Although the oldest, Suresh, seems to be on good terms with his mother, Norma is tearful and sullen, and Suminda does poorly in school, picks quarrels, and otherwise seems withdrawn from the world. Still, at the end of the film, we see Josephine once again leave her three children in Sri Lanka to return to Isadora in Athens. For Josephine can either live with her children in desperate poverty or make money by living apart from them. Unlike her affluent First World employers, she cannot both live with her family and support it.

Thanks to the process we loosely call "globalization," women are on the move as never before in history. In images familiar to the West from television commercials for credit cards, cell phones, and airlines, female executives jet about the world, phoning home from luxury hotels and reuniting with eager children in airports. But we hear much less about a far more prodigious flow of female labor and energy: the increasing migration of millions of women from poor countries to rich ones, where they serve as nannies, maids, and sometimes sex workers. In the absence of help from male partners, many women have succeeded in tough "male world" careers only by turning over the care of their children, elderly parents, and homes to women from the Third World. This is the female underside of globalization, whereby millions of Josephines from poor countries in the south migrate to do the "women's work" of the north—work that affluent women are no longer able or willing to do. These migrant workers often leave their own children in the care of grandmothers, sisters, and sisters-in-law. Sometimes a young daughter is drawn out of school to care for her younger siblings.

This pattern of female migration reflects what could be called a worldwide gender revolution. In both rich and poor countries, fewer families can rely solely on a male breadwinner. In the United States, the earning power of most men has declined since 1970, and many women have gone out to "make up the difference." By one recent estimate, women were the sole, primary, or coequal earners in more than half of American families.[1] So the question arises: Who will take care of the children, the sick, the elderly? Who will make dinner and clean house?

While the European or American woman commutes to work an average twenty-eight minutes a day, many nannies from the Philippines, Sri Lanka, and India cross the globe to get to their jobs. Some female migrants from the Third World do find something like "liberation," or at least the chance to become independent breadwinners and to improve their children's material lives. Other, less fortunate migrant women end up in the control of criminal employers—their passports stolen, their mobility blocked, forced to work without pay in brothels or to provide sex along with cleaning and child-care services in affluent homes. But even in more typical cases, where benign employers pay wages on time, Third World migrant women achieve their success only by assuming the cast-off domestic roles of middle- and high-income women in the First World—roles that have been previously rejected, of course, by men. And their "commute" entails a cost we have yet to fully comprehend.

The migration of women from the Third World to do "women's work" in affluent countries has so far received little media attention—for reasons that are easy enough to guess. First, many, though by no means all, of the new female migrant workers are women of color, and therefore subject to the racial "discounting" routinely experienced by, say, Algerians in France, Mexicans in the United States and Asians in the United Kingdom. Add to racism the private "indoor" nature of so much of the new migrants' work. Unlike factory workers, who congregate in large numbers, or taxi drivers, who are visible on the street, nannies and maids are often hidden away, one or two at a time, behind closed doors in private homes. Because of the illegal nature of their work, most sex workers are even further concealed from public view.

At least in the case of nannies and maids, another factor contributes to the invisibility of migrant women and their work—one that, for their affluent employers, touches closer to home. The Western culture of individualism, which finds extreme expression in the United States, militates against acknowledging help or human interdependency of nearly any kind. Thus, in the time-pressed upper middle class, servants are no longer displayed as status symbols, decked out in white caps and aprons, but often remain in the background, or disappear when company comes. Furthermore, affluent careerwomen increasingly earn their status not through leisure, as they might have a century ago, but by apparently "doing it all"—producing a full-time career, thriving children, a contented spouse, and a well-managed home. In order to preserve this illusion, domestic workers and nannies make the house hotel-room perfect, feed and bathe the children, cook and clean up—and then magically fade from sight.

The lifestyles of the First World are made possible by a global transfer of the services associated with a wife's traditional role—child care, homemaking, and sex—from poor countries to rich ones. To generalize and perhaps oversimplify: in an earlier phase of imperialism, northern countries extracted natural resources and agricultural products—rubber, metals, and sugar, for example—from lands they conquered and colonized. Today, while still relying on Third World countries for agricultural and industrial labor, the wealthy countries also seek to extract something harder to measure and quantify, something that can look very much like love. Nannies like Josephine bring the distant families that employ them real maternal affection, no doubt enhanced by the heartbreaking absence of their own children in the poor countries they leave behind. Similarly, women who migrate from country to country to work as maids bring not only their muscle power but an attentiveness to detail and to the human relationships in the household that might otherwise have been invested in their own families. Sex workers offer the simulation of sexual and romantic love, or at least transient sexual companionship. It is as if the wealthy parts of the world are running short on precious emotional and sexual resources and have had to turn to poorer regions for fresh supplies.

There are plenty of historical precedents for this globalization of traditional female services. In the ancient Middle East, the women of populations defeated in war were routinely enslaved and hauled off to serve as household workers and concubines for the victors. Among the Africans brought to North America as slaves in the sixteenth through nineteenth centuries, about a third were women and children, and many of those women were pressed to be concubines, domestic servants, or both. Nineteenth-century Irishwomen—along with many rural Englishwomen— migrated to English towns and cities to work as domestics in the homes of the growing upper middle class. Services thought to be innately feminine—child care, housework, and sex—often win little recognition or pay. But they have always been sufficiently in demand to transport over long distances if necessary. What is new today is the sheer number of female migrants and the very long distances they travel. Immigration statistics show huge numbers of women in motion, typically from poor countries to rich. Although the gross statistics give little clue as to the jobs women eventually take, there are reasons to infer that much of their work is "caring work," performed either in private homes or in institutional settings such as hospitals, hospices, child-care centers, and nursing homes.

The statistics are, in many ways, frustrating. We have information on legal migrants but not on illegal migrants, who, experts tell us, travel in equal if not greater numbers. Furthermore, many Third World countries lack data for past years, which makes it hard to trace trends over time; or they use varying methods of gathering information, which makes it hard to compare one country with another. Nevertheless, the trend is clear enough for some scholars, including Stephen Castles, Mark Miller, and Janet Momsen, to speak of a "feminization of migration."[2] From 1950 to 1970, for example, men predominated in labor migration to northern Europe from Turkey, Greece, and North Africa. Since then, women have been replacing men. In 1946, women were fewer than 3 percent of the Algerians and Moroccans living in France; by 1990, they were more than 40 percent.[3] Overall, half of the world's 120 million legal and illegal migrants are now believed to be women.

Patterns of international migration vary from region to region, but women migrants from a surprising number of sending countries actually outnumber men, sometimes by a wide margin. For example, in the 1990s, women made up over half of Filipino migrants to all countries and 84 percent of Sri Lankan migrants to the Middle East.[4] Indeed, by 1993 statistics, Sri Lankan women such as Josephine vastly outnumbered Sri Lankan men as migrant workers who'd left for Saudi Arabia, Kuwait, Lebanon, Oman, Bahrain, Jordan, and Qatar, as well as to all countries of the Far East, Africa, and Asia.[5] About half of the migrants leaving Mexico, India, Korea, Malaysia, Cyprus, and Swaziland to work elsewhere are also women. Throughout the 1990s women outnumbered men among migrants to the United States, Canada, Sweden, the United Kingdom, Argentina, and Israel.[6]

Most women, like men, migrate from the south to the north and from poor countries to rich ones. Typically, migrants go to the nearest comparatively rich country, preferably one whose language they speak or whose religion and culture they share. There are also local migratory flows: from northern to southern Thailand, for instance, or from East Germany to West. But of the regional or cross-regional flows, four stand out. One goes from Southeast Asia to the oil-rich Middle and Far East—from Bangladesh, Indonesia, the Philippines, and Sri Lanka to Bahrain, Oman, Kuwait, Saudi Arabia, Hong Kong, Malaysia, and Singapore. Another stream of migration goes from the former Soviet bloc to western Europe—from Russia, Romania, Bulgaria, and Albania to Scandinavia, Germany, France, Spain, Portugal, and England. A third goes from south to north in the Americas, including the stream from Mexico to the United States, which scholars say is the longest-running labor migration in the world. A fourth stream moves from Africa to various parts of Europe. France receives many female migrants from Morocco, Tunisia, and Algeria. Italy receives female workers from Ethiopia, Eritrea, and Cape Verde.

Female migrants overwhelmingly take up work as maids or domestics. As women have become an ever greater proportion of migrant workers, receiving countries reflect a dramatic influx of foreign-born domestics. In the United States, African-American women, who accounted for 60 percent of domestics in the 1940s, have been largely replaced by Latinas, many of them recent migrants from Mexico and Central America. In England, Asian migrant women have displaced the Irish and Portuguese domestics of the past. In French cities, North African women have replaced rural French girls. In western Germany, Turks and women from the former East Germany have replaced rural native-born women. Foreign females from countries outside the European Union made up only 6 percent of all domestic workers in 1984. By 1987, the percentage had jumped to 52, with most coming from the Philippines, Sri Lanka, Thailand, Argentina, Colombia, Brazil, El Salvador, and Peru.[7]

The governments of some sending countries actively encourage women to migrate in search of domestic jobs, reasoning that migrant women are more likely than their male counterparts to send their hard-earned wages to their families rather than spending the money on themselves. In general, women send home anywhere from half to nearly all of what they earn. These remittances have a significant impact on the lives of children, parents, siblings, and wider networks of kin—as well as on cash-strapped Third World governments. Thus, before Josephine left for Athens, a program sponsored by the Sri Lankan government taught her how to use a microwave oven, a vacuum cleaner, and an electric mixer. As she awaited her flight, a song piped into the airport departure lounge extolled the opportunity to earn money abroad. The songwriter was in the pay of the Sri Lanka Bureau of Foreign Employment, an office devised to encourage women to migrate. The lyrics say:

After much hardship, such difficult times

How lucky I am to work in a foreign land.

As the gold gathers so do many greedy flies.

But our good government protects us from them.

After much hardship, such difficult times,

How lucky I am to work in a foreign land.

I promise to return home with treasures for everyone.

Why this transfer of women's traditional services from poor to rich parts of the world? The reasons are, in a crude way, easy to guess. Women in Western countries have increasingly taken on paid work, and hence need other—paid domestics and caretakers for children and elderly people—to replace them.[8] For their part, women in poor countries have an obvious incentive to migrate: relative and absolute poverty. The "care deficit" that has emerged in the wealthier countries as women enter the workforce *pulls* migrants from the Third World and postcommunist nations; poverty *pushes* them.

In broad outline, this explanation holds true. Throughout western Europe, Taiwan, and Japan, but above all in the United States, England, and Sweden, women's employment has increased dramatically since the 1970s. In the United States, for example, the proportion of women in paid work rose from 15 percent of mothers of children six and under in 1950 to 65 percent today. Women now make up 46 percent of the U.S. labor force. Three-quarters of mothers of children eighteen and under and nearly two-thirds of mothers of children age one and younger now work for pay. Furthermore, according to a recent International Labor Organization study, working Americans averaged longer hours at work in the late 1990s than they did in the 1970s. By some measures, the number of hours spent at work have increased more for women than for men, and especially for women in managerial and professional jobs.

Meanwhile, over the last thirty years, as the rich countries have grown much richer, the poor countries have become—in both absolute and relative terms—poorer. Global inequalities in wages are particularly striking. In Hong Kong, for instance, the wages of a Filipina domestic are about fifteen times the amount she could make as a schoolteacher back in the Philippines. In addition, poor countries turning to the IMF or World Bank for loans are often forced to undertake measures of so-called structural adjustment, with disastrous results for the poor and especially for poor women and children. To qualify for loans, governments are usually required to devalue their currencies, which turns the hard currencies of rich countries into gold and the soft currencies of poor countries into straw. Structural adjustment programs also call for cuts in support for "noncompetitive industries," and for the reduction of public services such as health care and food subsidies for the poor. Citizens of poor countries, women as well as men, thus have a strong incentive to seek work in more fortunate parts of the world.

But it would be a mistake to attribute the globalization of women's work to a simple synergy of needs among women—one group, in the affluent countries, needing help and the other, in poor countries, needing jobs. For one thing, this formulation fails to account for the marked failure of First World governments to meet the needs created by its women's entry into the workforce. The downsized American—and to a lesser degree, western European—welfare state has become a "deadbeat dad." Unlike the rest of the industrialized world, the United States does not offer public child care for working mothers, nor does it ensure paid family and medical leave. Moreover, a series of state tax revolts in the 1980s reduced the number of hours public libraries were open and slashed school-enrichment and after-school programs. Europe did not experience anything comparable. Still, tens of millions of western European women are in the work-force who were not before—and there has been no proportionate expansion in public services.

Secondly, any view of the globalization of domestic work as simply an arrangement among women completely omits the role of men. Numerous stud-ies, including some of our own, have shown that as American women took on paid employment, the men in their families did little to increase their contribu-tion to the work of the home. For example, only one out of every five men among the working couples whom Hochschild interviewed for *The Second Shift* in the 1980s shared the work at home, and later studies suggest that while work-ing mothers are doing somewhat less housework than their counterparts twenty years ago, most men are doing only a little more.[9] With divorce, men frequently abdicate their child-care responsibilities to their ex-wives. In most cultures of the First World outside the United States, powerful traditions even more firmly dis-courage husbands from doing "women's work." So, strictly speaking, the pres-ence of immigrant nannies does not enable affluent women to enter the workforce; it enables affluent *men* to continue avoiding the second shift.

The men in wealthier countries are also, of course, directly responsible for the demand for immigrant sex workers—as well as for the sexual abuse of many migrant women who work as domestics. Why, we wondered, is there a particu-lar demand for "imported" sexual partners? Part of the answer may lie in the fact that new immigrants often take up the least desirable work, and, thanks to the AIDS epidemic, prostitution has become a job that ever fewer women deliber-ately choose. But perhaps some of this demand … grows out of the erotic lure of the "exotic." Immigrant women may seem desirable sexual partners for the same reason that First World employers believe them to be especially gifted as care-givers: they are thought to embody the traditional feminine qualities of nurtur-ance, docility, and eagerness to please. Some men feel nostalgic for these qualities, which they associate with a bygone way of life. Even as many wage-earning Western women assimilate to the competitive culture of "male" work and ask respect for making it in a man's world, some men seek in the "exotic Orient" or "hot-blooded tropics" a woman from the imagined past.

Of course, not all sex workers migrate voluntarily. An alarming number of women and girls are trafficked by smugglers and sold into bondage. Because traf-ficking is illegal and secret, the numbers are hard to know with any certainty. Kevin Bales estimates that in Thailand alone, a country of 60 million, half a mil-lion to a million women are prostitutes, and one out of every twenty of these is enslaved.[10] … [M]any of these women are daughters whom northern hill-tribe families have sold to brothels in the cities of the south. Believing the promises of jobs and money, some begin the voyage willingly, only to discover days later that the "arrangers" are traffickers who steal their passports, define them as debtors, and enslave them as prostitutes. Other women and girls are kidnapped, or sold by their impoverished families, and then trafficked to brothels. Even worse fates befall women from neighboring Laos and Burma, who flee crushing poverty and repression at home only to fall into the hands of Thai slave traders.

If the factors that pull migrant women workers to affluent countries are not as simple as they at first appear, neither are the factors that push them. Certainly relative poverty plays a major role, but, interestingly, migrant women often do not come from the poorest classes of their societies.[11] In fact, they are typically more affluent and better educated than male migrants. Many female migrants from the Philippines and Mexico, for example, have high school or college diplomas and have held middle-class—albeit low-paid—jobs back home. One study of Mexican migrants suggests that the trend is toward increasingly better-educated female migrants. Thirty years ago, most Mexican-born maids in the United States had been poorly educated maids in Mexico. Now a majority have high school degrees and have held clerical, retail, or professional jobs before leaving for the United States.[12] Such women are likely to be enterprising and adventurous enough to resist the social pressures to stay home and accept their lot in life.

Noneconomic factors—or at least factors that are not immediately and directly economic—also influence a woman's decision to emigrate. By migrating, a woman may escape the expectation that she care for elderly family members, relinquish her paycheck to a husband or father, or defer to an abusive husband. Migration may also be a practical response to a failed marriage and the need to provide for children without male help. In the Philippines, ... Rhacel Salazar Parreñas tells us, migration is sometimes called a "Philippine divorce." And there are forces at work that may be making the men of poor countries less desirable as husbands. Male unemployment runs high in the countries that supply female domestics to the First World. Unable to make a living, these men often grow demoralized and cease contributing to their families in other ways. Many female migrants ... tell of unemployed husbands who drink or gamble their remittances away. Notes one study of Sri Lankan women working as maids in the Persian Gulf: "It is not unusual ... for the women to find upon their return that their Gulf wages by and large have been squandered on alcohol, gambling and other dubious undertakings while they were away."[13]

To an extent then, the globalization of child care and housework brings the ambitious and independent women of the world together: the career-oriented upper-middle-class woman of an affluent nation and the striving woman from a crumbling Third World or postcommunist economy. Only it does not bring them together in the way that second-wave feminists in affluent countries once liked to imagine—as sisters and allies struggling to achieve common goals. Instead, they come together as mistress and maid, employer and employee, across a great divide of privilege and opportunity.

This trend toward global redivision of women's traditional work throws new light on the entire process of globalization. Conventionally, it is the poorer countries that are thought to be dependent on the richer ones—a dependency symbolized by the huge debt they owe to global financial institutions. What we explore … however, is a dependency that works in the other direction, and it is a dependency of a particularly intimate kind. Increasingly often, as affluent and middle-class families in the First World come to depend on migrants from poorer regions to provide child care, homemaking, and sexual services, a global relationship arises that in some ways mirrors the traditional relationship between the sexes. The First World takes on a role like that of the old-fashioned male in the family—pampered, entitled, unable to cook, clean, or find his socks. Poor countries take on a role like that of the traditional woman within the family—patient, nurturing, and self-denying. A division of labor feminists critiqued when it was "local" has now, metaphorically speaking, gone global.

To press this metaphor a bit further, the resulting relationship is by no means a "marriage," in the sense of being openly acknowledged. In fact, it is striking how invisible the globalization of women's work remains, how little it is noted or discussed in the First World. Trend spotters have had almost nothing to say about the fact that increasing numbers of affluent First World children and elderly persons are tended by immigrant care workers or live in homes cleaned by immigrant maids. Even the political groups we might expect to be concerned about this trend—antiglobalization and feminist activists—often seem to have noticed only the most extravagant abuses, such as trafficking and female enslavement. So if a metaphorically gendered relationship has developed between rich and poor countries, it is less like a marriage and more like a secret affair.

But it is a "secret affair" conducted in plain view of the children. Little Isadora and the other children of the First World raised by "two mommies" may be learning more than their ABCs from a loving surrogate parent. In their own living rooms, they are learning a vast and tragic global politics.[14] Children see. But they also learn how to disregard what they see. They learn how adults make the visible invisible. That is their "early childhood education."

ENDNOTES

1. See Ellen Galinsky and Dana Friedman, *Women: The New Providers,* Whirlpool Foundation Study, Part 1 (New York Families and Work Institute, 1995), p. 37.

2. Special thanks to Roberta Espinoza …In addition to material directly cited, this introduction draws from the following works: Kathleen M. Adams and Sara Dickey, eds., *Home and Hegemony: Domestic Service and Identity Politics in South and Southeast Asia* (Ann Arbor: University of Michigan Press, 2000); Floya Anthias and Gabriella Lazaridis, eds., *Gender and Migration in Southern Europe: Women on the Move* (Oxford and New York: Berg, 2000); Stephen Castles and Mark J. Miller, *The Age of Migration: International Population Movements in the Modern World* (New York and London: The Guilford Press, 1998); Noeleen Heyzer, Geertje Lycklama à Nijehold, and Nedra Weerakoon, eds., *The Trade in Domestic Workers: Causes, Mechanisms, and Consequences of International Migration* (London: Zed Books, 1994); Eleanore Kofman, Annie Phizacklea, Parvati Raghuram, and Rosemary Sales, *Gender and International Migration in Europe: Employment, Welfare, and Politics* (New York and London: Routledge, 2000); Douglas S. Massey, Joaquin Arango, Graeme Hugo, All

Kouaouci, Adela Pellegrino, and J. Edward Taylor, *Worlds in Motion: Understanding International Migration at the End of the Millennium* (Oxford: Clarendon Press, 1999); Janet Henshall Momsen, ed., *Gender, Migration, and Domestic Service* (London: Routledge, 1999); Katie Willis and Brenda Yeoh, eds., *Gender and Immigration* (London: Edward Elgar Publishers, 2000).

3. Illegal migrants are said to make up anywhere from 60 percent (as in Sri Lanka) to 87 percent (as in Indonesia) of all migrants. In Singapore in 1994, 95 percent of Filipino overseas contract workers lacked work permits from the Philippine government. The official figures based on legal migration therefore severely underestimate the number of migrants. See Momsen, 1999, p. 7.

4. Momsen, 1999, p. 9.

5. Sri Lanka Bureau of Foreign Employment, 1994, as cited in G. Gunatilleke, "The Economic, Demographic, Sociocultural and Political Setting for Emigration from Sri Lanka," *International Migration*, vol. 23 (3/4), 1995, pp. 667–98.

6. Anthias and Lazandis, 2000; Heyzer, Nijehold, and Weerakoon, 1994, pp. 4–27; Momsen, 1999, p. 21; "Wistat: Women's Indicators and Statistics Database," version 3, CD-ROM (United Nations, Department for Economic and Social Information and Policy Analysis, Statistical Division, 1994).

7. Geovanna Campani, "Labor Markets and Family Networks: Filipino Women in Italy," in Hedwig Rudolph and Mirjana Morokvasic, eds., *Bridging States and Markets: International Migration in the Early 1990s* (Berlin: Edition Sigma, 1993), p. 206.

8. This "new" source of the Western demand for nannies, maids, child-care, and elder-care workers does not, of course, account for the more status-oriented demand in the Persian Gulf states, where most affluent women don't work outside the home.

9. For information on male work at home during the 1990s, see Arlie Russell Hochschild and Anne Machung, *The Second Shift: Working Parents and the Revolution at Home* (New York: Avon, 1997), p. 277.

10. Kevin Bales, *Disposable People: New Slavery in the Global Economy* (Berkeley: University of California Press, 1999), p. 43.

11. Andrea Tyree and Katharine M. Donato, "A Demographic Overview of the International Migration of Women," in *International Migration: The Female Experience*, eds. Rita Simon and Caroline Bretell (Totowa, N.J.: Rowman & Allanheld, 1986), p. 29. Indeed, many immigrant maids and nannies are more educated than the people they work for.

12. Momsen, 1999, pp. 10, 73.

13. Grete Brochmann, *Middle East Avenue: Female Migration from Sri Lanka to the Gulf* (Boulder, Colo.: Westview Press, 1993), pp. 179, 215

14. On this point, thanks to Raka Ray, Sociology Department at the University of California, Berkeley.

23

The Care Crisis in the Philippines

Children and Transnational Families in the New Global Economy

RHACEL SALAZAR PARREÑAS

This study provides a close look at how global migration affects careworkers and their families. As Parreñas describes, when parents are forced to leave their children behind in the Philippines in order to provide them with economic security, their absence extracts tremendous family costs.

A growing crisis of care troubles the world's most developed nations. Even as demand for care has increased, its supply has dwindled. The result is a care deficit, to which women from the Philippines have responded in force. Roughly two-thirds of Filipino migrant workers are women, and their exodus, usually to fill domestic jobs, has generated tremendous social change in the Philippines. When female migrants are mothers, they leave behind their own children, usually in the care of other women. Many Filipino children now grow up in divided households, where geographic separation places children under serious emotional strain. And yet it is impossible to overlook the significance of migrant labor to the Philippine economy. Some 34 to 54 percent of the Filipino population is sustained by remittances from migrant workers.

Women in the Philippines, just like their counterparts in postindustrial nations, suffer from a "stalled revolution." Local gender ideology remains a few steps behind the economic reality, which has produced numerous female-headed, transnational households. Consequently, a far greater degree of anxiety attends the quality of family life for the dependents of migrant mothers than for those of migrant fathers. The dominant gender ideology, after all, holds that a woman's rightful place is in the home, and the households of migrant mothers present a challenge to this view. In response, government officials and journalists denounce migrating mothers, claiming that they have caused the Filipino family to deteriorate, children to be abandoned, and a crisis of care to take root in the Philippines. To end this crisis, critics admonish, these mothers must return.

SOURCE: From Rhacel Salazar Parreñas, "The Care Crisis in the Philippines: Children and Transnational Families in the New Global Economy," pp. 39–54, in *Global Woman: Nannies, Maids, and Sex Workers in the New Economy*, Barbara Ehrenreich and Arlie Russell Hochschild (eds.), (New York Metropolitan, 2003). Reprinted with permission.

Indeed, in May 1995, Philippine president Fidel Ramos called for initiatives to keep migrant mothers at home. He declared, "We are not against overseas employment of Filipino women. We are against overseas employment at the cost of family solidarity." Migration, Ramos strongly implied, is morally accept-able only when it is undertaken by single, childless women.

The Philippine media reinforce this position by consistently publishing sensa-tionalist reports on the suffering of children in transnational families. These reports tend to vilify migrant mothers, suggesting that their children face more profound problems than do those of migrant fathers; and despite the fact that most of the children in question are left with relatives, journalists tend to refer to them as having been "abandoned." One article reports, "A child's sense of loss appears to be greater when it is the mother who leaves to work abroad." Others link the emigration of mothers to the inadequate child care and unstable family life that eventually lead such children to "drugs, gambling, and drinking." Writes one columnist, "Incest and rapes within blood relatives are alarmingly on the rise not only within Metro Manila but also in the provinces. There are some indications that the absence of mothers who have become OCWs [overseas contract workers] has something to do with the situation." The same columnist elsewhere expresses the popular view that the children of migrants become a burden on the larger society: "Guidance counselors and social welfare agencies can show grim statistics on how many chil-dren have turned into liabilities to our society because of absentee parents."

From January to July 2000, I conducted sixty-nine in-depth interviews with young adults who grew up in transnational households in the Philippines. Almost none of these children have yet reunited with their migrant parents. I interviewed thirty children with migrant mothers, twenty-six with migrant fathers, and thirteen with two migrant parents. The children I spoke to certainly had endured emotional hardships; but contrary to the media's dark presentation, they did not all experience their mothers' migration as abandonment. The hard-ships in their lives were frequently diminished when they received support from extended families and communities, when they enjoyed open communication with their migrant parents, and when they clearly understood the limited finan-cial options that led their parents to migrate in the first place.

To call for the return of migrant mothers is to ignore the fact that the Philippines has grown increasingly dependent on their remittances. To acknowledge this reality could lead the Philippines toward a more egalitarian gender ideology. Casting blame on migrant mothers, however, serves only to divert the society's attention away from these children's needs, finally aggravating their difficulties by stigmatizing their families' choices.

The Philippine media has certainly sensationalized the issue of child welfare in migrating families, but that should not obscure the fact that the Philippines faces a genuine care crisis. Care is now the country's primary export. Remittances—mostly from migrant domestic workers—constitute the economy's largest source of foreign currency, totaling almost $7 billion in 1999. With limited choices in the Philippines, women migrate to help sustain their families financially, but the price is very high. Both mothers and children suffer from family separation, even under the best of circumstances.

ant mothers who work as nannies often face the painful prospect of
caring r other people's children while being unable to tend to their own.
One such mother in Rome, Rosemarie Samaniego, describes this predicament:

> When the girl that I take care of calls her mother "Mama," my heart
> jumps all the time because my children also call me "Mama." I feel the
> gap caused by our physical separation especially in the morning, when
> I pack [her] lunch, because that's what I used to do for my children....
> I used to do that very same thing for them. I begin thinking that at this
> hour I should be taking care of my very own children and not someone
> else's, someone who is not related to me in any way, shape, or form....
> The work that I do here is done for my family, but the problem is they
> are not close to me but are far away in the Philippines. Sometimes, you
> feel the separation and you start to cry. Some days, I just start crying
> while I am sweeping the floor because I am thinking about my children
> in the Philippines. Sometimes, when I receive a letter from my children
> telling me that they are sick, I look up out the window and ask the
> Lord to look after them and make sure they get better even without me
> around to care after them. [*Starts crying.*] If I had wings, I would fly
> home to my children. Just for a moment, to see my children and take
> care of their needs, help them, then fly back over here to continue
> my work.

The children of migrant workers also suffer an incalculable loss when a parent
disappears overseas. As Ellen Seneriches, a twenty-one-year-old daughter of a
domestic worker in New York, says:

> There are times when you want to talk to her, but she is not there. That
> is really hard, very difficult.... There are times when I want to call her,
> speak to her, cry to her, and I cannot. It is difficult. The only thing that
> I can do is write to her. And I cannot cry through the e-mails and
> sometimes I just want to cry on her shoulder.

Children like Ellen, who was only ten years old when her mother left for
New York, often repress their longings to reunite with their mothers. Knowing
that their families have few financial options, they are left with no choice but to
put their emotional needs aside. Often, they do so knowing that their mothers'
care and attention have been diverted to other children. When I asked her how
she felt about her mother's wards in New York, Ellen responded:

> Very jealous. I am very, very jealous. There was even a time when she
> told the children she was caring for that they are very lucky that she was
> taking care of them, while her children back in the Philippines don't
> even have a mom to take care of them. It's pathetic, but it's true. We
> were left alone by ourselves and we had to be responsible at a very
> young age without a mother. Can you imagine?

Children like Ellen do experience emotional stress when they grow up in
transnational households. But it is worth emphasizing that many migrant mothers

attempt to sustain ties with their children, and their children often recognize and appreciate these efforts. Although her mother, undocumented in the United States, has not returned once to the Philippines in twelve years, Ellen does not doubt that she has struggled to remain close to her children despite the distance. In fact, although Ellen lives only three hours away from her father, she feels closer to and communicates more frequently with her mother. Says Ellen:

> I realize that my mother loves us very much. Even if she is far away, she would send us her love. She would make us feel like she really loved us. She would do this by always being there. She would just assure us that whenever we have problems to just call her and tell her. [*Pauses.*] And so I know that it has been more difficult for her than other mothers. She has had to do extra work because she is so far away from us.

Like Ellen's mother, who managed to "be there" despite a vast distance, other migrant mothers do not necessarily "abandon" their traditional duty of nurturing their families. Rather, they provide emotional care and guidance from afar. Ellen even credits her mother for her success in school. Now a second-year medical school student, Ellen graduated at the top of her class in both high school and college. She says that the constant, open communication she shares with her mother provided the key to her success. She reflects:

> We communicate as often as we can, like twice or thrice a week through e-mails. Then she would call us every week. And it is very expensive, I know.... My mother and I have a very open relationship. We are like best friends. She would give me advice whenever I had problems.... She understands everything I do. She understands why I would act this or that way. She knows me really well. And she is also transparent to me. She always knows when I have problems, and like-wise I know when she does. I am closer to her than to my father.

Ellen is clearly not the abandoned child or social liability the Philippine media describe. She not only benefits from sufficient parental support—from both her geographically distant mother and her nearby father—but also exceeds the bar of excellence in schooling. Her story indicates that children of migrant parents can overcome the emotional strains of transnational family life, and that they can enjoy sufficient family support, even from their geographically distant parent.

Of course, her good fortune is not universal. But it does raise questions about how children withstand such geographical strains; whether and how they maintain solid ties with their distant parents; and what circumstances lead some children to feel that those ties have weakened or given out. The Philippine media tend to equate the absence of a child's biological mother with abandonment, which leads to the assumption that all such children, lacking familial support, will become social liabilities. But I found that positive surrogate parental figures and open communication with the migrant parent, along with acknowledgment of the migrant parent's contribution to the collective mobility of the family, allay many of the emotional insecurities that arise from transnational

household arrangements. Children who lack these resources have greater difficulty adjusting.

Extensive research bears out this observation. The Scalabrini Migration Center, a nongovernmental organization for migration research in the Philippines,
surveyed 709 elementary-school-age Filipino children in 2000, comparing the
experiences of those with a father absent, a mother absent, both parents absent,
and both parents present. While the researchers observed that parental absence
does prompt feelings of abandonment and loneliness among children, they concluded that "it does not necessarily become an occasion for laziness and
unruliness." Rather, if the extended family supports the child and makes him
or her aware of the material benefits migration brings, the child may actually
be spurred toward greater self-reliance and ambition, despite continued longings
for family unity.

Jeek Pereno's life has been defined by those longings. At twenty-five, he is a
merchandiser for a large department store in the Philippines. His mother more
than adequately provided for her children, managing with her meager wages first
as a domestic worker and then as a nurse's aide, to send them $200 a month and
even to purchase a house in a fairly exclusive neighborhood in the city center.
But Jeek still feels abandoned and insecure in his mother's affection; he believes
that growing up without his parents robbed him of the discipline he needed.
Like other children of migrant workers, Jeek does not feel that his faraway
mother's financial support has been enough. Instead, he wishes she had offered
him more guidance, concern, and emotional care.

Jeek was eight years old when his parents relocated to New York and left
him, along with his three brothers, in the care of their aunt. Eight years later,
Jeek's father passed away, and two of his brothers (the oldest and youngest)
joined their mother in New York. Visa complications have prevented Jeek and
his other brother from following—but their mother has not once returned to
visit them in the Philippines. When I expressed surprise at this, Jeek solemnly
replied: "Never. It will cost too much, she said."

Years of separation breed unfamiliarity among family members, and Jeek
does not have the emotional security of knowing that his mother has genuinely
tried to lessen that estrangement. For Jeek, only a visit could shore up this security after seventeen years of separation. His mother's weekly phone calls do not
suffice. And because he experiences his mother's absence as indifference, he does
not feel comfortable communicating with her openly about his unmet needs.
The result is repression, which in turn aggravates the resentment he feels. Jeek
told me:

> I talk to my mother once in a while. But what happens, whenever she
> asks how I am doing, I just say okay. It's not like I am really going to
> tell her that I have problems here…. It's not like she can do anything
> about my problems if I told her about them. Financial problems, yes she
> can help. But not the other problems, like emotional problems. She will
> try to give advice, but I am not very interested to talk to her about
> things like that…. Of course, you are still young, you don't really know

what is going to happen in the future. Before you realize that your parents left you, you can't do anything about it anymore. You are not in a position to tell them not to leave you. They should have not left us. [*Sobs*.]

I asked Jeek if his mother knew he felt this way. "No," he said, "she doesn't know." Asked if he received emotional support from anyone, Jeek replied, "As much as possible, if I can handle it, I try not to get emotional support from anyone. I just keep everything inside me."

Jeek feels that his mother not only abandoned him but failed to leave him with an adequate surrogate. His aunt had a family and children of her own. Jeek recalled, "While I do know that my aunt loves me and she took care of us to the best of her ability, I am not convinced that it was enough.... Because we were not disciplined enough. She let us do whatever we wanted to do." Jeek feels that his education suffered from this lack of discipline, and he greatly regrets not having concentrated on his studies. Having completed only a two-year vocational program in electronics, he doubts his competency to pursue a college degree. At twenty-five, he feels stuck, with only the limited option of turning from one low-paying job to another.

Children who, unlike Jeek, received good surrogate parenting managed to concentrate on their studies and in the end to fare much better. Rudy Montoya, a nineteen-year-old whose mother has done domestic work in Hong Kong for more than twelve years, credits his mother's brother for helping him succeed in high school:

My uncle is the most influential person in my life. Well, he is in Saudi Arabia now.... He would tell me that my mother loves me and not to resent her, and that whatever happens, I should write her. He would encourage me and he would tell me to trust the Lord. And then, I remember in high school, he would push me to study. I learned a lot from him in high school. Showing his love for me, he would help me with my school work.... The time that I spent with my uncle was short, but he is the person who helped me grow up to be a better person.

Unlike Jeek's aunt, Rudy's uncle did not have a family of his own. He was able to devote more time to Rudy, instilling discipline in his young charge as well as reassuring him that his mother, who is the sole income provider for her family, did not abandon him. Although his mother has returned to visit him only twice—once when he was in the fourth grade and again two years later—Rudy, who is now a college student, sees his mother as a "good provider" who has made tremendous sacrifices for his sake. This knowledge offers him emotional security, as well as a strong feeling of gratitude. When I asked him about the importance of education, he replied, "I haven't given anything back to my mother for the sacrifices that she has made for me. The least I could do for her is graduate, so that I can find a good job, so that eventually I will be able to help her out, too."

Many children resolve the emotional insecurity of being left by their parents the way that Rudy has: by viewing migration as a sacrifice to be repaid by adult children. Children who believe that their migrant mothers are struggling for the sake of the family's collective mobility, rather than leaving to live the "good life," are less likely to feel abandoned and more likely to accept their mothers' efforts to sustain close relationships from a distance. One such child is Theresa Bascara, an eighteen-year-old college student whose mother has worked as a domestic in Hong Kong since 1984. As she puts it, "[My inspiration is] my mother, because she is the one suffering over there. So the least I can give back to her is doing well in school."

For Ellen Seneriches, the image of her suffering mother compels her to reciprocate. She explained:

> Especially after my mother left, I became more motivated to study harder. I did because my mother was sacrificing a lot and I had to compensate for how hard it is to be away from your children and then crying a lot at night, not knowing what we are doing. She would tell us in voice tapes. She would send us voice tapes every month, twice a month, and we would hear her cry in these tapes.

Having witnessed her mother's suffering even from a distance, Ellen can acknowledge the sacrifices her mother has made and the hardships she has endured in order to be a "good provider" for her family. This knowledge assuaged the resentment Ellen frequently felt when her mother first migrated.

Many of the children I interviewed harbored images of their mothers as martyrs, and they often found comfort in their mothers' grief over not being able to nurture them directly. The expectation among such children that they will continue to receive a significant part of their nurturing from their mothers, despite the distance, points to the conservative gender ideology most of them maintain. But whether or not they see their mothers as martyrs, children of migrant women feel best cared for when their mothers make consistent efforts to show parental concern from a distance. As Jeek's and Ellen's stories indicate, open communication with the migrant parent soothes feelings of abandonment; those who enjoy such open channels fare much better than those who lack them. Not only does communication ease children's emotional difficulties; it also fosters a sense of family unity, and it promotes the view that migration is a survival strategy that requires sacrifices from both children and parents for the good of the family.

For daughters of migrant mothers, such sacrifices commonly take the form of assuming some of their absent mothers' responsibilities, including the care of younger siblings. As Ellen told me:

> It was a strategy, and all of us had to sacrifice for it.... We all had to adjust, every day of our lives.... Imagine waking up without a mother calling you for breakfast. Then there would be no one to prepare the clothes for my brothers. We are all going to school.... I had to wake up earlier. I had to prepare their clothes. I had to wake them up and help

them prepare for school. Then I also had to help them with their homework at night. I had to tutor them.

Asked if she resented this extra work, Ellen replied, "No. I saw it as training, a training that helped me become a leader. It makes you more of a leader doing that every day. I guess that is an advantage to me, and to my siblings as well."

Ellen's effort to assist in the household's daily maintenance was another way she reciprocated for her mother's emotional and financial support. Viewing her added work as a positive life lesson, Ellen believes that these responsibilities enabled her to develop leadership skills. Notably, her high school selected her as its first ever female commander for its government-mandated military training corps.

Unlike Jeek, Ellen is secure in her mother's love. She feels that her mother has struggled to "be there"; Jeek feels that his has not. Hence, Ellen has managed to successfully adjust to her household arrangement, while Jeek has not. The continual open communication between Ellen and her mother has had ramifications for their entire family: in return for her mother's sacrifices, Ellen assumed the role of second mother to her younger siblings, visiting them every weekend during her college years in order to spend quality time with them.

In general, eldest daughters of migrant mothers assume substantial familial responsibilities, often becoming substitute mothers for their siblings. Similarly, eldest sons stand in for migrant fathers. Armando Martinez, a twenty-nine-year-old entrepreneur whose father worked in Dubai for six months while he was in high school, related his experiences:

> I became a father during those six months. It was like, ugghhh, I made the rules…. I was able to see that it was hard if your family is not complete, you feel that there is something missing…. It's because the major decisions, sometimes, I was not old enough for them. I was only a teenager, and I was not that strong in my convictions when it came to making decisions. It was like work that I should not have been responsible for. I still wanted to play. So it was an added burden on my side.

Even when there is a parent left behind, children of migrant workers tend to assume added familial responsibilities, and these responsibilities vary along gender lines. Nonetheless, the weight tends to fall most heavily on children of migrant mothers, who are often left to struggle with the lack of male responsibility for care work in the Philippines. While a great number of children with migrant fathers receive full-time care from stay-at-home mothers, those with migrant mothers do not receive the same amount of care. Their fathers are likely to hold full-time jobs, and they rarely have the time to assume the role of primary caregiver. Of thirty children of migrant mothers I interviewed, only four had stay-at-home fathers. Most fathers passed the caregiving responsibilities on to other relatives, many of whom, like Jeek's aunt, already had families of their own to care for and regarded the children of migrant relatives as an extra burden. Families of migrant fathers are less likely to rely on the care work of extended kin. Among my interviewees, thirteen of twenty-six children with migrant fathers lived with and were cared for primarily by their stay-at-home mothers.

Children of migrant mothers, unlike those of migrant fathers, have the added burden of accepting nontraditional gender roles in their families. The Scalabrini Migration Center reports that these children "tend to be more angry, confused, apathetic, and more afraid than other children." They are caught within an "ideological stall" in the societal acceptance of female-headed transnational households. Because her family does not fit the traditional nuclear household model, Theresa Bascara sees her family as "broken," even though she describes her relationship to her mother as "very close." She says, "A family, I can say, is only whole if your father is the one working and your mother is only staying at home. It's okay if your mother works too, but somewhere close to you."

Some children in transnational families adjust to their household arrangements with greater success than others do. Those who feel that their mothers strive to nurture them as well as to be good providers are more likely to be accepting. The support of extended kin, or perhaps a sense of public accountability for their welfare, also helps children combat feelings of abandonment. Likewise, a more gender-egalitarian value system enables children to appreciate their mothers as good providers, which in turn allows them to see their mothers' migrations as demonstrations of love.

Even if they are well-adjusted, however, children in transnational families still suffer the loss of family intimacy. They are often forced to compensate by accepting commodities, rather than affection, as the most tangible reassurance of their parents' love. By putting family intimacy on hold, children can only wait for the opportunity to spend quality time with their migrant parents. Even when that time comes, it can be painful. As Theresa related:

> When my mother is home, I just sit next to her. I stare at her face, to see the changes in her face, to see how she aged during the years that she was away from us. But when she is about to go back to Hong Kong, it's like my heart is going to burst. I would just cry and cry. I really can't explain the feeling. Sometimes, when my mother is home, preparing to leave for Hong Kong, I would just start crying, because I already start missing her. I ask myself, how many more years will it be until we see each other again?.... Telephone calls. That's not enough. You can't hug her, kiss her, feel her, everything. You can't feel her presence. It's just words that you have. What I want is to have my mother close to me, to see her grow older, and when she is sick, you are the one taking care of her and when you are sick, she is the one taking care of you.

Not surprisingly, when asked if they would leave their own children to take jobs as migrant workers, almost all of my respondents answered, "Never." When I asked why not, most said that they would never want their children to go through what they had gone through, or to be denied what they were denied, in their childhoods. Armando Martinez best summed up what children in transnational families lose when he said:

> You just cannot buy the times when your family is together. Isn't that right? Time together is something that money can neither buy nor

replace.... The first time your baby speaks, you are not there. Other people would experience that joy. And when your child graduates with honors, you are also not there.... Is that right? When your child wins a basketball game, no one will be there to ask him how his game went, how many points he made. Is that right? Your family loses, don't you think?

Children of transnational families repeatedly stress that they lack the pleasure and comfort of daily interaction with their parents. Nonetheless, these children do not necessarily become "delinquent," nor are their families necessarily broken, in the manner the Philippine media depicts. Armando mirrored the opinion of most of the children in my study when he defended transnational families: "Even if [parents] are far away, they are still there. I get that from basketball, specifically zone defense." [*He* laughed.] "If someone is not there, you just have to adjust. It's like a slight hindrance that you just have to adjust to. Then when they come back, you get a chance to recover. It's like that."

Recognizing that the family is an adaptive unit that responds to external forces, many children make do, even if doing so requires tremendous sacrifices. They give up intimacy and familiarity with their parents. Often, they attempt to make up for their migrant parents' hardships by maintaining close bonds across great distances, even though most of them feel that such bonds could never possibly draw their distant parent close enough. But their efforts are frequently sustained by the belief that such emotional sacrifices are not without meaning—that they are ultimately for the greater good of their families and their future. Jason Halili's mother provided care for elderly persons in Los Angeles for fifteen years. Jason, now twenty-one, reasons, "If she did not leave, I would not be here right now. So it was the hardest route to take, but at the same time, the best route to take."

Transnational families were not always equated with "broken homes" in the Philippine public discourse. Nor did labor migration emerge as a perceived threat to family life before the late 1980s, when the number of migrant women significantly increased. This suggests that changes to the gendered division of family labor may have as much as anything else to do with the Philippine care crisis.

The Philippine public simply assumes that the proliferation of female-headed transnational households will wreak havoc on the lives of children. The Scalabrini Migration Center explains that children of migrant mothers suffer more than those of migrant fathers because child rearing is "a role women are more adept at, are better prepared for, and pay more attention to." The center's study, like the Philippine media, recommends that mothers be kept from migrating. The researchers suggest that "economic programs should be targeted particularly toward the absorption of the female labor force, to facilitate the possibility for mothers to remain in the family." Yet the return migration of mothers is neither a plausible nor a desirable solution. Rather, it implicitly accepts gender inequities in the family, even as it ignores the economic pressures generated by globalization.

As national discourse on the care crisis in the Philippines vilifies migrant women, it also downplays the contributions these women make to the country's economy. Such hand-wringing merely offers the public an opportunity to discipline women morally and to resist reconstituting family life in a manner that reflects the country's increasing dependence on women's foreign remittances. This pattern is not exclusive to the Philippines. As Arjun Appadurai observes, globalization has commonly led to "ideas about gender and modernity that create large female work forces at the same time that cross-national ideologies of 'culture,' 'authenticity,' and national honor put increasing pressure on various communities to morally discipline working women."

The moral disciplining of women, however, hurts those who most need protection. It pathologizes the children of migrants, and it downplays the emotional difficulties that mothers like Rosemarie Samaniego face. Moreover, it ignores the struggles of migrant mothers who attempt to nurture their children from a distance. Vilifying migrant women as bad mothers promotes the view that the return to the nuclear family is the only viable solution to the emotional difficulties of children in transnational families. In so doing, it directs attention away from the special needs of children in transnational families—for instance, the need for community projects that would improve communication among far-flung family members, or for special school programs, the like of which did not exist at my field research site. It's also a strategy that sidelines the agency and adaptability of the children themselves.

To say that children are perfectly capable of adjusting to nontraditional households is not to say that they don't suffer hardships. But the overwhelming public support for keeping migrant mothers at home does have a negative impact on these children's adjustment. Implicit in such views is a rejection of the division of labor in families with migrant mothers, and the message such children receive is that their household arrangements are simply wrong. Moreover, calling for the return migration of women does not necessarily solve the problems plaguing families in the Philippines. Domestic violence and male infidelity, for instance—two social problems the government has never adequately addressed—would still threaten the well-being of children.

Without a doubt, the children of migrant Filipina domestic workers suffer from the extraction of care from the global south to the global north. The plight of these children is a timely and necessary concern for nongovernmental, governmental, and academic groups in the Philippines. Blaming migrant mothers, however, has not helped, and has even hurt, those whose relationships suffer most from the movement of care in the global economy. Advocates for children in transnational families should focus their attention not on calling for a return to the nuclear family but on trying to meet the special needs transnational families possess. One of those needs is for a reconstituted gender ideology in the Philippines; another is for the elimination of legislation that penalizes migrant families in the nations where they work.

If we want to secure quality care for the children of transnational families, gender egalitarian views of child rearing are essential. Such views can be fostered by recognizing the economic contributions women make to their families and by redefining motherhood to include providing for one's family. Gender should be recognized as a fluid social category, and masculinity should be redefined, as the larger society questions the biologically based assumption that only women have an aptitude to provide care. Government officials and the media could then

stop vilifying migrant women, redirecting their attention, instead, to men. They could question the lack of male accountability for care work, and they could demand that men, including migrant fathers, take more responsibility for the emotional welfare of their children.

The host societies of migrant Filipina domestic workers should also be held more accountable for their welfare and for that of their families. These women's work allows First World women to enter the paid labor force. As one Dutch employer states, "There are people who would look after children, but other things are more fun. Carers from other countries, if we can use their surplus carers, that's a solution."

Yet, as we've seen, one cannot simply assume that the care leaving disadvantaged nations is surplus care. What is a solution for rich nations creates a problem in poor nations. Mothers like Rosemarie Samaniego and children like Ellen Seneriches and Jeek Pereno bear the brunt of this problem, while the receiving countries and the employing families benefit.

Most receiving countries have yet to recognize the contributions of their migrant care workers. They have consistently ignored these workers' rights and limited their full incorporation into society. The wages of migrant workers are so low that they cannot afford to bring their own families to join them, or to regularly visit their children in the Philippines; relegated to the status of guest workers, they are restricted to the low-wage employment sector, and with very few exceptions, the migration of their spouses and children is also restricted. These arrangements work to the benefit of employers, since migrant care workers can give the best possible care for their employers' families when they are free of care-giving responsibilities to their own families. But there is a dire need to lobby for more inclusive policies, and for employers to develop a sense of accountability for their workers' children. After all, migrant workers significantly help their employers to reduce *their* families' care deficit.

24

Childbirth at the Global Crossroads

ARLIE RUSSELL HOCHSCHILD

*This article describes a new transnational exchange—women in the developing
world who are paid to bear other people's children. Commercial surrogacy, however,
is emotionally complicated for both surrogates and genetic parents, especially when
they come from different parts of the globe.*

The auto-rickshaw driver honks his way through the dusty chaos of Anand,
Gujarat, India, swerving around motorbikes, grunting trucks, and ancient
large-wheeled bullock-carts packed with bags of fodder. Both sides of the street
are lined with plastic trash and small piles of garbage on which untethered cows
feed. The driver turns off the pavement onto a narrow, pitted dirt road, slows to
circumvent a pair of black and white spotted goats, and stops outside a dusty
courtyard. To one side stands a modest white building with a sign that reads, in
English and Gujarati, "Akanksha Clinic."

Two dozen dainty Indian women's sandals, toes pointed forward, are lined
along the front porch. For it is with bare feet that one enters a clinic housing
what may be the world's largest group of gestational surrogates—women who
rent their wombs to incubate the fertilized eggs from clients from around the
globe. Since India declared commercial surrogacy legal in 2002, some 350 assisted
reproductive technology (ART) clinics have opened their doors. Surrogacy is
now a burgeoning part of India's medical tourism industry, which is slated to add
$2 billion to the nation's gross domestic product by 2012. Advertisements describe
India as a "global doctor" offering First World skill at Third World prices, short
waits, privacy, and—important in the case of surrogacy—absence of red tape. To
encourage this lucrative trend, the Indian government gives tax breaks to private
hospitals treating overseas patients and lowers import duties on medical supplies.

In his 2007 book, *Supercapitalism*, Robert B. Reich argues that while industrial and clerical jobs could be outsourced to cheaper labor pools abroad, service
jobs would stay in America. But Reich didn't count on First World clients flying
to the global South to find low-cost retirement care or reproductive services.
The Akanksha clinic is just one point on an ever-widening two-lane global
highway that connects poor nations in the Southern Hemisphere to rich nations

SOURCE: Reprinted with permission from Arlie Hochschild, "Childbirth at the Global
Crossroads," *The American Prospect*: October 2009, Volume 20, Issue 8. http://www.
prospect.org. *The American Prospect*, 1710 Rhode Island Avenue, NW, 12th Floor,
Washington, DC 20036. All rights reserved.

in the Northern Hemisphere, and poorer countries of Eastern Europe to richer ones in the West. A Filipina nanny heads north to care for an American child. A Sri Lankan maid cleans a house in Singapore. A Ukrainian nurse's aide carries lunch trays in a Swedish hospital. Marx's iconic male, stationary industrial worker has been replaced by a new icon: the female, mobile service worker.

We have grown used to the idea of a migrant worker caring for our children and even to the idea of hopping an overseas flight for surgery. As global service work grows increasingly personal, surrogacy is the latest expression of this trend. Nowadays, a wealthy person can purchase it all—the egg, the sperm, and time in the womb. "A childless couple gains a child. A poor woman earns money. What could be the problem?" asks Dr. Nayna Patel, Akanksha's founder and director.

But despite Patel's view of commercial surrogacy as a straightforward equation, it's far more complicated for both the surrogates and the genetic parents. Like nannies or nurses, surrogates perform "emotional labor" to suppress feelings that could interfere with doing their job. Parents must decide how close they are willing (or able) to get to the woman who will give birth to their child.

As science and global capitalism gallop forward, they kick up difficult questions about emotional attachment. What, if anything, is too sacred to sell?

I follow a kindly embryologist, Harsha Bhadarka, to an upstairs office of the clinic to talk with two surrogates whom I will call Geeta and Saroj. (Aditya Ghosh, a journalist with the *Hindustan Times*, has kindly offered to join me.) The room is small, and the two surrogate mothers enter the room nodding shyly. Both live on the second floor of the clinic, but most of its 24 residents live in one of two hostels for the duration of their pregnancy. The women are brought nutritious food on tin trays, injected with iron (a common deficiency), and supervised away from prying in-laws, curious older children, and lonely husbands with whom they are allowed no visits home or sex.

Geeta, a 22-year-old, light-skinned, green-eyed beauty, is the mother of three daughters, one of whom is sitting quietly and wide-eyed on her lap. To be accepted as a surrogate, Akanksha requires a woman to be a healthy, married mother. As one doctor explains, "If she has children of her own, she'll be less tempted to attach herself to the baby."

"How did you decide to become a surrogate?" I ask.

"It was my husband's idea," Geeta replies. "He makes *pav bhaji* [a vegetable dish] during the day and serves food in the evening [at a street-side fast-food shop]. He heard about surrogacy from a customer at his shop, a Muslim like us. The man told my husband, 'It's a good thing to do,' and then I came to madam [Dr. Patel] and offered to try. We can't live on my husband's earnings, and we had no hope of educating our daughters."

Geeta says she has only briefly met the parents whose genes her baby carries. "They're from far away. I don't know where," she says. "They're Caucasian, so the baby will come out white." The money she has been promised, including a monthly stipend to cover vitamins and medications, is wired to a bank account that Patel has opened in Geeta's name. "I keep myself from getting too attached," she says. "Whenever I start to think about the baby inside me, I turn my attention to my own daughter. Here she is." She bounces the child on her lap. "That way, I manage."

Seated next to Geeta is Saroj, a heavy-set, dark woman with intense, curious eyes, and, after a while, an easy smile. Like other Hindu surrogates at Akanksha, she wears *sindoor* (a red powder applied to the part in her hair) and *mangalsutra* (a necklace with a gold pendant), both symbols of marriage. She is, she tells us, the mother of three children and the wife of a vegetable street vendor. She gave birth to a surrogate child a year and three months ago and is waiting to see if a second implantation has taken. The genetic parents are from Bangalore, India. (It is estimated that half the clients seeking surrogacy from Indian ART clinics are Indian and the other half, foreign. Of the foreign clients, roughly half are American.) Saroj, too, knows almost nothing about her clients. "They came, saw me, and left," she says.

Given her husband's wages, 1,260 rupees (or $25) a month, Saroj turned to surrogacy so she could move to a rain-proof house and feed her family well. Yet she faced the dilemma of all rural surrogates: being suspected of adultery—a cause for shunning or worse. I ask the women whether the money they earn has improved their social standing. For the first time the two women laugh out loud and talk to each other excitedly. "My father-in-law is dead, and my mother-in-law lives separately from us, and at first I hid it from her," Saroj says. "But when she found out, she said she felt blessed to have a daughter-in-law like me because I've given more money to the family than her son could. But some friends ask me why I am putting myself through all this. I tell them, 'It's my own choice.'"

Since Dr. Patel began offering surrogacy services in 2004, 232 surrogates have given birth at Akanksha. A 2007 study of 42 Akanksha surrogates found that nearly half described themselves as housewives and the rest were a mix of domestic, service, and manual laborers. Hindu, Muslim, and Christian, most had seventh- to 12th-grade educations, six were illiterate, and one—who turned to surrogacy to pay for a small son's heart surgery—had a bachelor's degree. Each surrogate negotiates a different sum: one surrogate carrying twins for an Indian couple discovered she was being paid less (about $3,600) than a surrogate in the next bed who was carrying one baby for an American couple for about $5,600.

Observers fear that a lack of regulation could spark a price war for surrogacy—Thailand underselling India, Cambodia underselling Thailand, and so on—with countries slowly under-cutting fees and legal protections for surrogates along the way. It could happen. Right now international surrogacy is a highly complex legal patchwork. Surrogacy is banned in China and much of Europe. It is legal but regulated in New Zealand and Great Britain. Only 17 of the United States have laws on the books; it is legal in Florida and banned in New York.

In India, commercial surrogacy is legal but unregulated, although a 135-page regulatory law, long in the works, will be sent to Parliament later this year. Even if the law is passed, however, some argue it would do little to improve life for women such as Geeta and Saroj. For example, it specifies that the doctor, not the surrogate, has the right to decide on any "fetal reduction" (an abortion). Moreover, most Indian federal laws are considered "advisory" to powerful state governments, and courts—where a failure to enforce such laws might be challenged—are back-logged for years, often decades. Dr. B. N. Chakravarty, the Calcutta-based chair of the surrogacy law drafting committee, says that the growth of the industry is "inevitable," but it needs regulating. Even if the law were written to protect surrogates and then actually enforced, it would do nothing to address the crushing poverty that often presses Indian women to "choose" surrogacy in the first place.

For N. B. Sarojini, director of the Delhi-based Sama Resource Group for Women and Health, a nonprofit feminist research institute, the problem is one of distorted priorities. "The ART clinics are posing themselves as the answer to an illusory 'crisis' of infertility," she says. "Two decades back, a couple might consider themselves 'infertile' after trying for five years to conceive. Then it moved to four years. Now couples rush to ARTs after one or two. Why not put the cultural spotlight on *alternatives*? Why not urge childless women to adopt orphans? And what, after all, is wrong with remaining childless?"

But Dr. Patel, a striking woman in an emerald green sari and with black hair flowing down her back, sees for-profit surrogacy as a "win-win" for the clinic, the surrogate, and the genetic parents. She also sees no problem with running the clinic like a business, seeking to increase inventory, safeguard quality, and improve efficiency. That means producing more babies, monitoring surrogates' diet and sexual contact, and assuring a smooth, emotion-free exchange of baby for money. (For every dollar that goes to the surrogates, observers estimate, three go to the clinic.) In Akanksha's hostel, women sleep on cots, nine to a room, for nine months. Their young children sleep with them; older children do not stay in the hostel. The women exercise inside the hostel, rarely leaving it and then only with permission. Patel also advises surrogates to limit contact with clients. Staying detached from the genetic parents, she says, helps surrogate mothers give up their babies and get on with their lives—and maybe with the next surrogacy. This ideal of the de-personalized pregnancy is eerily reminiscent of Aldous Huxley's 1932 dystopian novel *Brave New World*, in which babies are emotionlessly mass-produced in the Central London Hatchery.

Patel's business may seem coldly efficient, but it also has a touch of Mother Teresa. Akanksha residents are offered daily English classes and weekly lessons in computer use. Patel arranges for film screenings and gives out school backpacks and pencil boxes to surrogates' children. She hopes to attract donations from grateful clients to help pay children's school fees as well. "For me this is a mission," Patel says.

In light of appalling government neglect of a population totally untouched by India's recent economic boom, this charity sounds wonderful. But is it wonderful enough to cancel out concerns about the factory?

After leaving Anand, I head to Dr. Nandita Palshetkar's office in Mumbai. With Alifiya Khan, another journalist from the *Hindustan Times*, I meet with Leela, a lively 28-year-old who gave birth to a baby for Indian clients about six months ago. Like Geeta and Saroj, Leela had been desperate for money, but her experience of pregnancy was utterly different. On the day I meet her, she is dressed in a pink sari, hair drawn back from her olive-skinned face into a long black braid. She leans forward, smiling broadly, eager to talk about her baby, his genetic parents, and her feelings about being a surrogate mother.

At age 20, Leela married a fellow worker at a Mumbai-based company canteen. "I didn't know he was alcoholic until after we married," she says.

"My husband ran up a $7,000 debt with the moneylender who sent agents to pressure him to repay it.... We couldn't stop the moneylender from hounding us. I decided to act. I heard from my sister-in-law that I could get money for donating my eggs, and I did that twice. When I came back to do it a third time, madam [Dr. Palshetkar] told me I could earn more as a surrogate."

Was she able to pay off the debt? Leela lowers her head: "Half of it."

She ate better food during her paid pregnancy than during her other pregnancies and delivered the baby in a better hospital than the one where she delivered her own children. Unlike others I spoke with, Leela openly bonded with her baby. "I am the baby's *real* mother," she says. "I carried him. I felt him kick. I prayed for him. At seven months I held a celebration for him. I saw his legs and hands on the sonogram. I suffered the pain of birth."

The baby's genetic parents, Indians from a nearby affluent suburb, kindly reached out to Leela. The genetic mother "sees me as her little sister, and I see her as my big sister," Leela says. "They check in with me every month, even now, and call me the baby's 'auntie.' I said 'yes' and they brought him to my house, but I was disappointed to see he was long and fair, not like me. Still, to this day, I feel I have three children." A friendship of sorts arose between the two mothers, although Leela's doctor, like Patel, discouraged it. "I deleted their phone number from my list because madam told me it's not a good thing to keep contact for long," she says.

In a November 2008 *New York Times Magazine* article titled "Her Body, My Baby," American journalist Alex Kuczynski describes searching through profiles of available surrogates. "None were living in poverty," she writes. Cathy, the woman she eventually chose to carry her son, was a college-educated substitute teacher, a gifted pianist, and fellow fan of Barack Obama. They shared a land, a language, a level of education, a political bent—coming together to create a baby didn't seem like such a giant leap. But when the surrogate and genetic mother come from different corners of the globe—when one is an Indian woman who bails monsoon rains from her mud-floor hut and the other is an American woman who drives an SUV and vacations at ski resorts—the gap is more like a chasm. And as one childless American friend (rendered infertile through a defective Dalkon Shield intrauterine device) told me, "If I had hired a surrogate, I'm not sure how close I'd want to be to her. How open can you keep your heart when it's broken? Sometimes it's better not to touch unhealed wounds." A code of detachment seems almost necessary to circumvent the divide.

But detachment isn't so easy in practice. Even if you can separate the genetic parents from the surrogate, you cannot separate the surrogate from her womb. One surrogate mother told the sociologist Amrita Pande, "It's my blood, even if it's their genes." Psychologists tell us that a baby in utero recognizes the sound of its mother's voice. Surrogates I spoke with seemed to be struggling to detach. One said, "I try to think of my womb as a carrier." Another said, "I try not to think about it." Is the bond between mother and child fixed by nature or is it a culturally inspired fantasy we yearn to be true?

I asked Dr. Chakravarty if he thought that some children born of surrogacy would one day fly to India in search of their "womb mothers." (The proposed regulation requires parents to reveal to an inquiring child the fact of surrogacy, though not the identity of the surrogate.) "Yes," he said. But chances are such an 18-year-old would not find her womb mother. Instead, she might come to realize she had been made a whole person by uniting parts drawn from tragically unequal worlds.

In a larger sense, so are we all. Person to person, family to family, the First World is linked to the Third World through the food we eat, the clothes we wear, and the care we receive. That Filipina nanny who cares for an American child leaves her own children in the care of her mother and another nanny. In turn, that nanny leaves her younger children in the care of an eldest daughter. First World genetic parents pay a Third World woman to carry their embryo. The surrogate's husband cares for their older children. The worlds of rich and poor are invisibly bound through chains of care.

Before we leave the Akanksha clinic in Anand, the gentle embryologist, Bhadarka, remains across the table from Aditya and me after Geeta and Saroj have left the room. I ask Bhadarka if the clinic offers psychological counseling to the surrogates. "We explain the scientific process," she answers, "and they already know what they're getting into." Then she moves her hands across the table and adds softly, "In the end, a mother is a mother, isn't that true? In the birthing room there is the surrogate, the doctor, the nurse, the nurse's aide, and often the genetic mother. Sometimes we all cry."

25

The Feminization of Mexican Agriculture

JOHN ROSS

Ross examines farming's gender changes in the context of global economic conditions affecting Mexico. His article shows how men's out-migration alters the lifestyles of the women who stay behind.

When I first settled into this tiny Purepecha Indian village high in the Meseta Tarasca of west-central Michoacan state 50 years ago, few women tilled the land. Tending the "milpa" (corn patch) was strictly a man's work. The men ploughed the fields and planted in the spring and the wives and daughters would help to weed ("barbechar") and glean in the harvest—but it was the men who strapped on the "tchundi" basket as they moved up and down the rows, snapping off the big ears of maiz to be sold in the markets of neighboring cities.

While the men lorded it over the corn patch, women had dominion over the home and the children. They cared for the kids and the chickens and prepared the meals. At mid-day, they wrapped up fresh, warm tortillas in colorful "servietas" and carried them out to the fields to feed their husbands.

Only two women in Tanaco actually worked their own "parcelas" (plots). Dona Teresa Garcia had a handful of fields scattered up and down the valley she had inherited from her murdered husband and many sons to work them and although she was known to get her hands dirty, she was more an overseer and administrator. Slight and sprightly, Tere delighted in a full storehouse and was proudest of her purple and red and blue pinto corn she grew from her cache of grandfather seeds.

Nana Eloisa on the other hand was a mountain of a woman who ploughed the rocky valley soil at the foot of volcanic mountains and lush pine forests—when she didn't have an ox or the wherewithal to rent one, Eloisa was known to harness up the plough and pull it herself. Nana Eloisa had no husband although men sometimes hid in her long serge skirts. Unlike Dona Teresa who preferred to negotiate offstage with the men who ruled the community, Eloisa, who was equipped with a stentorian voice, often spoke up at assemblies of the

SOURCE: From John Ross, "The Feminization of Mexican Agriculture," *Counterpunch*. Used with permission.

"comuneros" (indigenous landholders). The neighbors talked about her in awed whispers.

Times have changed up in the Meseta—and changed again. In the 1980s, as the first of five neo-liberal regimes took hold far away in Mexico City, the Purepechas who never strayed far from the Meseta unlike their mestizo neighbors in Tangancicuaro and Gomez Farias who first began trekking north a hundred years ago, plunged into the immigration stream with a vengeance. Fathers and sons went off to find their fortunes in El Norte and many never came back.

The women were left in charge of the house and the milpa both, a double workday ("doble jornada"). Their husbands would send home the "remisas" (money orders) with instructions on where and how much corn to plant. Any cash left over was destined to pay off loans for the "coyotes" who charged thousands of pesos to get the men across the border.

Often the women would hire "peones" and "jornaleros" to do the field-work but others worked the milpas on their own. Gradually the women began to make their own decisions about their husbands' land. Many stepped out of the traditional long Purepecha skirts and literally and figuratively put on the "pantalones."

Women outweigh men in Mexico 53,000,000 to 50,000,000 according to the 2005 half census. Although many are still tied to the home, women now comprise 40 percent of the workforce. In the rural sector where 28 percent of the population continues to subsist, the stats are even more skewed. One estimate is that 18 million women are now the primary workers on the land—but only 4.5 million actually have title to it. Title allows them membership and voice and vote in the ejido (villages that are designated rural production units) and community, access to agricultural credits, and full agrarian rights. But women landholders are often relegated to servant stature in the ejido assemblies where only 2.5 percent serve as officials of the 28,000 communal farms so designated by the Secretary of Agriculture.

Although many women farmers or "campesinas" join mixed-gender farmers' organizations like the PRI party–run National Confederation of Campesinos (CNC) or the more left UNORCA and El Barzan, the dismaying disparity in their recognition as producers have motivated the women to form their own groupings such as the Ecological Campesinas of the Sierra of Petatlan Guerrero and the CONOC (National Council of Women Farmers' Organizations).

But whether within the male-dominated farmers' centrals or those of their own making, equal recognition has been slow in coming for the campesinas. Although agricultural budgets put together by the Secretary of Agriculture (SAGARPA) and the Secretary of Social Development (SEDESO) appear to allocate 42 percent of their resources to women, the numbers are deceiving—most of the money designated for women farmers is assistencial aid drawn down from the "Oportunidades" poverty program.

Other monies are assigned to crafts collectives such as the ceramicists of Ocumicho just over the mountain from Tanaco where the women throw the much-in-demand pots and the men bring the wood to keep the ovens fired up. Funds for micro-projects such as keeping chickens are available to women farmers but as Blanca Rubio writes in the left daily *La Jornada*, the campesinas would rather be recognized as producers of maiz than for their ancillary talents.

In addition to the gender of farming, the gender of out-migration from feeder states like Michoacan, Jalisco, Guanajuato, Zacatecas, and more indigenous Chiapas and Oaxaca, has changed radically. Once upon a time only men headed for El Norte and the potentially mortal consequences of this dangerous migration but womens' numbers in the flow north have tripled in the last decade as neo-liberal agrarian policies imposed from Mexico City have devastated the "campo" and the bottom has fallen out of Mexican agriculture.

Under presidents Carlos Salinas and Ernesto Zedillo (1988–2000), the Constitution was mutilated to allow the privatization of communally held land, grain distribution was handed over to transnationals like the Cargill Corporation, guaranteed prices were scrapped, and credit for poor farmers dried up. Vicente Fox and Felipe Calderon (2000–2010), presidents chosen from the right-wing PAN party, have hastened the demise of the agricultural sector.

The coffin nail was the 1994 North American Free Trade Agreement. Every year since, millions of tons of cheap U.S. and Canadian corn swamp Mexico forcing small-hold campesinos and campesinas out of business. A Carnegie Endowment investigation into the impacts of NAFTA on poor Mexican farmers published on the tenth anniversary of the trade treaty calculated that 1.8 million farmers had abandoned their milpas in NAFTA's first decade—since each farm family represents five Mexicans, the real number of expulsees comes in close to 10,000,000, at least half of them women.

One consequence is that women now swim in the migration stream in dramatically increased numbers. Sisters follow their brothers north and wives their husbands, leaving the children at home with the grandmothers. A third of the households in Tanaco and just down the valley in Cucucho have no mother or father at home.

For those women who stay behind, lifestyles have changed. Families have abandoned or sold off their milpas and the remisas from El Norte (which decreased 20 percent in recession-ridden 2009) are now invested in building up the house, laying cement floors and hooking up electricity lines. Women open "changaros," storefronts where they sell knickknacks and snacks to their neighbors.

Women farmers who still till their parcelas now have to work a triple workday ("triple jornada") just to make ends meet, finding jobs outside of the community as domestics or factory workers, taking care of the house and the kids and the chickens, and tending to the milpa. When the husbands do come home, the once rigidly defined roles of men and women in the Mexican countryside have been irreversibly altered. Men are not the sole breadwinners now and decisions must be taken together. Left to their own devices to survive, the campesinas have become empowered. They have feminized agriculture.

The feminization of the Mexican campo is a bright light in a dismal prospectus thinks the much-respected agrarian analyst Armando Bartra. Gender articulates how farmers approach the land, Bartra writes. Men wrest the crops from the soil. They plant to achieve bigger and better harvests and resort to chemical fertilizers and pesticides and genetically modified seed to speed up the bounty. They pin their hopes on the market, Bartra underscores, "and the market has no future" for small farmers.

By way of contrast, women are more in sync with the land. They don't till the soil for profit as much as to keep their families well nourished. They are committed to auto-sufficiency first and do not poison the land upon which they grow their family's food with chemicals. The feminization of farming, Bartra concludes, is "the only salvation for Mexican agriculture."

26

Masculinities and Globalization

R. W. CONNELL

Noted scholar R. W. Connell has written extensively about multiple forms of
masculinity. In this essay, he examines key strands in "the world's gender order"
and shows how men and masculinities are being reconfigured by transnational
power relations.

THE WORLD GENDER ORDER

Masculinities do not first exist and then come into contact with femininities;
they are produced together, in the process that constitutes a gender order.
Accordingly, to understand the masculinities on a world scale, we must first have
a concept of the globalization of gender.

This is one of the most difficult points in current gender analysis because the
very conception is counterintuitive. We are so accustomed to thinking of gender
as the attribute of an individual, even as an unusually intimate attribute, that it
requires a considerable wrench to think of gender on the vast scale of global society. Most relevant discussions, such as the literature on women and development,
fudge the issue. They treat the entities that extend internationally (markets, corporations, intergovernmental programs, etc.) as ungendered in principle—but
affecting unequally gendered recipients of aid in practice, because of bad policies.
Such conceptions reproduce the familiar liberal-feminist view of the state as in
principle gender-neutral, though empirically dominated by men.

But if we recognize that very large scale institutions such as the state are
themselves gendered, in quite precise and specifiable ways (Connell 1990b),
and if we recognize that international relations, international trade, and global
markets are inherently an arena of gender formation and gender politics (Enloe
1990), then we can recognize the existence of a world gender order. The term
can be defined as the structure of relationships that interconnect the gender
regimes of institutions, and the gender orders of local society, on a world scale.
That is, however, only a definition. The substantive questions remain: what is

the shape of that structure, how tightly are its elements linked, how has it arisen historically, what is its trajectory into the future?

Current business and media talk about globalization pictures a homogenizing process sweeping across the world, driven by new technologies, producing vast unfettered global markets in which all participate on equal terms. This is a misleading image. As Hirst and Thompson (1996) show, the global economy is highly unequal and the current degree of homogenization is often overestimated. Multinational corporations based in the three major economic powers (the United States, European Union, and Japan) are the major economic actors worldwide.

The structure bears the marks of its history. Modern global society was historically produced, as Wallerstein (1974) argued, by the economic and political expansion of European states from the fifteenth century on and by the creation of colonial empires. It is in this process that we find the roots of the modern world gender order. Imperialismwas, from the start, a gendered process. Its first phase, colonial conquest and settlement, was carried out by gender-segregated forces, and it resulted in massive disruption of indigenous gender orders. In its second phase, the stabilization of colonial societies, new gender divisions of labor were produced in plantation economies and colonial cities, while gender ideologies were linked with racial hierarchies and the cultural defense of empire. The third phase, marked by political decolonization, economic neocolonialism, and the current growth of world markets and structures of financial control, has seen gender divisions of labor remade on a massive scale in the "global factory" (Fuentes and Ehrenreich 1983), as well as the spread of gendered violence alongside Western military technology.

The result of this history is a partially integrated, highly unequal and turbulent world society, in which gender relations are partly but unevenly linked on a global scale. The unevenness becomes clear when different substructures of gender (Connell 1987; Walby 1990) are examined separately.

The Division of Labor

A characteristic feature of colonial and neocolonial economies was the restructuring of local production systems to produce a male wage worker–female domestic worker couple (Mies 1986). This need not produce a "housewife" in the Western suburban sense, for instance, where the wage work involved migration to plantations or mines (Moodie 1994). But it has generally produced the identification of masculinity with the public realm and the money economy and of femininity with domesticity, which is a core feature of the modern European gender system (Holter 1997).

Power Relations

The colonial and postcolonial world has tended to break down purdah systems of patriarchy in the name of modernization, if not of women's emancipation (Kandiyoti 1994). At the same time, the creation of a westernized public realm has seen the growth of large-scale organizations in the form of the state and corporations, which in the great majority of cases are culturally masculinized and controlled by men. In *comprador* capitalism, however, the power of local elites depends on their relations with the metropolitan powers, so the hegemonic masculinities of neocolonial societies are uneasily poised between local and global cultures.

Emotional Relations

Both religious and cultural missionary activity has corroded indigenous homosexual and cross-gender practice, such as the native American *berdache* and the Chinese "passion of the cut sleeve" (Hinsch 1990). Recently developed Western models of romantic heterosexual love as the basis for marriage and of gay identity as the main alternative have now circulated globally—though as Airman (1996) observes, they do not simply displace indigenous models, but interact with them in extremely complex ways.

SYMBOLIZATION

Mass media, especially electronic media, in most parts of the world follow North American and European models and relay a great deal of metropolitan content; gender imagery is an important part of what is circulated. A striking example is the reproduction of a North American imagery of femininity by Xuxa, the blonde television superstar in Brazil (Simpson 1993). In counterpoint, exotic gender imagery has been used in the marketing strategies of newly industrializing countries (e.g., airline advertising from Southeast Asia)—a tactic based on the longstanding combination of the exotic and the erotic in the colonial imagination (Jolly 1997).

Clearly, the world gender order is not simply an extension of a traditional European–American gender order. That gender order was changed by colonialism, and elements from other cultures now circulate globally. Yet in no sense do they mix on equal terms, to produce a United Colours of Benetton gender order. The culture and institutions of the North Atlantic countries are hegemonic within the emergent world system. This is crucial for understanding the kinds of masculinities produced within it.

THE REPOSITIONING OF MEN AND THE RECONSTITUTION OF MASCULINITIES

The positioning of men and the constitution of masculinities may be analyzed at any of the levels at which gender practice is configured: in relation to the body, in personal life, and in collective social practice. At each level, we need to consider how the processes of globalization influence configurations of gender.

Men's bodies are positioned in the gender order, and enter the gender process, through body reflexive practices in which bodies are both objects and agents (Connell 1995)—including sexuality, violence, and labor. The conditions of such practice include where one is and who is available for interaction. So it is a fact of considerable importance for gender relations that the global social order distributes and redistributes bodies, through migration, and through political controls over movement and interaction.

The creation of empire was the original "elite migration," though in certain cases mass migration followed. Through settler colonialism, something close to the gender order of Western Europe was reassembled in North America and in Australia. Labor migration within the colonial systems was a means by which gender practices were spread, but also a means by which they were reconstructed, since labor migration was itself a gendered process—as we have seen in relation to the gender division of labor. Migration from the colonized world to the metropole became (except for Japan) a mass process in the decades after World War II. There is also migration within the periphery, such as the creation of a very large immigrant labor force, mostly from other Muslim countries, in the oil-producing Gulf States.

These relocations of bodies create the possibility of hybridization in gender imagery, sexuality, and other forms of practice. The movement is not always toward synthesis, however, as the race/ethnic hierarchies of colonialism have been recreated in new contexts, including the politics of the metropole. Ethnic and racial conflict has been growing in importance in recent years, and as Klein (1997) and Tillner (1997) argue, this is a fruitful context for the production of masculinities oriented toward domination and violence. Even without the context of violence, there can be an intimate interweaving of the formation of masculinity with the formation of ethnic identity, as seen in the study by Poynting, Noble, and Tabar (1997) of Lebanese youths in the Anglo-dominant culture of Australia.

At the level of personal life as well as in relation to bodies, the making of masculinities is shaped by global forces. In some cases, the link is indirect, such as the working-class Australian men caught in a situation of structural unemployment (Connell 1995), which arises from Australia's changing position in the global economy. In other cases, the link is obvious, such as the executives of multinational corporations and the financial sector servicing international trade. The requirements of a career in international business set up strong pressures on domestic life: almost all multinational executives are men, and the assumption in business magazines and advertising directed toward them is that they will have dependent wives running their homes and bringing up their children.

At the level of collective practice, masculinities are reconstituted by the remaking of gender meanings and the reshaping of the institutional contexts of practice. Let us consider each in turn.

The growth of global mass media, especially electronic media, is an obvious "vector" for the globalization of gender. Popular entertainment circulates stereotyped gender images, deliberately made attractive for marketing purposes. The example of Xuxa in Brazil has already been mentioned. International news media are also controlled or strongly influenced from the metropole and circulate Western definitions of authoritative masculinity, criminality, desirable femininity, and so on. But there are limits to the power of global mass communications. Some local centers of mass entertainment differ from the Hollywood model, such as the Indian popular film industry centered in Bombay. Further, media research emphasizes that audiences are highly selective in their reception of media messages, and we must allow for popular recognition of the fantasy in mass entertainment. Just as economic globalization can be exaggerated, the creation of a global culture is a more turbulent and uneven process than is often assumed (Featherstone 1995).

More important, I would argue, is a process that began long before electronic media existed, the export of institutions. Gendered institutions not only circulate definitions of masculinity (and femininity), as sex role theory notes. The functioning of gendered institutions, creating specific conditions for social practice, calls into existence specific patterns of practice. Thus, certain patterns of collective violence are embedded in the organization and culture of a Western-style army, which are different from the patterns of precolonial violence. Certain patterns of calculative egocentrism are embedded in the working of a stock market; certain patterns of rule following and domination are embedded in a bureaucracy.

Now, the colonial and postcolonial world saw the installation in the periphery, on a very large scale, of a range of institutions on the North Atlantic model: armies, states, bureaucracies, corporations, capital markets, labor markets, schools, law courts, transport systems. These are gendered institutions and their functioning has directly reconstituted masculinities in the periphery. This has not necessarily meant photocopies of European masculinities. Rather, pressures for change are set up that are inherent in the institutional form.

To the extent that particular institutions become dominant in world society, the patterns of masculinity embedded in them may become global standards. Masculine dress is an interesting indicator: almost every political leader in the world now wears the uniform of the Western business executive. The more common pattern, however, is not the complete displacement of local patterns but the articulation of the local gender order with the gender regime of global-model institutions. Case studies such as Hollway's (1994) account of bureaucracy in Tanzania illustrate the point; there, domestic patriarchy articulated with masculine authority in the state in ways that subverted the government's formal commitment to equal opportunity for women.

We should not expect the overall structure of gender relations on a world scale simply to mirror patterns known on the smaller scale. In the most vital of respects, there is continuity. The world gender order is unquestionably patriarchal, in the sense that it privileges men over women. There is a patriarchal dividend for men arising from unequal wages, unequal labor force participation, and a highly unequal structure of ownership, as well as cultural and sexual privileging. This has been extensively documented by feminist work on women's situation globally (e.g., Taylor 1985), though its implications for masculinity have mostly been ignored. The conditions thus exist for the production of a hegemonic masculinity on a world scale, that is to say, a dominant form of masculinity that embodies, organizes, and legitimates men's domination in the gender order as a whole.

The conditions of globalization, which involve the interaction of many local gender orders, certainly multiply the forms of masculinity in the global gender order. At the same time, the specific shape of globalization, concentrating economic and cultural power on an unprecedented scale, provides new resources for dominance by particular groups of men. This dominance may become institutionalized in a pattern of masculinity that becomes, to some degree, standardized across localities. I will call such patterns *globalizing masculinities*, and it is among them, rather than narrowly within the metropole, that we are likely to find candidates for hegemony in the world gender order.

GLOBALIZING MASCULINITIES

In this section, I will offer a sketch of major forms of globalizing masculinity in the three historical phases identified above in the discussion of globalization.

Masculinities of Conquest and Settlement

The creation of the imperial social order involved peculiar conditions for the gender practices of men. Colonial conquest itself was mainly carried out by segregated groups of men—soldiers, sailors, traders, administrators, and a good many who were all these by turn (such as the Rum Corps in early New South Wales, Australia). They were drawn from the more segregated occupations and milieux in the metropole, and it is likely that the men drawn into colonization tended to be the more rootless. Certainly the process of conquest could produce frontier masculinities that combined the occupational culture of these groups with an unusual level of violence and egocentric individualism. The vehement contemporary debate about the genocidal violence of the Spanish conquistadors—who in fifty years completely exterminated the population of Hispaniola—points to this pattern (Bitterli 1989).

The political history of empire is full of evidence of the tenuous control over the frontier exercised by the state—the Spanish monarchs unable to rein in the conquistadors, the governors in Sydney unable to hold back the squatters and in Capetown unable to hold back the Boers, gold rushes breaking boundaries everywhere, even an independent republic set up by escaped slaves in Brazil. The point probably applies to other forms of social control too, such as customary controls on men's sexuality. Extensive sexual exploitation of indigenous women was a common feature of conquest. In certain circumstances, frontier masculinities might be reproduced as a local cultural tradition long after the frontier had passed, such as the gauchos of southern South America, the cowboys of the western United States.

In other circumstances, however, the frontier of conquest and exploitation was replaced by a frontier of settlement. Sex ratios in the colonizing population changed, as women arrived and locally born generations succeeded. A shift back toward the family patterns of the metropole was likely. As Cain and Hopkins (1993) have shown for the British empire, the ruling group in the colonial worlds as a whole was an extension of the dominant class in the metropole, the landed gentry, and tended to reproduce its social customs and ideology. The creation of a settler masculinity might be the goal of state policy, as it seems to have been in late nineteenth-century New Zealand, as part of a general process of pacification and the creation of an agricultural social order (Phillips 1987). Or it might be undertaken through institutions created by settler groups, such as the elite schools in Natal studied by Morrell (1994).

The impact of colonialism on the construction of masculinity among the colonized is much less documented, but there is every reason to think it was severe. Conquest and settlement disrupted all the structures of indigenous society, whether or not this was intended by the colonizing powers (Bitierli 1989).Indigenous gender orders were no exception. Their disruption could result from the pulverization of indigenous communities (as in the seizure of land in eastern North America and southeastern Australia), through gendered labor migration (as in gold mining with Black labor in South Africa; see Moodie 1994), to ideological attacks on local gender arrangements (as in the missionary assault on the berdache tradition in North America; see Williams 1986). The varied course of resistance to colonization is also likely to have affected the making of masculinities. This is clear in the region of Natal in South Africa, where sustained resistance to colonization by the Zulu kingdom was a key to the mobilization of ethnic/national masculine identities in the twentieth century (Morrell 1996).

Masculinities of Empire

The imperial social order created a hierarchy of masculinities, as it created a hierarchy of communities and races. The colonizers distinguished "more manly" from "less manly" groups among their subjects. In British India, for instance, Bengali men were supposed effeminate while Pathans and Sikhs were regarded as strong and warlike. Similar distinctions were made in South Africa between Hottentots and Zulus, in North America between Iroquois, Sioux, and Cheyenne on one side, and southern and southwestern tribes on the other.

At the same time, the emerging imagery of gender difference in European culture provided general symbols of superiority and inferiority. Within the imperial "poetics of war" (MacDonald 1994), the conqueror was virile, while the colonized were dirty, sexualized, and effeminate or childlike. In many colonial situations, indigenous men were called "boys" by the colonizers (e.g., in Zimbabwe; see Shire 1994). Sinha's (1995) interesting study of the language of political controversy in India in the 1880s and 1890s shows how the images of "manly Englishman" and "effeminate Bengali" were deployed to uphold colonial privilege and contain movements for change. In the late nineteenth century, racial barriers in colonial societies were hardening rather than weakening, and gender ideology tended to fuse with racism in forms that the twentieth century has never untangled.

The power relations of empire meant that indigenous gender orders were generally under pressure from the colonizers, rather than the other way around. But the colonizers too might change. The barriers of late colonial racism were not only to prevent pollution from below but also to forestall "going native," a well-recognized possibility—the starting point, for instance, of Kipling's famous novel *Kim* ([1901] 1987). The pressures, opportunities, and profits of empire might also work changes in gender arrangements among the colonizers, for instance, the division of labor in households with a large supply of indigenous workers as domestic servants (Bulbeck 1992). Empire might also affect the gender order of the metropole itself by changing gender ideologies, divisions of labor, and the nature of the metropolitan state. For instance, empire figured prominently as a source of masculine imagery in Britain, in the Boy Scouts, and in the cult of Lawrence of Arabia (Dawson 1991). Here we see examples of an important principle: the interplay of gender dynamics between different parts of the world order.

The world of empire created two very different settings for the modernization of masculinities. In the periphery, the forcible restructuring of economics and workforces tended to individualize, on one hand, and rationalize, on the other. A widespread result was masculinities in which the rational calculation of self-interest was the key to action, emphasizing the European gender contrast of rational man/irrational woman. The specific form might be local—for instance, the Japanese "salaryman," a type first recognized in the 1910s, was specific to the Japanese context of large, stable industrial conglomerates (Kinmonth 1981). But the result generally was masculinities defined around economic action, with both workers and entrepreneurs increasingly adapted to emerging market economies.

In the metropole, the accumulation of wealth made possible a specialization of leadership in the dominant classes, and struggles for hegemony in which masculinities organized around domination or violence were split from masculinities organized around expertise. The class compromises that allowed the development of the welfare state in Europe and North America were paralleled by gender compromises—gender reform movements (most notably the women's suffrage movement) contesting the legal privileges of men and forcing concessions from the state. In this context, agendas of reform in masculinity emerged: the temperance movement, compassionate marriage, homosexual rights movements, leading eventually to the pursuit of androgyny in "men's liberation" in the 1970s (Kimmel and Mosmiller 1992). Not all reconstructions of masculinity, however, emphasized tolerance or moved toward androgyny. The vehement masculinity politics of fascism, for instance, emphasized dominance and difference and glorified violence, a pattern still found in contemporary racist movements (Tillner 1997).

Masculinities of Postcolonialism and Neoliberalism

The process of decolonization disrupted the gender hierarchies of the colonial order and, where armed struggle was involved, might have involved a deliberate cultivation of masculine hardness and violence (as in South Africa; see Xaba

1997). Some activists and theorists of liberation struggles celebrated this, as a necessary response to colonial violence and emasculation; women in liberation struggles were perhaps less impressed. However one evaluates the process, one of the consequences of decolonization was another round of disruptions of community-based gender orders and another step in the reorientation of masculinities toward national and international contexts.

Nearly half a century after the main wave of decolonization, the old hierarchies persist in new shapes. With the collapse of Soviet communism, the decline of postcolonial socialism, and the ascendancy of the new right in Europe and North America, world politics is more and more organized around the needs of transnational capital and the creation of global markets.

The neoliberal agenda has little to say, explicitly, about gender: it speaks a gender-neutral language of "markets," "individuals," and "choice." But the world in which neoliberalism is ascendant is still a gendered world, and neoliberalism has an implicit gender politics. The "individual" of neoliberal theory has in general the attributes and interests of a male entrepreneur, the attack on the welfare state generally weakens the position of women, while the increasingly unregulated power of transnational corporations places strategic power in the hands of particular groups of men. It is not surprising, then, that the installation of capitalism in Eastern Europe and the former Soviet Union has been accompanied by a reassertion of dominating masculinities and, in some situations, a sharp worsening in the social position of women.

We might propose, then, that the hegemonic form of masculinity in the current world gender order is the masculinity associated with those who control its dominant institutions: the business executives who operate in global markets, and the political executives who interact (and in many contexts, merge) with them. I will call this *transnational business masculinity*. This is not readily available for ethnographic study, but we can get some clues to its character from its reflections in management literature, business journalism, and corporate self-promotion, and from studies of local business elites (e.g., Donaldson 1997).

As a first approximation, I would suggest this is a masculinity marked by increasing egocentrism, very conditional loyalties (even to the corporation), and a declining sense of responsibility for others (except for purposes of image making). Gee, Hull and Lankshear (1996), studying recent management textbooks, note the peculiar construction of the executive in "fast capitalism" as a person with no permanent commitments, except (in effect) to the idea of accumulation itself. Transnational business masculinity is characterized by a limited technical rationality (management theory), which is increasingly separate from science.

Transnational business masculinity differs from traditional bourgeois masculinity by its increasingly libertarian sexuality, with a growing tendency to commodity relations with women. Hotels catering to businessmen in most parts of the world now routinely offer pornographic videos, and in some parts of the world, there is a well-developed prostitution industry catering for international businessmen. Transnational business masculinity does not require bodily force, since the patriarchal dividend on which it rests is accumulated by impersonal, institutional means. But corporations increasingly use the exemplary bodies of elite sportsmen as a marketing tool (note the phenomenal growth of corporate "sponsorship" of sport in the last generation) and indirectly as a means of legitimation for the whole gender order.

MASCULINITY POLITICS ON A WORLD SCALE

Recognizing global society as an arena of masculinity formation allows us to pose new questions about masculinity politics. What social dynamics in the global arena give rise to masculinity politics, and what shape does global masculinity politics take?

The gradual creation of a world gender order has meant many local instabilities of gender. Gender instability is a familiar theme of poststructuralist theory, but this school of thought takes as a universal condition a situation that is historically specific. Instabilities range from the disruption of men's local cultural dominance as women move into the public realm and higher education, through the disruption of sexual identities that produce "queer" politics in the metropole, to the shifts in the urban intelligentsia that produced "the new sensitive man" and other images of gender change.

One response to such instabilities, on the part of groups whose power is challenged but still dominant, is to reaffirm *local* gender orthodoxies and hierarchies. A masculine fundamentalism is, accordingly, a common response in gender politics at present. A soft version, searching for an essential masculinity among myths and symbols, is offered by the mythopoetic men's movement in the United States and by the religious revivalists of the Promise Keepers (Messner 1997). A much harder version is found, in that country, in the right-wing militia movement brought to world attention by the Oklahoma City bombing (Gibson 1994), and in contemporary Afghanistan, if we can trust Western media reports, in the militant misogyny of the Taliban. It is no coincidence that in the two latter cases, hardline masculine fundamentalism goes together with a marked anti-internationalism. The world system—rightly enough—is seen as the source of pollution and disruption.

Not that the emerging global order is a hotbed of gender progressivism. Indeed, the neoliberal agenda for the reform of national and international economics involves closing down historic possibilities for gender reform. I have noted how it subverts the gender compromise represented by the metropolitan welfare state. It has also undermined the progressive-liberal agendas of sex role reform represented by affirmative action programs, anti-discrimination provisions, child care services, and the like. Right-wing parties and governments have been persistently cutting such programs, in the name of either individual liberties or global competitiveness. Through these means, the patriarchal dividend to men is defended or restored, without an *explicit* masculinity politics in the form of a mobilization of men.

Within the arenas of international relations, the international state, multinational corporations, and global markets, there is nevertheless a deployment of masculinities and a reasonably clear hegemony. The transnational business masculinity described above has had only one major competitor for hegemony in recent decades, the rigid, control-oriented masculinity of the military, and the military-style bureaucratic dictatorships of Stalinism. With the collapse of Stalinism and the end of the cold war, Big Brother (Orwell's famous parody of this form of masculinity) is a fading threat, and the more flexible, calculative, egocentric masculinity of the fast capitalist entrepreneur holds the world stage.

We must, however, recall two important conclusions of the ethnographic moment in masculinity research: that different forms of masculinity exist together and that hegemony is constantly subject to challenge. These are possibilities in the global arena too. Transnational business masculinity is not completely homogeneous; variations of it are embedded in different parts of the world system, which may not be completely compatible. We may distinguish a Confucian variant, based in East Asia, with a stronger commitment to hierarchy and social consensus, from a secularized Christian variant, based in North America, with more hedonism and individualism and greater tolerance for social conflict. In certain arenas, there is already conflict between the business and political leaderships embodying these forms of masculinity: initially over human rights versus Asian values, and more recently over the extent of trade and investment liberalization.

If these are contenders for hegemony, there is also the possibility of opposition to hegemony. The global circulation of "gay" identity (Airman 1996) is an important indication that nonhegemonic masculinities may operate in global arenas, and may even find a certain political articulation, in this case around human rights and AIDS prevention.

REFERENCES

Altman, Dennis. 1996. Rupture or continuity? The internationalisation of gay identities. *Social Text* 48 (3): 77–94.

Barrett, Frank J. 1996. The organizational construction of hegemonic masculinity: The case of the U.S. Navy. *Gender, Work and Organization* 3 (3): 129–42.

BauSteineMaenner, ed. 1996. *Kritische Maennerforschung* [Critical research on men]. Berlin: Argument.

Bitterli, Urs. 1989. *Cultures in Conflict: Encounters between European and Non-European Cultures, 1492–1800,* Stanford, CA: Stanford University Press.

Bolin, Anne. 1988. *In Search of Eve: Transexual Rites of Passage.* Westport, CT: Bergin & Garvey.

Bulbeck, Chilla. 1992. *Australian women in Papua New Guinea: Colonial passages 1920–1960.* Cambridge, U.K.: Cambridge University Press.

Cain, P. J., and A. G. Hopkins. 1993. *British Imperialism: Innovation and Expansion, 1688–1914.* New York: Longman.

Carrigan, Tim, Bob Connell, and John Lee. 1985. Toward a new sociology of masculinity. *Theory and Society* 14 (5): 551–604.

Chodorow, Nancy. 1994. *Femininities, Masculinities, Sexualities: Freud and Beyond.* Lexington: University Press of Kentucky.

Cockburn, Cynthia. 1983. *Brothers: Male dominance and technological change.* London: Pluto.

Cohen, Jon. 1991. NOMAS: Challenging male supremacy. *Changing Men* (Winter/ Spring): 45–46

Connell, R. W. 1987. *Gender and power.* Cambridge, MA: Polity.

———. 1990a. An iron man: The body and some contradictions of hegemonic masculinity. In *Sport, Men and the Gender Order: Critical Feminist Perspectives,* edited by Michael A. Messner and Donald F. Sabo, 83–95. Champaign, IL: Human Kinetics Books.

———. 1990b. The state, gender and sexual politics: Theory and appraisal. *Theory and Society* 19: 507–44.

———. 1992. A very straight gay: Masculinity, homosexual experience and the dynamics of gender. *American Sociological Review* 57 (6): 735–51.

———. 1995. *Masculinities.* Cambridge, MA: Polity.

———. 1996. Teaching the boys: New research on masculinity, and gender strategies for schools. *Teachers College Record* 98 (2): 206–35.

Cornwall, Andrea and Nancy Lindisfame, eds. 1994. *Dislocating masculinity: Comparative ethnographies.* London: Routledge.

Dawson, Graham. 1991. The blond Bedouin: Lawrence of Arabia, imperial adventure and the imagining of English-British masculinity. In *Manful assertions: Masculinities in Britain since* 1800, edited by Michael Roper and John Tosh, 113–44. London: Routledge.

Donaldson, Mike. 1991. *Time of our lives: Labour and love in the working class.* Sydney: Allen & Unwin.

———. 1997. *Growing up very rich: The masculinity of the hegemonic.* Paper presented at the conference "Masculinities: Renegotiating Genders," June, University of Wollongong, Australia.

Enloe, Cynthia. 1990. *Bananas, Beaches and Bases: Making Feminist Sense of International Politics.* Berkeley: University of California Press.

Featherstone, Mike. 1995. *Undoing culture: Globalization, postmodernism and identity.* London: Sage.

Foley, Douglas E. 1990. *Learning Capitalist Culture: Deep in the Heart of Tejas.* Philadelphia: University of Pennsylvania Press.

Fuentes, Annette, and Barbara Ehrenreich. 1983. *Women in the Global Factory.* Boston: South End.

Gee, James Paul, Glynda Hull, and Colin Lankshear. 1996. *The new work order: Behind the language of the new capitalism.* Sydney: Allen & Unwin.

Gender Equality Ombudsman. 1997. The father's quota. Information sheet on parental leave entitlements, Oslo.

Gibson, J. William. 1994. *Warrior Dreams: Paramilitary Culture in Post-Vietnam America.* New York: Hill and Wang.

Hagemann-White, Carol, and Maria S. Rerrich, eds. 1988. *FrauenMaennerBilder* (Women, Imaging, Men). Bielefeld: AJZ–Verlag.

Hearn, Jeff. 1987. *The gender of oppression: Men, masculinity and the critique of Marxism.* Brighton, U.K.: Wheatsheaf.

Herdt, Gilbert H. 1981. *Guardians of the Flutes: Idioms of Masculinity.* New York: McGraw-Hill.

_____. ed. 1984. *Ritualized Homosexuality in Melanesia.* Berkeley: University of California Press.

Heward, Christine. 1988. *Making a man of him: Parents and their sons' education at an English public school 1929–1950.* London: Routledge.

Hinsch, Bret. 1990. *Passions of the Cut Sleeve: The Male Homosexual Tradition in China.* Berkeley: University of California Press.

Hirst, Paul, and Grahame Thompson. 1996. *Globalization in Question: The International Economy and the Possibilities of Governance.* Cambridge, MA: Polity.

Hollstein, Walter. 1992. *Machen Sie Platz, mein Herr! Teilen statt Herrschen* [Sharing instead of dominating]. Hamburg: Rowohlt.

Hollway, Wendy. 1994. Separation, integration and difference: Contradictions in a gender regime. In *Power/gender: Social relations in theory and practice,* edited by H. Lorraine Radtke and Henderikus Stam, 247–69. London: Sage.

Holter, Oystein Gullvag. 1997. *Gender, patriarchy and capitalism: A social forms analysis.* Ph.D. diss., University of Oslo, Faculty of Social Science.

Hondagneu-Sotelo, Pierrette, and Michael A. Messner. 1994. Gender displays and men's power: The "new man" and the Mexican immigrant man. In *Theorizing Masculinities,* edited by Harry Brod and Michael Kaufman, 200–218. Twin Oaks, CA: Sage.

Ito Kimio. 1993. *Oiokorashisa-no-yukue* [Directions for masculinities]. Tokyo: Shinyo-sha.

Jolly, Margaret. 1997. From Point Venus to Bali Ha'i: Eroticism and exoticism in representations of the Pacific. In *Sites of Desire, Economies of Pleasure: Sexualities in Asia and the Pacific,* edited by Lenore Manderson and Margaret Jolly, 99–122. Chicago: University of Chicago Press.

Kandiyoti, Deniz. 1994. The paradoxes of masculinity: Some thoughts on segregated societies. In *Dislocating masculinity: Comparative ethnographies,* edited by Andrea Cornwall and Nancy Lindisfarne, 197–213. London: Routledge.

Kaufman, Michael. 1997. Working with men and boys to challenge sexism and end men's violence. Paper presented at UNESCO expert group meeting on Male Roles and Masculinities in the Perspective of a Culture of Peace, September, Oslo.

Kimmel, Michael S. 1987. Rethinking "masculinity": New directions in research. In *Changing Men: New Directions in Research on Men and Masculinity,* edited by Michael S. Kimmel, 9–24. Newbury Park, CA: Sage.

_____. 1996. *Manhood in America: A Cultural History.* New York: Free Press, Michael S. Kimmel, and Thomas P. Mosmiller, eds. 1992. *Against the Tide: Pro-feminist Men in the United States, 1776–1990,* a documentary history. Boston: Beacon.

Kindler, Heinz. 1993. *Maske(r)ade: Jungen und Maennerarbeit fuer die Pratis* [Work with youth and men]. Neuling: Schwaebisch Gmuend und Tuebingen.

Kinmonth, Earl H. 1981. *The Self-made Man in Meiji Japanese Thought: From Samurai to Salary Man.* Berkeley: University of California Press.

Kipling, Rudyard. [1901] 1987. *Kim.* London: Penguin.

Klein, Alan M. 1993. *Little Big Men: Bodybuilding Subculture and Gender Construction.* Albany: State University of New York Press.

Klein, Uta. 1997. *Our best boys: The making of masculinity in Israeli society*. Paper presented at UNESCO expert group meeting on Male Roles and Masculinities in the Perspectives of a Culture of Peace, September, Oslo.

Lewes, Kenneth. 1988. *The Psychoanalytic Theory of Male Homosexuality*. New York: Simon & Schuster.

MacDonald, Robert H. 1994. *The language of empire: Myths and metaphors of popular imperialism, 1880–1918*. Manchester, U.K.: Manchester University Press.

McElhinny, Bonnie. 1994. An economy of affect: Objectivity, masculinity and the gendering of police work. In *Dislocating masculinity: Comparative ethnographies*, edited by Andrea Cornwall and Nancy Lindisfarne, 159–71. London: Routledge.

McKay, Jim, and Debbie Huber. 1992: Anchoring media images of technology and sport. *Women's Studies International Forum* 15 (2): 205–18.

Messerschmidt, James W. 1997. *Crime as Structured Action: Gender, Race, Class, and Crime in the Making*. Thousand Oaks, CA: Sage.

Messner, Michael A. 1992. *Power at Play: Sports and the Problem of Masculinity*. Boston: Beacon.

_____. 1997. *The Politics of Masculinities: Men in Movements*. Thousand Oaks, CA: Sage.

Metz-Goeckel, Sigrid, and Ursula Mueller. 1986. *Der Mann: Die Brigitte-Studie* [The male]. Beltz: Weinheim & Basel.

Mies, Maria. 1986. *Patriarchy and accumulation on a world scale: Women in the international division of labour*. London: Zed.

Moodie, T. Dunbar. 1994. *Going for gold: Men, mines, and migration*. Johannesburg: Witwatersand University Press.

Morrell, Robert. 1994. Boys, gangs, and the making of masculinity in the White secondary schools of Natal, 1880–1930. *Masculinities* 2 (2): 56–82.

_____. ed. 1996. *Political economy and identities in KwaZulu-Natal: Historical and social perspectives*. Durban, Natal: Indicator Press.

Nakamura, Akira. 1994. *Watashi-no Danseigaku* [My men's studies]. Tokyo: Kindaibugeisha.

Oftung, Knut, ed. 1994. *Menns bilder og bilder av menn* [Images of men]. Oslo: Likestillingsradet.

Phillips, Jock. 1987. *A man's country? The image of the Pakeha male, a history*. Auckland: Penguin.

Poynting, S., G. Noble, and P. Tabar. 1997. *"Intersections" of masculinity and ethnicity: A study of male Lebanese immigrant youth in Western Sydney*. Paper presented at the conference "Masculinities: Renegotiating Genders," June, University of Wollongong, Australia.

Roper, Michael. 1991. Yesterday's model: Product fetishism and the British company man, 1945–85. In *Manful assertions: Masculinities in Britain since 1800*, edited by Michael Roper and John Tosh, 190–211. London: Routledge.

Schwalbe, Michael. 1996. *Unlocking the Iron Cage: The Men's Movement, Gender Politics, and American Culture*. New York: Oxford University Press.

Segal, Lynne. 1997. *Slow motion: Changing masculinities, changing men*. 2nd ed. London: Virago.

Seidler, Victor J. 1991. *Achilles heel reader: Men, sexual politics and socialism*. London: Routledge.

Shire, Chenjerai. 1994. Men don't go to the moon: Language, space and masculinities in Zimbabwe. In *Dislocating masculinity: Comparative ethnographies,* edited by Andrea Cornwall and Nancy Lindisfarne, 147–58. London: Routledge.

Simpson, Amelia. 1993. *Xuxa: The Mega-marketing of gender, Race and Modernity.* Philadelphia: Temple University Press.

Sinha, Mrinalini. 1995. *Colonial masculinity: The manly Englishman and the effeminate Bengali in the late nineteenth century.* Manchester, U.K.: Manchester University Press.

Taylor, Debbie. 1985. Women: An analysis. In *Women: A world report,* 1–98. London: Methuen.

Theberge, Nancy. 1991. Reflections on the body in the sociology of sport. *Quest* 43: 123–34.

Thorne, Barrie. 1993. *Gender play: Girls and boys in school.* New Brunswick, NJ: Rutgers University Press.

Tillner, Georg. 1997. Masculinity and xenophobia. Paper presented at UNESCO meeting on Male Roles and Masculinities in the Perspective of a Culture of Peace. September, Oslo.

Tomsen, Stephen. 1997. A top night: Social protest, masculinity and the culture of drinking violence. *British Journal of Criminology* 37 (1): 90–103.

Tosh, John. 1991. Domesticity and manliness in the Victorian middle class: The family of Edward White Benson. In *Manful assertions: Masculinities in Britain since 1800,* edited by Michael Roper and John Tosh, 44–73. London: Routledge.

United Nations Educational, Scientific and Cultural Organization (UNESCO). 1997. *Male roles and masculinities in the perspective of a culture of peace: Report of expert group meeting, Oslo, 24–28 September 1997.* Paris: Women and a Culture of Peace Programme, Culture of Peace Unit, UNESCO.

Walby, Sylvia. 1990. *Theorizing patriarchy.* Oxford, U.K.: Blackwell.

Walker, James C. 1988. *Louts and legends: Male youth culture in an inner-city school.* Sydney: Allen & Unwin.

Wallerstein, Immanuel. 1974. *The Modern World-System: Capitalist Agriculture and the Origins of the European World-Economy in the Sixteenth Century.* New York: Academic Press.

Whitson, David. 1990. Sport in the social construction of masculinity. In *Sport, Men, and the Gender Order: Critical Feminist Perspectives,* edited by Michael A. Messner and Donald F. Sabo, 19–29. Champaign, IL: Human Kinetics Books.

Widersprueche. 1995. Special Issue: Maennlichkeiten. Vol. 56/57.

Williams, Walter L. 1986. *The Spirit and the Flesh: Sexual Diversity in American Indian Culture.* Boston: Beacon.

Xaba, Thokozani. 1997. Masculinity in a transitional society: The rise and fall of the "young lions." Paper presented at the conference "Masculinities in Southern Africa," June, University of Natal-Durban, Durban.

REFLECTION QUESTIONS FOR CHAPTER 7

1. What is wrong with a gender-neutral analysis of globalization?
2. How does research on maids, nannies, and sex workers help us to see globalization in a new light?
3. Explain how commercial surrogacy is a new form of global servicework.
4. Why has Mexican agriculture become feminized? What are the implications of the change for women themselves?
5. How does Connell's discussion of "hegemonic masculinity" illustrate his argument that although globalization implies homogenization, it produces many unequal forms of masculinity?
6. Does globalization reproduce the worst tendencies of gender inequality, or can it provide opportunities for women to leave the worst excesses of patriarchy behind? Provide examples to support your point of view.

Chapter 8

The Globalization of Social Problems

Globalization, as represented by transnational flows of people, information, and commerce, has intensified social problems both globally and locally. Crime networks control the global trade in illicit drugs, the black market in guns and explosives, and trafficking in sex workers, sweatshop workers, and domestic servants. Transnational corporations and banks are sometimes involved in money laundering, which evades taxes and indirectly supports criminal activities. Transnational corporations also exploit workers and degrade the environment in low-wage and developing countries. Of the many social problems exacerbated by globalization, we focus here on four: environmental degradation, swift moving diseases, the movement of guns and drugs between the United States and Mexico, and the trafficking of prostitutes.

To examine the issue of environmental degradation, we have selected three essays. The first looks at the consequences of the 2010 oil spill in the Gulf of Mexico. The spill site was owned by BP, a British corporation, and run jointly by Transocean, a Swiss corporation, and a U.S.-based conglomerate, Halliburton. This article by John Ross focuses on the spill's damage to the sea turtles of Mexico. In the second selection, Elizabeth Becker shows how the annual transnational travel of about a billion tourists threatens the planet. The next selection investigates the environmental consequences of throwing away outmoded electronic technology. This electronic waste is loaded with toxic materials. It is mostly discarded by the richest nations and relocated in the poorest countries where is has dire effects on the environment and public health.

The next article describes the way in which communicable diseases spread easily and quickly in an increasingly globalized world. People are exposed to swine flu, for example, because of massive international travel, unsanitary food processing, and the transportation of food over great distances without adequate health safeguards.

The last two selections examine two quite different aspects of global crime syndicates. James McKinley's article illustrates the symbiotic relationship between U.S. gun laws and the firepower of Mexican drug cartels. The United States is the world's largest user of illicit drugs, most of which are either grown or manufactured in Mexico or travel through Mexico on their way to the United States. The various drug cartels in Mexico used armed force to protect their interests and territories, and the primary source of their guns is the United States. The final article in this chapter directs attention to the "New Slavery," which includes more than 27 million slaves in the world today. The "New Slavery," like slavery in other times, entails the exploitation of human beings for profit using violence or the threat of violence. But today's forms of slavery differ from those practiced in the past in that, typically, slavery is no longer a lifelong condition, as the slave is freed[1] or discarded after he or she is no longer useful or has paid off his or her debt. We focus here on sex slaves, prostitutes who are bought and sold and then become debt slaves who, in order to obtain their freedom, must repay the amount their new owners paid for them.

ENDNOTE

1.   Ken Bales, 1999. *Disposable People: New Slavery in the Global Economy*. Berkeley: University of California Press; Kevin Bales, Zoe Trodd, and Alex Kent Williamson, *Modern Slavery: The Secret World of 27 Million People*, Oxford, U.K.: Oneworld, 2009; and John Bowe, *Nobodies: Modern American Slave Labor and the Dark Side of the New Global Economy*. New York: Random House, 2007.

27

Saving Turtle Island

JOHN ROSS

The BP explosion on its Gulf of Mexico deep sea oil rig in 2010 killed workers and was the largest environmental disaster in U.S. history, spewing giant plumes of oil mixed with methane into the ocean for months. The environmental damage was most damaging to the states of Louisiana, Mississippi, Alabama, and Florida. In time, the disaster will reach Cuba and other Caribbean nations to the east and Mexico to the west. This article examines the disastrous effects on the turtles of Caribbean Mexico. The oil spill has also had effects on domestic Mexican politics, where President Calderon seeks to privatize the government owned petroleum industry (PEMEX), a policy promoted by Halliburton, the oil company's primary subcontractor.

The turtles of Caribbean Mexico are an ancient race. Their ancestors paddled with dinosaurs and prehistoric fish. Kemp's Ridley turtles were burying their eggs in Gulf Coast sanctuaries countless millenniums before the Olmecs, Mexico's matrix civilization, installed their mysterious giant heads on the Veracruz plain. The presence of turtles in indigenous iconography is evidenced by artifacts displayed in anthropological museums in Mexico City and Jalapa Veracruz. Twentieth-century naturalists recorded "arribos" ("arrivals") of tens of thousands of Kemp's Ridley females at Rancho Nuevo beach Tamaulipas; with few exceptions, Kemp's Ridleys (named for an amateur turtle-ologist and the smallest and rarest of all sea turtles) nest only at Rancho Nuevo and Padre Island Texas.

But for Gulf waters, turtles are like canaries in the coalmines. The 1979 blowout of Ixtoc 1, a Mexican National Petroleum Company (PEMEX) platform off the southern state of Tabasco, gushed uncontrollably for nine months. Some 3,000,000 barrels were spewed into the Gulf of Mexico, fouling beaches and nesting grounds. The Rancho Nuevo arribos shrank below 4000. Although Mexican Kemp Ridleys have staged a modest comeback (the population is now calculated at 8000), the April 20th explosion of a British Petroleum deep sea drilling rig on the Macondo Prospect (with apologies to Gabriel Garcia Marquez) 130 miles southeast of New Orleans could spell doomsday for these primordial creatures.

SOURCE: John Ross, "Saving Turtle Island," *Counterpunch* (June 4–6, 2010). Used with permission.

Across the Gulf, Mexican authorities are watching this travesty unfold with furrowed brows. The blow-out of the "Deepwater Horizon" platform that killed 11 and wounded 17 workers is now the largest oil spill in U.S. history, almost doubling the size of the Exxon Valdez fiasco in Alaskan waters (10,000,000 gallons) and threatening biblical devastation of Caribbean wildlife from Mexico to Cuba. Already, Gulf Coast fishing grounds have been shut down, shrimp and oyster beds contaminated, colonies of marine mammals such as dolphins and manatees are menaced, and bird life, particularly brown pelicans, is at extreme risk. One hundred and fifty dead Kemp Ridley sea turtles were counted during the first 20 days of the catastrophe.

The good news—at least for Mexico—is that deep-water oil plumes have been caught up in loop currents that threaten environmental mayhem as far east as the Florida Keys and Communist Cuba, but will not touch home. The bad news is that come August when the hurricane season blows in (2010 is being touted as a record year for tropical hurricanes with 15 giant storms headed for the Caribbean and the Gulf of Mexico), those currents will shift dramatically south towards Mexico—even now deep water "cyclones" are sweeping gobs of oil towards Veracruz and Tamaulipas turtle breeding grounds, and Mexico's Environmental Secretary, Rafael Elvira, is preparing to file suit against BP, whose earnings in 2009 were $16.6 Billion USD.

BP's efforts to plug the leak with everything from old tires to tons of mud, robot submarines and never-before-tested "domes" have met with serial failure. A slant drill to relieve pressure on the undersea gusher will not be in place until August, when the currents turn towards Mexico. Kemp Ridleys nest from April through August.

President Felipe Calderon's brow is further corrugated by the prospect that the mammoth BP spill will torpedo his pledge to privatize (he calls it "modernize") both Mexico's oil industry and PEMEX, the national petroleum consortium. The explosion of the Deepwater Horizon, a joint venture between BP, Halliburton, and Transocean (controlled by a Swiss holding company), has certainly slowed if not slain Calderon's plans to contract similar transnationals for deep sea drilling on Mexico's slice of the Gulf.

According to U.S. Department of Energy evaluations, Mexico has only nine years of proven reserves left before it becomes a net oil importer. Major offshore wells like Cantarell in the Sound of Campeche are played out, and no new land-based deposits have been located— rummaging through the remains of the old Chicontepec field in Veracruz (Halliburton is an important subcontractor) has yielded meager results.

One joke making the rounds has Calderon delighted by the BP spill, because it will bring more oil to Mexican waters.

In the vision of Big Oil, Mexico's only hope for economic survival lies in its "aguas profundas," or deep waters, five miles down in the Gulf. Of course, only Big Oil has the technology to get at these riches. According to the transnationals, PEMEX must be reformed and partner up with them ("an association of capitals") for a percentage of the take. So-called "risk contracts" are currently barred by the Mexican Constitution.

Following orders from his backers (Halliburton, the number one PEMEX subcontractor, was a generous contributor to Calderon's fraud-tarred 2006 election victory), the Mexican president submitted "energy reform" legislation to Congress in 2008 that laid out a "strategic alliance" with Big Oil and "flexibilization" of PEMEX, opening the state company to private investment and risk contracts. The Calderon media machine cranked up an infomercial campaign depicting an azure Caribbean under which Mexico's true wealth lay buried. "The Treasure of Mexico" was repeatedly shown at prime time on this distant neighbor nation's two-headed television monopoly, Televisa and TV Azteca.

Mexico is fast running out of oil, the President warned to make his point. Deep sea drilling is the only option. "Energy reform" was put on congressional fast track.

By seeking to privatize Mexico's petroleum industry, Felipe Calderon is swimming against global currents. World-class producers like Russia and Saudi Arabia are consolidating their state-run oil companies, Glasprom and Aramco, rather than selling them off to the private sector.

Petroleum is a volatile liquid in the Mexican mix. Oil and sovereignty have been joined at the hip ever since depression-era president Lazaro Cardenas expropriated and nationalized the industry in 1938 from Anglo and American owners—the so-called "Seven Sisters"—when they defied the Mexican Supreme Court during an oil workers' strike, and those opposed to Calderon's scheme went into hullabaloo mode to push back his privatization legislation.

Ex-left presidential candidate Andres Manuel Lopez Obrador, from whom many Mexicans believe Calderon stole the 2006 election, organized his social base and the "Adelitas," women partisans who dressed up as "soldaderas" or female fighters in the Mexican revolution, donned sombreros and long skirts, toy carbines and bandaleros of fake bullets crisscrossed across their breasts, and encircled the Mexican Senate. Inside both houses of Congress,Lopez Obrador's colleagues seized the podiums and paralyzed all legislative activity for ten days.

The stand-off eventually resulted in a series of nationally televised debates over the next four months during which energy experts, academics, Big Oil reps, PEMEX honchos, lawyers, leftists, senators, deputies, impresarios, and even a poet or two argued about the privatization proposal. The debates were carried live on a big screen in the great Zocalo plaza, where hundreds of outraged citizens gathered every afternoon to cuss out the privatizers.

By autumn 2008, a compromise was struck between Calderon's PAN party and the former ruling PRI, which still holds a majority in both houses. Anti–Lopez Obrador elements within the left-center PRD also signed off on the deal that delineated hundreds of exploration tracts in Mexican deep sea waters, but put a hold on transnational participation and risk contracts. The compromise did not please the transnationals, but Calderon okayed it reluctantly and was preparing fresh legislation to assuage their concerns when the Deepwater Horizon blew out at the bottom of the Gulf, putting the kibosh on Big Oil's pipedreams.

The struggle to stop the privatization of PEMEX is symbolic and illusory. Thirty-one out of the company's 41 divisions are, in effect, subcontracted out to the likes of BP and Halliburton; most contracts are concentrated in the PEP or exploration and perforation sector. Ironically, players like BP, the biggest producer in the Gulf of Mexico today, and Shell are reincarnations of British interests that dominated petroleum production in Veracruz before expropriation—Royal Dutch Shell evolved from Lord Cowdry's (Weetman Pierson) Aguila Oil. Moreover, EXXON is reported to be dickering for BP (which now incorporates Amoco and Atlantic-Richfield), a merger that would restore John D. Rockefeller's Standard Oil, taken down by trustbusters in 1911. Standard Oil's James Doheny and Pierson ruled Mexican oilfields before 1938 and once threatened to secede and form their own "Republic of the Gulf of Mexico."

The U.S. and Mexico dispute a pair of potentially abundant fields in the deep waters of the Gulf. Designated "Donas," the eastern polygon is triangulated between the Yucatan, New Orleans, and Cuba. The much-larger (16,000 square kilometers) western polygon sits between Tamaulipas and Texas. Mexico's share of the western "Dona" (62 percent) purportedly holds up to 34,000,000,000 barrels, twice current reserves.

Preliminary delineation of the Donas was agreed upon by Washington and Mexico City in 2000 and deep sea drilling is set to begin as early as next year. Chevron and Shell have reportedly already won contracts to work the U.S. sites. But Mexico does not have the technology to get at its "treasure" and Houston oil guru George Baker confirms that it will be another decade before PEMEX comes into possession of the tools to drill baby drill at such depths.

Advocates for continued state control of Mexico's oil like Professor Fabio Barbosa of the National Autonomous University (UNAM) rebut the claim that PEMEX cannot drill deep, citing development of the Nab platform in mile-deep waters off Yucatan (the Dona reserves are thought to be three to five miles down in the Gulf).

In a recent *El Universal* op-ed, Barbosa recalled then–BP vice president Cris Sladen's warning to a 2006 oil conference in Veracruz that Mexico would go belly-up if it didn't dissolve PEMEX and let the latest version of the Seven Sisters handle the deep sea exploration and drilling.

Closer to the bottom of the food chain, the voices of the turtles are not heard in this debate between privatizers and nationalists. Deep sea drilling presages unprecedented carnage for their already exhausted species. BP itself has an unblemished record of intended genocide—its Arctic projects threaten protected bowhead whales in the Beaufort Sea and a 900,000 gallon spill in Prudhoe Bay in 2000, plus its plans to trash the Alaska National Wildlife put dozens of species, from Polar bears and caribou to the Arctic tern, the longest-flying migratory bird on Planet Earth, on the brink of extinction.

In an exhibition of unbridled cynicism, BP greenwashes its tarnished image with full-page *New York Times* professions of its concerns for the environment and by handling out conservation awards and grants. So far as is known, no Kemp's Ridley sea turtle has ever won one.

The indigenous peoples of the Pacific Northwest liken the American continent to the back of a turtle—humans are allowed to live on it but must do so in harmony with the planet. "Turtle Island" is the translation of the name of the place where we live in several Indian languages, a designation that once lent its name to Gary Snyder's Pulitzer Prize–winning poems imploring environmental respect and salvation.

But the poet's metaphors do not carry much weight in the boardroom. BP and its cronies in corporate crime and capitalist greed have put Turtle Island at the top their hit list.

28

Don't Go There: The Whole World Has the Travel Bug. And It's Ravaging the Planet

ELIZABETH BECKER

Washington Post journalist Elizabeth Becker directs attention to the ways that global tourism threatens the environment. Tourists have huge carbon footprints, as they use fossil fuels for transport. Moreover, they leave in their wake depleted natural resources, spoiled natural habitat, and mountains of refuse. This exposé goes beyond the environmental impacts of tourism to reveal its other negative consequences. Local culture is undermined by tourism, the poor are uprooted so that luxury vacation houses and spas can be built, and tourism even contributes to human rights violations, most notably the exploitation of young girls and boys for sex tourism.

Did you manage to find someplace for your vacation this summer where you could get away from it all and immerse yourself in nature, or whatever it is that you like to do with a free week or two?

I didn't think so.

It's getting harder and harder. The world has shrunk—and the tourist legions have exploded. The streets of Paris and Venice are so crowded that you can barely move. Cruise ships are filling harbors and disgorging hordes of day trippers the world over. Towering hotels rise in ever-greater numbers along once pristine and empty beaches.

Thanks to globalization and cheap transportation, there aren't many places where you can travel today to avoid the masses of adventure or relaxation-seekers who seem to alight at every conceivable site. I used to love going back to my old haunt in a Himalayan hill station where, as a student in India in 1970, I climbed those steep, silent paths and watched langur monkeys swinging in the trees outside my window. No longer. Now, Mussoorie is chock-a-block with tourist lodges, garbage and noise; the monkeys are fleeing.

SOURCE: From Elizabeth Becker, "Don't Go There: The Whole World Has the Travel Bug. And It's Ravaging the Planet," *The Washington Post National Weekly Edition* (September 8–14, 2008), p. 27. Used with permission.

This problem goes far beyond a veteran traveler's complaint that things aren't the way they used to be, or annoyance at sharing the Eiffel Tower or the Taj Mahal with thousands of other photo-snapping tourists loudly asking questions in languages the locals don't understand. What's happening today is of another magnitude.

The places we love are rapidly disappearing. Global tourism today is not only a major industry—it's nothing short of a planet-threatening plague. It's polluting land and sea, destroying wildlife and natural habitat and depleting energy and natural resources. From Asia to Africa, look-alike resorts and spas are replacing and undermining local culture, and the international quest for vacation houses is forcing local residents out of their homes. It's giving rise to official corruption, wealth inequities and heedless competition. It's even contributing to human rights violations, especially through the scourge of sex tourism.

Look at Cambodia. The monumental temples at Angkor and the beaches on the Gulf of Thailand have made that country a choice destination, especially for Asians, who spent $1 billion there last year. But the foundations of those celebrated temples are in danger of sinking as the 856,000 tourists who every year crowd into Siem Reap, the nearby town of 85,000, drain the surrounding water table.

Meanwhile, Cambodia's well-connected elite has moved to cash in on the bonanza, conspiring with police and the courts to evict peasants from their rural landscape, which is being transformed by high-end resorts catering to wealthy visitors. Cambodia's League for the Promotion and Defense of Human Rights is compiling files that bulge with photographs of thatched-roof houses being burned down while police restrain their traumatized owners. And at night along the riverfront in the capitol of Phnom Penh, the sight of aging Western men holding hands with Cambodian girls young enough to be their granddaughters is ugly evidence of the rampant sex-tourism trade.

All this came as a shock to me. I've been writing about Cambodia for more than 35 years, but I never considered tourism there a serious subject. But when I went back last November, I couldn't avoid the issue. In three short years, tourism had transformed the country. In every interview, the conversation wandered toward tourism, its potential and its abuses. When I went up to Siem Reap, I found the great hall temple of Angkor as crowded, as a colleague said, as Filene's Basement during a sale. Forget tapping into any sense of the divine.

I began researching the global tourism industry and why journalists have allowed it to fly under the radar. Newspapers, the Web and the airwaves are filled with stories celebrating travel; few examine the effects of mass tourism. As Nancy Newhouse, the former *New York Times* travel editor, told me: "We never did the ten worst [places to visit], only the ten best."

Most people can't imagine that tourism could be a global menace. Even the word "tourism" sounds lightweight. And travel has always been surrounded by an aura of romance. For centuries, beginning with the first tourists on holy pilgrimages, travel has been about adventure and discovery and escape from the pressures of daily life.

It wasn't until the end of the 20th century that tourism was added to the list of industries measured in the U.S. gross domestic product. And the results were a revelation: about $1.2 trillion of the $13 trillion U.S. economy is derived from tourism.

Tourism has become the stealth industry of the global era. According to the United Nations, the international tourist count in 1960, at the dawn of the modern era of air travel, was 25 million. By 1970, the figure was up to 165 million. Last year, about 898 million people traveled the globe, and the international tourism industry earned $7 trillion. (And those figures don't include people who vacation in their own countries.)

The U.N. World Tourism Organization was established as a special agency five years ago with the twin goals of keeping track of the tourism industry and figuring out how poor countries, in particular, can take advantage of the tourist boom without causing their own ruin. Geoffrey Lipman, the assistant secretary general of the new organization, has spent his life studying the industry. "Tourism," he told me, "is arguably the largest cluster of industrial sectors in the world" and needs to be included in any international discussions about eliminating poverty or protecting the environment. If properly conducted—maintaining respect for a country's environment and culture, providing local jobs and a market for local goods—tourism, the United Nations believes, is easily the best way for a poor nation to earn foreign currency.

There are several promising examples of this philosophy at work. The non-profit British National Trust offers tourist rentals in restored cottages and historic mansions and then uses the money to buy more land and properties to preserve and protect. The African nation of Namibia, meanwhile, has created what it calls "community-based tourism," which manages more than 25 million acres of wildlife preserves, opening much of the land to tourism—hunting or photo safaris, birding and white-water rafting—that employs local residents and has dramatically reduced poaching.

Most of the tourism industry, however, is heading in the opposite direction. Tourism is now responsible for 5 percent of the world's pollution, according to a recent study. Cruise ships are one of the biggest culprits. These floating hotels create three times more pollution per passenger mile than airplanes. Years of cruises have helped spoil the water of the Caribbean, which, according to the United Nations, absorbs half the waste dumped in the world's oceans. Now these ships are venturing into already fragile polar waters. Last year, Norway banned all cruise ships from visiting its region of the Arctic Circle.

Beach erosion has been swift. After the South Asian tsunami in 2004, fishermen were told to move their homes away from the beaches, but luxury hotel chains with clout were allowed to rebuild near the water's edge. In the United States, the upswing in violent hurricanes hasn't put a dent in the number of vacation homes being built by the sea. "Essentially every tropical island is in danger," the National Geographic Society's Jonathan Tourtellot told me.

In poorer nations, unregulated tourist developments have put unbearable strains on scant resources, especially water. High-end tourists often waste more water in a day with multiple daily showers and toilet flushes than some local families use in a month.

Then there's the fear that over time, major tourist destinations will become virtual ghost towns. Residents of Venice went on strike last spring to block licenses for more hotels; the city of canals is now so expensive that many locals have been pushed out, helping cut the permanent population nearly in half. This summer, the British government issued a report on rural living that included a serious warning that the rich were buying so many vacation or second homes in the countryside that many local residents couldn't afford to live in their villages anymore.

But of all the ills brought on by mass travel, none is as odious as sex tourism. The once-hidden trade is now open and global, with ever-younger girls and boys being forced into prostitution. The Department of Justice estimates that sex tourism provides from 2 to 14 percent of the gross national incomes of countries such as Thailand, Indonesia and the Philippines.

The United States has taken a lead in attempts to eliminate sex tourism, but otherwise, it has stayed out of the tourism debate, mostly viewing tourism as a private matter. Now, however, says Isabel Hill, director of the Commerce Department's Office of Travel and Tourism Industries, the questions raised by mass tourism have become too large to ignore. She hopes that the United States, like so many European countries, will "recognize our limitations and how we have to regulate our resources."

Still, there probably won't be a U.S. secretary for tourism and the environment anytime soon. But don't be surprised if the next international agreement on climate change mentions the role of tourism, or if some countries start regulating tourism along with the environment, because the two go hand-in-hand.

In fact, you'd better hope that they do—if you ever again want to find that cool vacation spot where you can get away from it all.

29

Global Trade of Electronic Waste

GARY M. KROLL
RICHARD H. ROBBINS

This essay, from a book on the environmental impact of globalization, focuses on the electronic waste (e.g., computers, cell phones, television sets) that is discarded annually, mostly in the developed countries, and usually shipped to Asian and African countries for inexpensive, labor-intensive recycling and disposal. This electronic waste contains many toxic materials, which harm those in direct contact with them. People are also negatively impacted by the pollution of the air, land, and water that takes place as a result of electronic waste in landfills.

For many of us, the cookie cutter definition of "modernity" and "progress" often focuses on technology, and specifically the electronic gadgets that increasingly fill our offices, homes, and pockets. These devices are the archetype type of artifice; that is, they appear as if they have been beamed to earth the far reaches of space. They thus represent an almost complete break with the natural world, but this is only the image. Our electronic ephemera are powered by fossil fuels (mostly coal), they contain petroleum-based plastics, and they possess a minilab of heavy metals. These are natural resources—little pieces of nature. When the technology is new, it looks like it will last forever. But all commodities have finite lives, and most commodities are destined to become waste. According to one study "an estimated 50 million metric tons of e-waste replete with toxic materials are generated annually as consumers replace used electronics such as computers and mobile phones with the latest models" ("Feature Focus" 2007: 47).

But globalization breathes new life into discarded electronic devices, usually as garbage that is legally bought and sold in the market. This is a particularly pernicious dynamic for poor people who sometimes accept payment to turn their backyards into havens for e-waste. In this chapter Elizabeth Grossman reports how this process often threatens environmental and human health in Asia, Africa, and Eastern Europe among poor people who have gone into the business of making money off discarded gadgetry.

SOURCE: From Gary M. Kroll and Richard H. Robbins, "Global Trade of Electronic Waste." In *World in Motion: The Environmental Reader*, Gary M. Kroll and Richard H. Robbins (eds.), Lanham, MD: Alta Mira Press, 2009. Used with permission.

ELIZABETH GROSSMAN, "WHERE COMPUTERS GO TO DIE—AND KILL" (2006)

A parade of trucks piled with worn-out computers and electronic equipment pulls away from container ships docked at the port of Taizhou in the Zhejiang province of southeastern China. A short distance inland, the trucks dump their loads in what looks like an enormous parking lot. Pools of dark oily liquid seep from under the mounds of junked machinery. The equipment comes mostly from the United States, Europe, and Japan.

For years, developed countries have been exporting tons of electronic waste to China for inexpensive, labor-intensive recycling and disposal. Since 2000, it's been illegal to import electronic waste into China for this kind of environmentally unsound recycling. But tons of debris are smuggled in with legitimate imports, corruption is common among local officials, and China's appetite for scrap is so enormous that the shipments just keep on coming.

In Taizhou's outdoor workshops, people bang apart the computers and toss bits of metal into brick furnaces that look like chimneys. Split open, the electronics release a stew of toxic materials—among them beryllium, cadmium, lead, mercury, and flame retardants—that can accumulate in the blood and disrupt the body's hormonal balance. Exposed to heat or allowed to degrade, the plastics in electronics can break down into organic pollutants that cause a host of health problems, including cancer. Wearing no protective clothing, workers roast circuit boards in big, uncovered wok-like pans to melt plastics and collect valuable metals. Other workers sluice open basins of acid over semiconductors to remove their gold, tossing the waste into nearby streams. Typical wages for this work are about $2 to $4 a day.

Jim Puckett, director of Basel Action Network, an environmental advocacy organization that tracks hazardous waste, filmed these Dickensian scenes in 2004. "The volume of junk was amazing," he says. "It was arriving twenty-four hours a day and there was so much scrap that one truck was loaded every two minutes." Nothing has changed in two years. "China is still getting the stuff," Puckett tells me in March 2006. In fact, he says, the trend in China now is "to push the ugly stuff out of sight into the rural areas."

The conditions in Taizhou are particularly distressing to Puckett because they underscore what he sees as a persistent failure by the U.S. government to stop the dumping of millions of used computers, TVs, cell phones, and other electronics in the world's developing regions, including those in China, India, Malaysia, the Philippines, Vietnam, Eastern Europe, and Africa.

Because high-tech electronics contain hundreds of materials packed into small spaces, they are difficult and expensive to recycle. Eager to minimize costs and maximize profits, many recyclers ship large quantities of used electronics to countries where labor is cheap and environmental regulations lax. U.S. recyclers and watchdog groups like Basel Action Network estimate that 50 percent or more of U.S. used computers, cell phones, and TVs sent to recyclers are shipped overseas for recycling to places like Taizhou or Lagos, Nigeria, as permitted by federal law. Much of this obsolete equipment ends up as toxic waste, with hazardous components exposed, burned, or allowed to degrade in landfills.

BAN first called widespread attention to the problem in 2002, when it released *Exporting Harm*, a documentary that revealed the appalling damage caused by electronic waste in China. In the southern Chinese village of Guiyu, many of the workers who dismantle high-tech electronics live only steps from their jobs. Their children wander over piles of burned wires and splash in puddles by the banks of rivers that have become dumping grounds for discarded computer parts. The pollution has been so severe that Guiyu's water supply has been undrinkable since the mid-1990s. Water samples taken in 2005 found levels of lead and other metals four hundred to six hundred times what international standards consider safe.

In the summer of 2005, Puckett investigated Lagos, another port bursting with what he calls the effluent of the affluent. "It appears that about five hundred loads of computer equipment are arriving in Lagos each month," he says. Ostensibly sent for resale in Nigeria's rapidly growing market for high-tech electronics, as much as 75 percent of the incoming equipment is unusable, Puckett discovered. As a result, huge quantities are simply dumped.

Photographs taken by BAN in Lagos show scrapped electronics lying in wetlands, along roadsides, being examined by curious children and burning in uncontained landfills. Seared, broken monitors and CPUs are nested in weeds, serving as perches for lizards, chickens, and goats. One mound of computer junk towers at least six feet high. Puckett found identification tags showing that some of the junked equipment originally belonged to the U.S. Army Corps of Engineers, the Illinois Department of Human Services, the Kansas Department of Aging, the State of Massachusetts, the Michigan Department of Natural Resources, the City of Houston, school districts, hospitals, banks, and numerous businesses, including IBM and Intel.

Under the Basel Convention, an international agreement designed to curtail trade in hazardous waste, none of this dumping should be happening. Leaded CRT glass, mercury switches, parts containing heavy metals, and other elements of computer scrap are considered hazardous waste under Basel and cannot be exported for disposal. Electronics can be exported for reuse, repair and—under certain conditions—recycling, creating a gray area into which millions of tons of obsolete electronics have fallen.

The United States is the only industrialized nation not to have ratified the Basel Convention, which would prevent it from trading in hazardous waste. The United States also has no federal laws that prohibit the export of toxic e-waste, nor has it signed the Basel Ban, a 1995 amendment to the convention that prohibits export of hazardous waste from Organization of Economic Cooperation and Development member countries to non–OECD countries—essentially from wealthy to poorer nations. While this policy is intended to spur reuse and recycling, it also makes it difficult to curtail the kind of shipments BAN found in Lagos.

Despite a growing awareness of e-waste's hazards, the U.S. government, says Puckett, has done nothing in the past several years to stem the flow of

e-trash. Given the Bush administration's reluctance to enact or support regulations that interfere with what it considers free trade and the difficulty of monitoring e-waste exports, the shipments continue. "Follow the material, and
you'll find the vast majority of e-waste is still going overseas," says Robert
Houghton, president of Redemtech, a company that handles electronics recycling for a number of Fortune 500 companies, including Kaiser Permanente.
As Puckett says, "Exploiting low-wage countries as a dumping ground is winning the day."

Over a billion computers are now in use worldwide—over 200 million in
the United States, which has the world's highest per capita concentration of PCs.
The average life span of an American computer is about three to five years, and
some 30 million become obsolete each year. According to the International
Association of Electronics Recyclers, approximately 3 billion pieces of consumer
electronics will be scrapped by 2010. Overall, high-tech electronics are the
fastest-growing part of the municipal waste stream in the United States and
Europe.

The EPA estimates that only about 10 percent of all obsolete consumer electronics are recycled. The rest are stored somewhere, passed on to second users, or
simply tossed in the trash. The EPA's most recent estimate is that over 2 million
tons of e-waste end up in U.S. landfills each year. As Jim Fisher of *Salon* reported
in 2000, a toxic stew from discarded computers leaches into groundwater surrounding landfills.

Current design, particularly of equipment now entering the waste stream,
makes separating the dozens of materials in electronics labor intensive. "Almost
every piece of equipment is different," says Greg Sampson of Earth Protection
Services, a national electronics recycler. The process almost always involves manual labor and, once the electronics are dismantled, sophisticated machinery is
required to safely separate and process metals and plastics.

The fragile CRTs with leaded glass used in traditional desktop monitors and
TV screens pose a particular recycling challenge. Metals are the easiest materials
to recycle and the most valuable. Circuit boards typically contain gold, silver, and
other precious metals. Plastics are the peskiest, as many different kinds may be
used in a single piece of equipment and markets for recycled plastics are far less
established than those for scrap metals.

E-Scrap News, a recycling industry trade magazine, features about 950 e-scrap
processors in its North American database—a list that doesn't include nonprofits
or reuse organizations. And not all electronics recyclers offer the same services.
Some dismantle the equipment and recover materials themselves. But many simply collect equipment and do initial disassembly, then contract with others for
materials recovery.

According to the International Association of Electronics Recyclers, this
business now generates about $700 million annually in the United States and is
increasing steadily. Most recyclers charge fees to process equipment. But essentially profits come from the sale of materials recovered or by selling equipment or
components to those who will do so. There's also a speculative aspect to the
business, especially when the scrap metal market is booming and the value of

recyclable circuit boards increasing. It reached an all-time high in January 2006 at $5,640 a ton.

Some recyclers—mostly smaller shops—acquire used equipment at surplus property auctions on eBay or other such resale outlets, then resell equipment whole or in parts by the pound to what Houghton calls "materials brokers" and "chop shops." One batch of equipment may end up being sold to a series of brokers before it reaches a materials processor, and much of what these brokers deal in ends up overseas where costs are lowest. "If a company is buying your electronic scrap or untested equipment," rather than charging for this service, "it's highly likely that it's going overseas," says Sarah Westervelt of BAN.

In 2000, *Salon's* Fisher noted that U.S. computer manufacturers bucked the European trend of instigating convenient buy-back programs for used computers—a resistance that continues today. Since 2000, the Silicon Valley Toxics Coalition, an environmental group, has maintained a "report card" of computer makers' environmental progress in recycling and manufacturing. In its most recent report card, it notes that the "most alarming trends in the electronics industry in the United States continue to be staunch opposition to producer take-back programs."

Currently there is no consistent, industry-wide, or government program to certify or license electronics recyclers. As a result, says Houghton, "It's extremely difficult to peel back the union far enough to find out where the equipment goes. It may change hands two, three or four times before it leaves the country." And, he explains, "The cost of shipping a forty-foot container full of computers, relative to the value of the equipment," even at scrap prices, "is pretty low." With dealers from China to Eastern Europe and Africa ready to buy used electronics for scrap or reuse, and U.S. domestic transportation and recycling costs high, it's actually more profitable to load up a container and send it to Nigeria or Taizhou than it is to process equipment at home.

So traveling the seas in the shadows of legitimate high-tech exports are huge containers that may hold as many as a thousand used computers. They're loaded on ships at East Coast and Gulf Coast ports for Atlantic crossings, or at European ports, including Felixstowe, Le Havre, and Rotterdam, arriving in West Africa by way of Spain. Others cross the Mediterranean from Israel and Dubai or travel Asian Pacific routes from the United States, Japan, Taiwan, and Korea.

Compounding the difficulty of tracking an individual computer is the fact that several different companies—including freight consolidators at both exporting and importing ports, some located in countries distant from both buyers and sellers—are responsible for moving these goods. A recycler in Texas may well be unaware of who is unloading or receiving his goods in China or Africa. Many international freight shippers make it easy to track a whole container—just punch the number into their web site—but information about who's shipping what is not public.

Even in Europe, where e-waste exports are regulated, illegal shipments slip through. "From our work, we have no doubt that there are improper shipments of waste," says Roy Watkinson of the U.K. Environment Agency, which in October of 2005 reported that 75 percent of the containers it had inspected that month contained some illegal waste, including e-scrap. A European group, IMPEL, a network of environmental regulators, has been monitoring this trade, and has found ships loaded with damaged computer equipment sailing out of Wales bound for Pakistan in containers marked "plastics."

According to accounts by Lai Yun of Greenpeace China and Mark Dallura of Chase Electronics in Philadelphia, and news reports from China, corruption is common among customs officials there. Dallura told the *Washington Post* in 2003 that the ships discarded computers to China via Taiwanese middlemen. "I sell it to [the Taiwanese] in Los Angeles and how they get it there is not my concern," Dallura said. "They pay the customs officials off. Everybody knows it. They show up with Mercedeses, rolls of hundred-dollar bills. This is not small-time. This is big-time stuff. There's a lot of money going on in this." Today, loads of e-scrap continue to enter the country despite the Chinese government's official crackdown on these imports.

In an attempt to find out how computers belonging to federal and state government agencies—including one from a Wisconsin school district—might end up in Lagos, Nigeria, I tried to get to the bottom of what happens to the half-million computers the federal government disposes of each year.

Much of the federal government's used but usable computer equipment (including cell phones) is placed with another government agency or donated to a school or community nonprofit (usually chosen and vetted by an individual agency office). The rest (the exact numbers are not known) goes to the General Services Administration—the agency that deals with the procurement, use, and disposal of government property—for public auction. State governments work similarly, usually through state surplus property offices or equivalent programs. No one I consulted had any estimate of how many computers state and local governments discard annually. What was clear is that the ultimate rate of significant quantities of government electronics is poorly documented.

Equipment left after these donations and sales is sent out for recycling. Some federal and state agencies choose their own recyclers. Some federal agencies send used computers to the recyclers awarded contracts under the EPA's electronics recycling program, called Recycling Electronics and Asset Disposition services. A number send equipment to the Federal Prison Industries' computer recycling facilities, which dismantle equipment and send parts on for materials recovery. Many state and local governments (and school districts) put their electronics recycling contracts out for bid, often choosing the company that charges the least to handle and process the equipment. This itself is a red flag. If there's no charge or prices are extremely low, especially for monitors, cautions Sampson of Earth Protection Services, "chances are high equipment is being recycled using cheap labor or by less than optimum methods."

What struck me about the GSA and other public auctions was the lack of oversight, both in terms of where used equipment might end up—potentially creating environmental hazards—and in terms of data security. BAN had scrapped hard drives that it purchased in Lagos analyzed by the Swiss firm NetMon, which found correspondence from staff at the World Bank and from Wisconsin's Child Protective Custody Agency, among others. As a result of chaotic recycling, "There's a definite concern for our security," says Eric Kar-ofsky, senior research analyst with AMR Research, a firm that analyzes business supply chains.

Recent GSA auctions have included computers belonging to the Census Bureau, the South Texas Veterans Health Care System, the Border Patrol, the Federal Aviation Administration, and the U.S. Department of Commerce. Any-one over eighteen from a country the United States does business with, who has a valid credit card, can buy at these auctions, many of which are conducted online. Auction participants are hard to identify as their bids are recorded only by user names, but it's unlikely that anyone is buying a load of seventy-five used CPUs for personal use. And there are thousands of waiting online buyers. In the United States, a laptop sells on eBay about every forty-five seconds, reports senior category manager Stephani Regalia, who helped launch eBay's ReThink program devoted to selling used electronics.

The GSA keeps records of who's bought equipment but does not track what happens to equipment that's been sold, nor does it ask buyers why they're pur-chasing the electronics. "Why would we?" asks a GSA staffer in Boston. The result is that at both the state and federal level, large quantities of electronics are purchased by brokers, auctioneers, and individual dealers who often sell the equipment for export.

For example, one company that has bid at GSA auctions, CTBI Co. of San Antonio, also works as the Morsi Corp. Mike Hancock, the company's proprie-tor, tells me that he sells working equipment to overseas buyers, including those in Indonesia. The scrap, he says, goes to China, Pakistan, and Canada, but another company handles those transactions, so he doesn't track things further. As far as he's concerned, none of his scrap has ended up in Nigeria. "I don't do business in Nigeria," Hancock says. "There are too many bad credit cards there."

One electronics recycler that does do business in Africa is Arizona-based ScrapComputer.com. The staff person I spoke to (who would not give me his name), in the company's Chicago office, says nothing ends up in landfills, and that working equipment is refurbished for schools or sold on eBay. But it also exports computers to India and China where, the staffer says, functional CRTs are remade into TVs. ScrapComputer also sends equipment—all working, I am told—to Malaysia and Egypt, and to West African countries including the Congo. Clearly, this is not the only company selling into Africa, but given the fluid nature of the business, it's extremely difficult to pin down which recyclers knowingly sell e-scrap with a blind eye to dumping and unsound recycling methods.

Still curious to know how a computer owned by Wisconsin's Wauwatosa School District ended up in Lagos, I tracked down the office, SWAP (Surplus with a Purpose), that handles used computers for Wisconsin school districts. Business manager Tim Sell tells me that SWAP—part of the University of Wisconsin—accounts for everything it handles. He says equipment not refur-bished for donations or placed in state offices goes to the Wisconsin State Cor-rections Department's computer recycling facilities, which refurbish and recycle used computers.

But he bemoans the legal loopholes that make e-scrap so hard to track. "Recyclers lie to us," he says, explaining that despite assurances, equipment and parts probably do end up being handled in ways SWAP would rather it did not. When I ask about the computer in Nigeria, Sell tells me he knows that individual customers buy equipment from SWAP and stockpile it for sale to bulk buyers either here or overseas, including those who buy to sell in Africa. With so many unknowns and loopholes in the current system of accounting for used electronics sent for recycling, "I don't know how you're going to stop these exports 100 percent," says Sell.

The United States may be one of the world's biggest consumers of high-tech electronics, but unlike the European Union or Japan, it has no national system for handling e-waste. Unless a state or local government prohibits it, it's currently legal to dump up to 220 pounds a month of e-waste, including CRTs and circuit boards, into local landfills. Several dozen states have introduced e-waste bills, and a handful of U.S. states—California, Maine, Maryland, Massachusetts, Minnesota, and Washington—have recently passed substantive e-waste bills, some of which bar CRTs from their landfills. E-waste bills have also been introduced in the House and Senate, but neither would create a national collection system.

The export of e-waste has been discussed in Congress but no legislation to regulate this trade has yet been introduced. Matt Gerien, press secretary to Rep. Mike Thompson (D–Calif.), who has cosponsored an e-waste bill in the House, says, "Ironically, what brought Representative Thompson to this issue are these export problems." But neither the bill that Rep. Thompson has cosponsored with Rep. Louise Slaughter (D–N.Y.), nor the one introduced by Sen. Ron Wyden (D–Ore.) and Sen. Jim Talent (R–Mo.) would deal with exports.

Meanwhile, says Laura Coughlan of the EPA's Office of Solid Waste, the Bush administration has drafted legislation that would allow the United States to ratify the Basel Convention but is waiting for final clearance for transmittal to Congress. And the Ban Amendment, which essentially prohibits send e-waste from wealthy to poorer countries, "has created issues for U.S. ratification of the convention," says Coughlan, who explains that no "U.S. administration has supported ratification of this amendment, and the U.S. government has been unable to reach consensus with domestic stakeholders."

Legislation in Europe has made electronics recycling mandatory throughout the European Union, as it is in Japan and some other countries. Companion legislation requires the elimination of certain toxics—among them lead, cadmium, and hexavalent chromium used in solder, batteries, inks, and paints—from electronic products, and given the global nature of the high-tech industry, these new materials standards could effectively become world standards. Many such changes have already been made and more are in the works, but the old equipment now being discarded remains laden with toxics.

As U.S. lawmakers, manufacturers, environmental advocates, waste haulers, and recyclers struggle to find a way to collect the nation's high-tech trash, Americans are left with what policymakers are fond of calling a patchwork of regulations and recycling options. This makes things as confusing for manufacturers as it does for consumers and recyclers. "At some point, the feds will have to step in and harmonize things," says Ted Smith of the Silicon Valley Toxics Coalition.

In 2005, the EPA held an electronics recycling summit. Among the issues participants grappled with, and on which there is no industry-wide or national policy, are certifying electronics recyclers and exporting electronic waste. Complaints were voiced about the difficulty of dealing with products designed with materials that make recycling complicated and expensive. But loudest of all were complaints that the United States had too many confusing and uncoordinated recycling efforts. A year later, a few more state laws regulating e-waste have been passed but little has been done to stop the steady stream of used computers, cell phones, and TVs that are ending up overseas, in dumps, polluting soil, water, and air.

REFERENCES

"Feature Focus: Environment and Globalization: Minimizing Risks, Seizing Opportunities." 2007. In *GeoYearBook: An Overview of Our Changing Environment. A Report from the United Nations Environmental Program*, www.unep.org/geo/yearbook/yb2007/PDF/6_Feature_Focus72dpi.pdf, p. 47.

30

U.S. Stymied as Guns Flow to Mexican Cartels

JAMES C. MCKINLEY, JR.

The Mexican drug cartels depend on U.S. guns, which are readily available in the U.S. because here guns are sold and possessed lawfully. Guns are purchased in the U.S. and smuggled to Mexico. The irony, of course, is that the United States has a policy to stop the supply of drugs into the country, yet its lax gun laws make it easier for drug cartels to supply drugs to the United States.

John Phillip Hernandez, a 24-year-old unemployed machinist who lived with his parents, walked into a giant sporting goods store here in July 2006, and plunked $2,600 in cash on a glass display counter. A few minutes later, Mr. Hernandez walked out with three military-style rifles.

One of those rifles was recovered seven months later in Acapulco, Mexico, where it had been used by drug cartel gunmen to attack the offices of the Guerrero State attorney general, court documents say. Four police officers and three secretaries were killed.

Although Mr. Hernandez was arrested last year as part of a gun-smuggling ring, most of the 22 others in the ring are still at large. Before their operation was discovered, the smugglers had transported what court documents described as at least 339 high-powered weapons to Mexico over a year and a half, federal agents said.

"There is no telling how long that group was operating before we caught on to them," said J. Dewey Webb, the agent in charge of the Houston division of the Bureau of Alcohol, Tobacco, Firearms and Explosives.

Noting there are about 1,500 licensed gun dealers in the Houston area, Mr. Webb added, "You can come to Houston and go to a different gun store every day for several months and never alert any one."

The case highlights a major obstacle facing the United States as it tries to meet a demand from Mexico to curb the flow of arms from the States to drug cartels. The federal system for tracking gun sales, crafted over the years to avoid infringements on Second Amendment rights, makes it difficult to spot suspicious

SOURCE: James C. McKinley, Jr., "U.S. Stymied as Guns Flow to Mexican Cartels," *New York Times* (April 14, 2009). http://www.nytimes.com/2009/04/15/us/15guns. html?_r=1+pageswanted=all. Used with permission.

trends quickly and to identify people buying for smugglers, law enforcement officials say.

As a result, in some states along the Southwest border where firearms are lightly regulated, gun smugglers can evade detection for months or years. In Texas, New Mexico and Arizona, dealers can sell an unlimited number of rifles to anyone with a driver's license and a clean criminal record without reporting the sales to the government.

At gun shows in these states, there is even less regulation. Private sellers, unlike licensed dealers, are not obligated to record the buyer's name, much less report the sale to the A.T.F.

Mexican officials have repeatedly asked the United States to clamp down on the flow of weapons and are likely to bring it up again with President Obama when he visits Mexico on Thursday.

Sending straw buyers into American stores, cartels have stocked up on semi-automatic AK-47 and AR-15 rifles, converting some to machine guns, investigators in both countries say. They have also bought .50 caliber rifles capable of stopping a car and Belgian pistols able to fire rifle rounds that will penetrate body armor.

Federal agents say about 90 percent of the 12,000 pistols and rifles the Mexican authorities recovered from drug dealers last year and asked to be traced came from dealers in the United States, most of them in Texas and Arizona.

The Mexican foreign minister, Patricia Espinosa, in talking with reporters recently, accused the United States of violating its international treaty obligations by allowing guns to flow into the hands of organized crime groups in Mexico.

But law enforcement officials on this side of the border say the legal hurdles to making cases against smugglers remain high.

"Guns are legal to possess in this country," said William J. Hoover, the assistant director for operations of the federal firearms agency. "If you stop me between the dealer and the border, I am still legal, because I can possess those guns."

To be sure, the A.T.F. and Immigration and Customs Enforcement have stepped up their efforts to stop smuggling over the last two years. Last year, some 200 indictments were handed up against straw buyers and gun smugglers, breaking up at least a dozen trafficking rings.

31

Diseases Travel Fast, but So Do Tools to Fight Them

KEVIN SULLIVAN

MARY JORDAN

The world's citizens are increasingly vulnerable to pandemics (worldwide epidemics). But, as journalists Kevin Sullivan and Mary Jordan report, these threats to health are resisted by globalized disease fighters. These health professionals monitor the spread of diseases, stockpile vaccines, and provide information through global networks.

Teenagers in New Zealand, honeymooners in Scotland, high-schoolers in New York and tourists in Israel all are sick from the same bug caught just days ago on trips to Mexico.

Their illnesses are the latest example of how diseases, from influenza to tuberculosis to cholera, are spreading ever more quickly in an increasingly globalized world. But so, too, are the tools necessary to combat outbreaks of disease: expertise, medicine, money and information.

"Things move incredibly fast; there has been an exponential rise in the numbers of people who move around the world," says Scott Dowell, a physician and head of global disease detection and response for the U.S. Centers for Disease Control and Prevention in Atlanta.

Although global pandemics are as old as history itself, diseases and the people who carry them have never been able to move so far, so fast. The number of international air travelers grew fivefold, to 824 million passengers per year, from 1980 to 2007.

Spanish explorer Hernán Cortés transported smallpox and measles to Mexico on long sea voyages in the 16th century. The current strain of swine flu leapt from Mexico to the far corners of the world on jumbo jets in a matter of hours.

"That makes it incredibly harder to manage these outbreaks," says Dowell, who is overseeing the CDC's assistance to Mexico on the swine flu case. In New Zealand, for example, officials were trying to track down all 350-plus passengers

SOURCE: From Kevin Sullivan and Mary Jordan, "Diseases Travel Fast, But So Do Tools to Fight Them," *The Washington Post National Weekly Edition* (May 4–10, 2009), p. 13. Used with permission.

who were on the same flight as the infected high-schoolers (Air New Zealand Flight 1 from Los Angeles to Auckland on April 25).

Although the world is more vulnerable to the rapid spread of disease, many experts say, it has never been more prepared.

Advances in the understanding of disease, stockpiling of vaccines and global networks of medical surveillance have better equipped health professionals to deal with outbreaks. Instant communications have allowed information on diseases to move faster than the bugs themselves. The swine flu page on the CDC's Web site lets users sign up for e-mail alerts, podcasts and news feeds; more than 36,000 people have signed up for CDC Twitter alerts.

"It's a very different world than it was even 10 years ago," says Robert F. Breiman, a physician and coordinator of the CDC's Global Disease Detection Division center in Nairobi.

The Nairobi center is one of six maintained by the CDC around the world; the others are in Egypt, Thailand, Kazakhstan, China and Guatemala. In each, CDC medical professionals work with local officials to detect disease outbreaks in the region and coordinate their responses.

The centers were established in the wake of the 2003 outbreak of severe acute respiratory syndrome, or SARS, which caused hundreds of deaths around the world.

Dowell says that CDC staffers, including infectious-disease specialists from the Guatemala office, have been dispatched to Mexico.

The CDC also works with the World Health Organization's Global Outbreak Alert and Response Network, which coordinates efforts among health officials around the globe.

"There is no such thing as a local outbreak," Michael Ryan, a physician who runs the WHO's global outbreak program, said in an interview before the epidemic in Mexico.

A report issued by a committee in the British House of Lords last year concluded that dramatic global population growth was a key factor in spreading infectious disease.

It said the world's population has risen from 2.5 billion in 1950 to more than 6 billion now and is projected to rise to 9 billion by 2050. And the population is growing fastest in many of the poorest countries with the biggest health problems.

Mass migrations have turned such places as Lagos, Nigeria, and Rio de Janeiro into megacities where over-crowding and poverty create ideal conditions for the spread of diseases such as dengue fever. An outbreak in Brazil killed at least 100 people last year.

Several thousand Bolivians have immigrated to Switzerland in recent years, and with them has come Chagas disease, a parasitic illness that can cause heart problems. Chagas, once confined to rural Latin America, also has spread with migrants to the United States and Asia.

The British report noted that rising populations are forcing people to push deeper into previously uninhabited land to live and grow crops. Going deeper

A look at what is known about the virus.

A Virus With Pandemic Potential

HOW THE FLU VIRUS MUTATES

The genetic machinery of the flu virus lacks a mechanism to correct errors when it replicates. As a result, it mutates at a high rate, allowing it to evade the body's natural defenses, vaccines and drugs.

Seasonal flu strains that circulate every winter generally have minor changes from those of the previous year. But people who have been exposed to flu in the past usually retain a measure of immunity.

Why is this virus killing healthy people?

A protein on the virus binds to receptors on healthy cells in the airways and lungs, causing the virus to open and release its RNA.

HOW THIS FLU VIRUS IS DIFFERENT

The pandemic threat arises from another trick of the flu virus, called genetic reassortment. When different strains infect the same host at the same time, it allows them to exchange whole sections of their genetic code.

▼ **Scientists think the current virus strain combines genetic material from pigs, birds and humans.** Segments from the three different viruses have created a reassorted virus that has not been seen before.

One theory is that the virus trigger's an excessively aggressive immune response that destroys the throat and lung tissue. Those with robust immune systems may be especially vulnerable.

The RNA moves to the cell nucleus, where it is incorporated in the cell's machinery, directing the cell to make copies of the virus.

Another protein on the virus punches a hole in the cell, killing it and releasing the replicated virus.

...either into the airway to find another cell to infect or ejected by a cough or sneeze and launched to find a new host.

How a flu virus works

The genetic code of the flu is contained in eight strands of RNA.

VIRUS

CELL

VIRUS

1

NUCLEUS

2

DEAD CELL

3

DEAD CELL

NEW VIRUS

4

Fighting the Flu

PREVENTION: VACCINE

ANTIBODIES

Vaccines teach the body's immune system to make antibodies to kill the virus. A weakened form of the virus is grown in hens' eggs, purified and killed with a chemical. Creating a new vaccine takes at least six months and recuires hundreds of millions of eggs.

INTERRUPTION: ANTIVIRAL DRUGS

Relenza and Tamiflu, both shown to be effective against this current virus, stop it from budding out of the cell if administered soon after symptoms appear. Antivirals can also be given to people in contact with an infected person to prevent the disease from spreading.

BY BRENNA MALONEY AND LAURA STANTON—THE WASHINGTON POST

SOURCES: Centers for Disease Control and Prevention, National Institute of Allergy and Infectious Diseases, World Health Organization

into jungles and forests has led to closer contact with wild animals, which can carry new pathogens that lead to previously unknown illnesses, the report said.

Such contact with animals, particularly primates, has often been documented in African outbreaks of Ebola, the acute hemorrhagic fever.

Climate change is creating droughts and floods where they were not common before, and that is also altering patterns and flows of disease. In warming temperatures, mosquitoes are migrating to new areas, carrying diseases once confined to the tropics.

Chikungunya, a tropical disease from Africa and Asia that causes severe joint pain, showed up in Italy in 2006 and has infected several hundred people there. A tourist who visited Kerala, India, is suspected to have carried the virus home. Then the Asian tiger mosquito, which is working its way north as temperatures rise, carried it from one infected person to the next.

Robin Weiss, a British virologist who has written about the globalization of disease, says instant global communication can rapidly spread helpful information, but it can also have downsides, such as needlessly alarming people with a flood of texts, e-mails and news bulletins.

"There are good and bad effects of globalization" on infectious disease, says Mario Raviglione, a top tuberculosis official at the WHO in Geneva.

Raviglione says the globalized economy means that money flows to poorer countries to help them improve health systems and fight high-profile diseases such as HIV/AIDS. But when the banking system falters in one country and drags down financial systems around the globe, that has a direct effect on how much families and governments can spend on fighting disease.

"We all live in a global village," Raviglione says.

32

Forced Prostitution

KEVIN BALES
ZOE TRODD
ALEX KENT WILLIAMSON

This excerpt focuses on one form of slavery—the international trafficking of women as prostitutes—and describes how "[s]laveholders use violence, starvation, and social isolation to retain control" over their slaves. Bales and his co-authors describe the extent of the practice in the West (Canada, the U.K., the U.S., and Western Europe) as well as in Africa and Asia.

FORCED PROSTITUTION IN THE WEST

The most well-publicized form of international trafficking is that of women into forced prostitution. At least half of international trafficking cases are for sexual exploitation and, while women account for fifty-six percent of forced economic exploitation, they account for ninety-eight percent of forced sexual exploitation. Many of these women accept job offers and sign phony contracts, beginning the journey into slavery. Told they owe money for their trip, they must work off the debt with clients. Slaveholders use violence, starvation, and social isolation to retain control. And there is often a double bind: not only of the brothel owner's restrictions, but the restrictions of a foreign country where they cannot speak the language, have no knowledge of their legal rights, and often fear the police.

Most are taken from poorer countries to richer countries, with many thousands arriving in Canada, the U.K., the U.S., and Western Europe. For example, women arrive in Canada from South Korea, Thailand, Cambodia, Malaysia, and Vietnam, and most are trafficked for commercial sexual exploitation. The Royal Canadian Mounted Police conservatively estimates that between 800 and 1200 people, most of them women, are victims of human trafficking in Canada each year. In the U.K., the Home Office recently announced that of the hundreds of people who had been trafficked into the country in just an eighteen-month period, seventy-five percent of them were women and girls.

SOURCE: From Kevin Bales, Zoe Trodd, and Alex Kent Williamson, "Forced Prostitution," in *Modern Slavery: The Secret World of 27 Million People,* Oxford, U.K.: Oneworld, 2009. Used with permission.

In the U.S., around eighty percent of the foreign-born individuals trafficked into the country each year are female, and seventy percent of these women end up in forced prostitution. Feeder countries include Albania, the Philippines, Thailand, Nigeria, and Mexico (many from the central region of Tlaxcala, a haven for slave traders). Major trafficking organizations are Asian criminal syndicates, Russian crime groups and syndicates, and loosely associated Latin American groups. Many women are lured from their home countries by false promises of legitimate employment in the U.S., and they are primarily trafficked in three ways: the illegal use of "legitimate" travel documents, imposter passports, and entry without inspection. The women are forced to work to pay off the debts imposed by their "smugglers"—debts ranging from $40,000 to $60,000 per person. They might perform 4000 acts of sexual intercourse each year to meet their quota, at $10 to $25 per act. In other situations, girls as young as twelve years old are forced to have sex seven days a week, with ten to fifteen people a day, and meet a quota of $500–$1000 a night.

Estimates of the number of people who are trafficked within and into Europe annually range from 100,000 to 500,000, and human trafficking is Europe's fastest-growing criminal activity. By some estimates, ninety percent of non-national women active in the sex industries of southeastern Europe have been trafficked. But most women are brought from the poorer eastern countries to the richer West: over two-thirds of trafficked women from former Soviet countries end up in Western Europe. The UN ranks Belgium, Germany, Greece, Italy, and the Netherlands "very high" as destination countries, and Albania, Bulgaria, Lithuania, and Romania "very high" as origin countries.

This process of trafficking from eastern to western countries began when socialism was dismantled in the USSR in 1991. "Transition countries" (nations that moved from socialism to capitalism) saw an explosion in the export of men, women, and children as slaves. The transition period in most of the ex-socialist countries has been marked by economic recession, hyperinflation, high unemployment, and armed conflict—prompting large numbers of refugees and economic migrants to seek entry to Western Europe. Traffickers make false promises of employment and corrupt border guards reportedly accept bribes to facilitate trafficking. As many as 100,000 women are now trafficked from and throughout the fifteen former Soviet countries annually and sold into international prostitution.

In Russia, tens of thousands of women are trafficked each year to over fifty countries for commercial sexual exploitation. One key overland corridor into the European Union is the "Eastern Route" through Poland. The women arrive in Central and Western Europe and also the Middle East. In Israel, women arrive from Russia, and also Ukraine, Moldova, Turkey, Uzbekistan, Lithuania, Belarus, Brazil, Colombia, and South Africa. Officially, the estimate of trafficked women each year in Israel is 3000, and in 2006, the UN ranked Israel "very high" as a destination country for trafficking.

The fall of socialism in 1991 also led to a rise in organized crime in Albania, which is currently one of the top ten countries of origin for sex trafficking. Albania suffered badly from the interruption of trade with countries of the

Eastern bloc after 1989. In 1991, its GDP was only half that of 1990 and half of all children were malnourished. Over the next ten years, an estimated 100,000 Albanian women and girls were trafficked. Today, more than sixty-five percent of Albanian sex-trafficking victims are minors at the time they are trafficked, and perhaps fifty percent of victims leave home under the false impression that they will be married or engaged to an Albanian or foreigner and live abroad. Another ten percent are kidnapped or forced into prostitution. The women and girls are commonly tortured if they do not comply.

From Armenia, women are trafficked to the UAE, where an estimated 10,000 women from Eastern Europe, sub-Saharan Africa, South and East Asia, Iraq, Iran, and Morocco are victims of sex trafficking. Armenian women are also taken to Greece and Turkey, where victims arrive from Ukraine and Moldova as well. Moldova is another main country of origin for the trafficking of women and children into European sexual exploitation. Again, the country's economic conditions fuel this trafficking. In 2000, the country's GDP was forty percent of its level in 1990. Moldova has the lowest average salary of countries in the former Soviet Union, and women's salaries are seventy to eighty percent of those of men. Unemployment remains extremely high, especially among women—who comprise around seventy percent of the unemployed—and more than half the population lives below the poverty line. Consequently, women are forced to look outside of the country for work and pimps take advantage of these migrants, though some victims are kidnapped. Thousands of women have been trafficked out of the country in recent years. Most Moldovan trafficking victims are taken to the Balkans and other destinations include Asia, Western Europe, and the Middle East.

The push for women to leave Ukraine is equally powerful. There, they account for up to ninety percent of the unemployed and are usually the first fired. Traffickers abduct an estimated 35,000 women from Ukraine each year, and in 1998 the Ukrainian Ministry of the Interior estimated that 400,000 Ukrainian women had been trafficked in the past decade, although Ukrainian researchers and NGOs believe the number is higher. Ukrainian consulates have brought 11,000 trafficking victims back to Ukraine, and the International Organisation for Migration (IOM) says it has helped more than 2100 Ukrainian victims since 2000, but adds that this is a small portion of the total number. Almost fifty countries serve as destination points throughout Europe and eastward. Germany is one of the most popular destinations in Europe for women trafficked from Ukraine, though victims also come from Africa (mainly Nigeria) and Asia (mainly Thailand).

FORCED PROSTITUTION IN AFRICA AND ASIA

While we have estimates for the number of trafficking victims in Europe and the U.S., the number of people who are trafficked around and out of African countries remains unknown. We can, however, identify some patterns. The UN notes that victims trafficked from African countries frequently arrive in Ivory Coast, Nigeria, and South Africa, and that those trafficked out of Africa arrive in the U.K., Italy, France, Belgium, the Netherlands, and Saudi Arabia. Trafficked women and children arrive in South Africa from Angola, Botswana, the Democratic Republic of Congo (DRC), Lesotho, Mozambique, Malawi, South Africa, Swaziland, Tanzania, Zimbabwe, and Zambia. Several major criminal groups in South Africa now traffic women: Bulgarian and Thai syndicates, the Russian and the Chinese Mafia, and African criminal organizations, mainly from West Africa. Women are trafficked through false promises of employment, marriage, or education, and some are simply abducted.

The UN ranks Nigeria particularly high as an origin country for human trafficking, and Benin, Ghana, and Morocco slightly less high. Around 45,000 Nigerian women have become victims of trafficking over the past fifteen years (two-thirds have gone to Europe and a third to the Gulf States). We know, too, that thousands of Ethiopian girls are trafficked into Lebanon each year for sexual exploitation, and that Zimbabwean women are forced into prostitution in the U.K. and the U.S. Zimbabwe's trafficking problem has worsened since 2005, when the Zimbabwean government began Operation Murambatsvina ("Clean-Up"), a campaign to forcibly clear slum areas. This has displaced hundreds of thousands of people, and left an estimated 223,000 children, especially girls, vulnerable to trafficking.

Yet more often than for sexual exploitation, individuals are trafficked from Africa for forced labor: thirty-five percent of reported trafficking cases in Africa are for forced labor, according to the UN. In Asia, however, only twenty percent of reported human trafficking cases are for forced labor, and the greater majority is for sexual exploitation. The IOM estimates that around 225,000 women and children are trafficked every year in Asia. In Southern Asia, Sri Lankan women are trafficked to Saudi Arabia, Kuwait, the UAE, Bahrain, and Qatar for sexual exploitation. And in Nepal, up to 12,000 women and children (many aged between nine and sixteen) are trafficked every year across the border into India. An estimated 200,000 trafficked Nepalese women and girls currently reside in Indian brothels and Nepal has an unknown number of internal sex trafficking victims as well.

India's National Crime Records Bureau reports that there are 9368 trafficked women and children in the country, though, [it] acknowledges that a sizeable number of crimes against women go unreported due to the attached social stigma. The Central Social Welfare Board of India has published much higher figures, for example estimating that up to one million women and children were being used in forced prostitution in six metropolitan cities in India, and that within this group, thirty percent had been enslaved when they were younger than eighteen.

In Southeastern Asia, Cambodia is ranked high as an origin country for trafficking by the UN. Prostitution escalated in Cambodia when the UN Transitional Authority Forces arrived in 1991. The number of prostitutes in the cities rose from 6000 in 1991 to 20,000 in 1992, and the Cambodian Women's Development Association claims that half of these women were trafficking victims. Women continue to be internally trafficked for sexual exploitation, usually from rural areas to the country's capital, Phnom Penh, and other secondary cities, and are also sent to Thailand and Malaysia for forced prostitution.

The UN ranks Thailand even higher as an origin country, and it is the only country in Asia to be ranked "very high" as an origin, transit, and destination country. Thai women and girls are lured from rural areas with the promise of work in restaurants or factories. The slave-recruiter offers the girls' parents an "advance" on their wages, but then they are enslaved and sold to a brothel. The brothel owners tell the girls that they must pay back their purchase price plus interest through prostitution, though they typically don't let them go until they can no longer be sold to men because of physical or mental breakdown. The brothels do have to feed the girls and keep them presentable, but if they become ill or injured, they are disposed of.

Thousands of women are also trafficked annually out of Thailand for sexual exploitation. The major destinations include Japan, Malaysia, Bahrain, Australia, Singapore, and the U.S. In Singapore, women and girls are also trafficked from Indonesia, Malaysia, the Philippines, Vietnam, and China. And in Japan, women are also trafficked from the Philippines, Russia, and Eastern Europe, and on a smaller scale from Colombia, Brazil, Mexico, Burma, and Indonesia.

The sale of sex in Japan is a vast industry. Japan is by far the richest country in Asia and the region's largest destination for trafficking—the only country in Eastern Asia that the UN ranks "very high" as a destination country for human trafficking. The U.S. State Department agrees. In its 2004 "Trafficking in Persons Report," Japan appeared in the same group as poor, war-torn countries like Serbia, Tajikistan, and Ivory Coast, and was in danger of being classified in the worst tier, where sanctions can apply. The Japanese population is one of the most law-abiding in the world, with a robbery rate of 1.3 per 100,000 people compared to a rate in the U.S. nearly 180 times greater. Yet slavery flourishes in Japan. Currently, the cost for sex in a "fashion massage" shop is $50 to $90. Moving down the ladder, sex with foreign women is sold on the street for $8 to $10. By 2001, the Japanese sex industry was thought to generate $20 billion per year and was known to be rapidly growing.

A conservative estimate of the total number of foreign women enslaved as prostitutes in Japan is 25,000. In addition, while the marriage rate of Japanese women to foreign men has remained fairly constant, marriages of foreign women to Japanese men have skyrocketed to more than 30,000 a year. An investigation by the Organization of American States explained that while many of these marriages are legitimate, some are known to be "phony marriages arranged by traffickers and used to move foreign women into the Japanese sex trade or the underground labor force."

Thousands of the foreign women in Japan are on overstayed visas and tourist visas, and as Nu explains, it was easy for her enslavers to manage the problem of overstayed visas:

> Sometimes the police would come in to check if there were visa-overstayers. The owner was mostly warned in advance by informants. Overstayers would be concealed, or heaped into a bus and hidden in a hotel close by in the mountains 'til the police left. At other times the bar would be closed for a day or two. There was also a time when only those with valid visas were produced before the police, and the police bribed.

According to government estimates, there were about 220,000 illegal residents who had overstayed their visas in the country in 2004, and hundreds of thousands more will arrive each year on tourist visas. Both categories will contain a significant number of women held against their will. But another part of Japan's slavery problem is its euphemistically titled "Entertainment Industry," which includes brothels, strip clubs, bathhouses and street prostitution. The government has a special "entertainer visa," supposedly given to singers and dancers that will be giving performances in theaters and nightclubs. If this were true, then Japan would have more professional entertainers than the rest of the world combined.

In reality, the visa is used to import large numbers of foreign women to meet the demands of Japanese men for sex and "entertainment." Between 1996 and 2003, the number of visas issued each year more than doubled (see table 8.1).

In 2003, approximately 80,000 "entertainers" came from the Philippines and, over the years, around 40,000 women have come from Latin America on entertainer visas. Under intense pressure from human rights groups and other countries, Japan finally agreed to better police the entertainer visa system from

T A B L E 8.1   Entry into Japan on "entertainer" visas

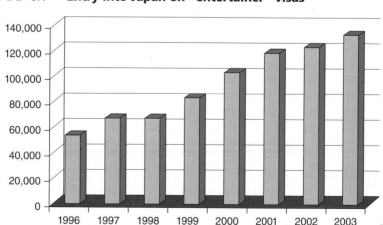

March 2005, but no figures have been released showing a fall in the number of "entertainers" brought to Japan.

REFLECTION QUESTIONS FOR CHAPTER 8

1. The massive BP oil spill in the Gulf of Mexico has enormous repercussions for the environment. The world, especially in developed nations, has an insatiable appetite for oil, causing the search for fossil fuels in ever more fragile environments. What are the alternatives to "drill, baby, drill"?

2. Almost a third of the world's population lives on less than $1 a day. Most live in rural areas and shanty towns in developing societies. Their poverty is a primary source of the "New Slaves" (sex workers, domestic workers, and sweatshop workers). How is poverty linked to the "New Slavery"? What can be done about this serious problem? What is the responsibility of the prosperous nations? Of transnational corporations? Of the nations of the developing world?

3. How can the Mexican drug cartels be diminished in power and consequence? How is the U.S. responsible for these thriving crime syndicates?

Chapter 9

Changing Global Structures
Resistance and Social Movements

The articles included in this book address issues and questions regarding the various manifestations of globalization: Does the presence of transnational corporations in the developing world raise the standard of living for the poor in those countries? Will transnational corporations rule the world? Are the political agencies of globalization—for example, the World Trade Organization and the International Monetary Fund—increasing the wealth gap between the rich and poor nations? Is globalization increasing threats to the environment? Is the future a homogenized Westernized world?

For many individuals and groups, the answers to these questions are in the negative. In opposition to the effects of globalization, they have organized social movements to resist and challenge globalization and to change its direction. Human agency and social movements have the potential to change the course of history. Social transformations do not always occur from the top (where the political and economic power is located) but sometimes from the bottom up (by the relatively powerless who join with others to wield a power greater than individuals could on their own).

Powerful social structures constrain what we do, but they can never control it entirely.[1] Human beings are not passive actors. Individuals acting alone or with others can shape, resist, challenge, and sometimes change the social structures that impinge on them. These actions constitute human agency.[2]

Individuals seeking change typically join with others for greater power to become part of social movements, "agencies of social transformation that emerge in response to certain social changes and conditions."[3] Many people have become part of social movements to, for example, resist global corporations, free trade, and transnational political/economic organizations. As *The Field Guide to the Global Economy* puts it,

> Millions of workers, consumers, environmentalists, religious activists, farmers, and women around the world are demanding their fair share of the fruits of the global economy. Their strategies are diverse. Some attempt to slow down aspects of globalization, while others aim to reshape its path in ways that promote democracy, equity, and sustainability. Campaigns operate on the local, national, and international levels.[4]

The first reading in this chapter is an excerpt from Jeremy Brecher, Tim Costello, and Brendan Smith's *Globalization from Below*. The authors address a wide range of efforts to stop, slow down, or change the course of globalization. They provide the blueprint, the rationale, and a brief history of how and why people have and are organizing at the grassroots to effect change "from the bottom up." The second part of their essay provides an understanding of how social movements arise and how they have the potential to overcome the politically and economically powerful.

The second reading also addresses grassroots processes in which individuals and groups are engaged. Sociologist Myra Marx Ferree turns our attention to transnational women's activism. In contrast to common thinking about globalization, which is gender-blind, Ferree shows how gender matters in social and political movements for change. Her article challenges other assumptions about women's activism, including the assumptions that women's movements are always driven by feminist goals and that feminist mobilization is restricted to women's organizations. Ferree explains how a complex variety of local and transnational settings provide opportunities for restructuring organizations in which women are marginalized by dominant institutions. Her discussion stresses how feminist goals are often combined with other pressures for change in the globalizing world and how such social and political movements become connected across national boundaries.

The third essay in this section addresses students' mobilization in global activism. Bhumika Muchhala provides an account of both his own radicalization in dissent movements and the history of the Students Against Sweatshops movement on U.S. campuses. Not only does he render an account of the movement's origin and structure, he also reflects on its place in the global struggle.

In the next reading, "The Rise of Food Democracy," Brian Halwell examines power relations in the global food system. From Nebraska to Norway, from Egypt to Zimbabwe, Halwell shows how small farms are reclaiming the "sovereignty" of small producers and local markets, with some notable effects at the level of the World Trade Organization and some of the world's largest food companies.

The final reading discusses community activism in Peru. Kelly Hearn describes local mobilization against giant oil corporations, their toxic legacies, and the injustices they leave behind.

ENDNOTES

1. Anthony Giddens. 1991. *Introduction to Sociology*. New York: W.W. Norton.
2. D. Stanley Eitzen and Maxine Baca Zinn. 2004. *In Conflict and Order: Understanding Society*, 10th ed. Boston: Allyn & Bacon, Chapter 18.
3. Robin Cohen and Paul Kennedy. 2000. *Global Sociology*. New York: New York University Press, p. 287.
4. Sarah Anderson and John Cavanagh, with Thea Lee. 2000. *Field Guide to the Global Economy*. New York: The New Press, p. 91.

33

Globalization and
Social Movements

JEREMY BRECHER
TIM COSTELLO
BRENDAN SMITH

*Globalization produces inequalities, which can also be used to mobilize for
change. Jeremy Brecher and his colleagues provide a broad outline of how people
in various grassroots settings around the world are organizing and connecting their
struggles.*

It is often said that globalization is inevitable and that there is no alternative.
But, in fact, the new global regime is highly vulnerable. It violates the interests
of the great majority of the world's people. It lacks political legitimacy. It is riven
with divisions and conflicting interests. It has the normal crisis-prone character of
capitalist systems, but few of the compensatory non-market institutions that
helped stabilize pre-globalization economies. And it has few means to control
its own tendency to destroy the natural environment on which it—and its
species—depend. These are the reasons that, as the *Financial Times* wrote, the
world had swung "from the triumph of global capitalism to its crisis in less than a
decade."[1]

GLOBALIZATION FROM BELOW

Just as the corporate and political elites are reaching across national borders to
further their agendas, people at the grassroots are connecting their struggles
around the world to impose their needs and interests on the global economy.
Globalization from above is generating a worldwide movement of resistance:
globalization from below.[2]

Throughout the 20th century, nationally based social movements have
placed limits on the downsides of capitalism. Workers and communities won

SOURCE: From Jeremy Brecher, Tim Costello, and Brendan Smith, *Globalization from
Below* South End Press. Cambridge, MA: 2000). Used with permission.

national economic regulation and protections ranging from environmental laws to labor unions and from public investment to progressive taxation.

Globalization outflanked both national movements and national economies. It caused a historic break in the institutions, traditions, and movements that had opposed unfettered capitalism since its inception. Not only Communism, but also social democracy, economic nationalism, trade unionism, and democratic government itself were rolled back by the neoliberal tide—and often found their own foundations crumbling from within in the face of forces they could not understand or control.

Nonetheless, the real problems of a system of unrestrained capitalism did not disappear. Globalization only intensified them. And so the impulses that had generated these counter-movements in the first place began to stir.

Like globalization from above, these counter-movements began from many diverse starting points, ranging from local campaigns against runaway plants to union organizing in poor countries, and from protection of indigenous peoples to resistance to corporate-engineered food. Their participants have come to the issues of globalization by way of many different itineraries. For example:

- Acid rain and global warming do not respect national borders. They have forced environmentalists around the world to recognize global ecological interdependence.[3] At the same time, environmentalists became increasingly conscious that the actions of global corporations and of institutions such as the World Bank destroyed local environments—symbolized by the destruction of the Amazon rain forest and India's Narmada Valley. While some argued that globalizing capitalism would actually promote environmentalism in the third world, environmentalists discovered that it was instead creating an environmental race to the bottom as countries lowered environmental standards to attract corporations. The WTO's anti-environmental rules—symbolized by its decision condemning a U.S. law for the protection of sea turtles—brought the environmental movement into direct confrontation with this central institution of globalization.

- In the 1970s, the world's poorer countries formed the G–77 and initiated a North–South Dialogue with the rich countries to formulate a New International Economic Order. When the rich countries withdrew from this effort in the 1980s and began instead to promote neoliberal policies coordinated through the IMF, World Bank, and WTO, most third world governments went along with their plans, albeit in many cases reluctantly. But networks of third world NGOs continued to develop an alternative agenda and to press it both on their own countries and on international institutions. Third world governments have recently begun to follow their lead. As the rich countries prepared their agenda for the 1999 Seattle WTO extravaganza, poor-country governments began to question whether they had benefitted from globalization. Encouraged by the global citizens' movement to halt any new round of WTO negotiations, third world delegations for the first time refused to go along with the rich countries' proposals until their own concerns were addressed, helping to bring the meeting down in shambles. Early in 2000, the G–77 held its first ever head-of-state-level meeting and proposed an alternative program that included debt relief, increased aid, access to technology, and a shift in economic decision making from the World Bank and IMF to the UN.[4]

- People in rich countries have a long history of compassionate assistance for poor countries—sometimes in alliance with religious proselytizing and colonialism. With the development of the third world debt crisis in the 1980s, however, many people of conscience in the first world became deeply concerned about the effect of crushing debts on third world people and began to demand cancellation of their debt. Many then went on to address the broader question of the devastating "structural adjustment" policies being imposed on the debtor countries by the IMF, World Bank, and rich countries.

- When negotiations started in 1986 for what became the WTO, critics argued that U.S. and other first world proposals would benefit agribusiness and transnational commodity traders, but would drive millions of small farmers in both the North and South off their farms. Advocates for small farmers around the world began holding regular counter-meetings at the negotiations and developed a global network to oppose the proposals. They provided much of the core for international opposition to the emerging WTO. What has been described as "the first really global demonstration," in December 1990, brought farmers from Europe, Japan, North America, Korea, Africa, and Latin America to Brussels—helping force the negotiations into deadlock.[5] Since then, small farmers have been at the forefront of opposition to WTO agricultural policies, efforts to turn seeds into private property, and genetically engineered organisms (GEOs).

- From World War II until the 1960s, the labor movement in the United States was a strong supporter of economic liberalization, both as an expression of its alliance with U.S. international policy and as a means to secure expanding markets for U.S.–made products. Faced with a massive loss of jobs in auto, steel, garment, and other industries in the 1970s, the labor movement increasingly campaigned for tariffs and other barriers to imports designed to "save American jobs." Over the 1990s, globalization made such economic nationalist strategies less and less credible. Organized labor increasingly moved toward demanding reform of the global economy as a whole, symbolized by demands for labor rights and environmental standards in international trade agreements to protect all the world's workers and communities from the race to the bottom. Its participation in the Seattle WTO protests represented a new page in U.S. labor history and was followed by the announcement of a long-term "Campaign for Global Fairness."

- The burgeoning identity-based movements of the late 20th century found that many identities did not respect national borders. The women's movement slogan "sisterhood is powerful" evolved into a consciousness that "sisterhood is global." A growing awareness of the global oppression of women led to a struggle to define women's rights as internationally protected human rights. Events surrounding the UN's 1995 Beijing women's conference brought large numbers of women in the United States to an awareness of the impact of IMF and World Bank–imposed structural adjustment austerity programs on women in poor countries, and their similarity to the implications of welfare reform for poor women in the United States. The fact that the great majority of those exploited in overseas factories were young women led to a growing concern about the global sweatshop.

- From the 1960s on, consumer movements in many countries had enshrined a wide range of protections in national laws and had developed effective legal techniques for imposing a degree of accountability on corporations. Consumer organizations—notably Ralph Nader's Public Citizen—discovered that trade agreements like NAFTA and the WTO were overriding high national standards for such things as food and product safety. They also realized that both neoliberal ideology and competition among countries for investment were tending to lower consumer protection standards all over the world. New consumer issues, such as the right of governments to regulate genetically engineered food, have steadily increased consumer concern over globalization.

- African American communities in the U.S. have been concerned with conditions in Africa from the mid–19th century to the struggle against South African apartheid. But the 1990s saw two specific concerns that brought attention to the global economy. The first was the devastation wreaked on African countries by international debt and the brutal structural adjustment conditionalities the IMF and World Bank imposed on African countries in exchange for helping them roll over their debts. The other was the struggle over the African trade bill (known to its critics as the "NAFTA for Africa" bill) that ostensibly opened U.S. markets to African exports but in fact imposed more stringent structural adjustment-type conditions while doing little to provide desperately needed debt relief. Many African American leaders, including a wide swath of black clergy, became involved in the Jubilee 2000 debt relief campaign and the fight against the "NAFTA for Africa" bill and for an alternative proposed by Rep. Jesse Jackson, Jr.

- Groups in Europe, Japan, and the U.S. that had been involved in support for development and popular movements in third world countries found those countries increasingly used as production platforms by global corporations. They began calling attention to the growth of sweatshops and pressuring companies like the Gap and Nike to establish acceptable labor and human rights conditions in their factories around the world. Their efforts gradually grew into an anti-sweatshop movement with strong labor and religious support and tens of thousands of active participants. In the U.S., college students took up the anti-sweatshop cause on hundreds of campuses, ultimately holding sit-ins on many campuses to force their colleges to ban the use of college logos on products not produced under acceptable labor conditions.

Many other people are following their own itineraries toward globalization from below. Some, such as activists in the human rights movement seeking to protect rights of people globally, or public health advocates trying to control tobacco companies and provide AIDS treatment for poor countries, are just as globalized as those described above. Some, such as activists in the immigrant networks spreading out around the world, are in some ways even more global and are challenging globalization from above by their very way of life. Some, like the tens of millions who have participated in nationally organized mass and general strikes and upheavals, are resisting the effects of globalization from above, even if (so far) they are doing so in a national framework.[6] Far more numerous still are the billions of people who are being adversely affected by globalization from above, but who have not yet found their own way to respond. Ultimately, their itineraries may be the most important of all.

Confluence

From diverse origins and through varied itineraries, these movements now find themselves starting to converge. Many of their participants are recognizing their commonalties and beginning to envision themselves as constructing a common movement.

This convergence is occurring because globalization is creating common interests that transcend both national and interest-group boundaries. As author and activist Vandana Shiva wrote in the wake of the Battle of Seattle,

> When labour joins hands with environmentalists, when farmers from the North and farmers from the South make a common commitment to say "no" to genetically engineered crops, they are not acting as special interests. They are defending the common interests and common rights of all people, everywhere. The divide and rule policy, which has attempted to pit consumers against farmers, the North against the South, labour against environmentalist has failed.[7]

Much of the convergence is negative: different groups find themselves facing the same global corporations, international institutions, and market-driven race to the bottom. But there is also a growing positive convergence around common values of democracy, environmental protection, community, economic justice, equality, and human solidarity.

Participants in this convergence have varied goals, but its unifying goal is to bring about sufficient democratic control over states, markets, and corporations to permit people and the planet to survive and begin to shape a viable future. This is a necessary condition for participants' diverse other goals.

Is this confluence a movement, or is it just a collection of separate movements? Perhaps it can most aptly be described as a movement in the early stages of construction. Within each of its components there are some people who see themselves as part of a global, multi-issue movement and others who do not. Those who do are often networked with their counterparts in other movements and other countries. Their numbers are increasing rapidly and they are playing a growing role within their movements and organizations. They are developing a shared vision. And they see themselves as constructing a common movement. It is this emerging movement that we refer to as globalization from below.

Globalization from below is certainly a movement with contradictions. Its participants have many conflicting interests. It includes many groups that previously defined themselves in part via negative reference to each other. It includes both rigidly institutionalized and wildly unstructured elements.

Globalization from below is developing in ways that help it cope with this diversity. It has embraced diversity as one of its central values, and asserts that cooperation need not presuppose uniformity. Its structure tends to be a network of networks, facilitating cooperation without demanding organizational centralization.

Older orientations toward charitable "us helping them" on the one hand, and narrow self-interest on the other, are still present; but there is also a new recognition of common interests in the face of globalization. Solidarity based on mutuality and common interest increasingly forms the basis for the relationships among different parts of the movement.

The movement is generally multi-issue, and even when participants focus on particular issues, they reflect a broader perspective. As Howard Zinn wrote of the Seattle WTO protests,

> In one crucial way it was a turning point in the history of movements
> of the recent decades—a departure from the single-issue focus of the
> Seabrook occupation of 1977, the nuclear-freeze gathering in Central
> Park in 1982, the great Washington events of the Million-Man March,
> [and] the Stand for Children [march].[8]

Globalization from below has now established itself as a global opposition, representing the interests of people and the environment worldwide. It has demonstrated that, even when governments around the world are dominated by corporate interests, the world's people can act to pursue their common interests.

Globalization from below grew both out of previous movements and out of their breakdown. There is much to be learned from the historical heritage of centuries of struggle to restrain or replace capitalism, and today's activists often draw on past values and practices in shaping their own. But it would be a mistake to simply treat this new movement as an extension of those that went before—or to attach it to their remnants.[9]

Globalization in all its facets presents new problems that the old movements failed to address. That is part of why they declined so radically. It also presents new opportunities that will be lost if the new wine is simply poured back into the old bottles. Besides, the historic break provides an invaluable opportunity to escape the dead hand of the past and to reground the movement to restrain global capital in the actual needs and conditions of people today.[10]

Globalization from below is now a permanent feature of the globalization epoch. Even if its current expressions were to fail, the movement would rise again, because it is rooted in a deep social reality: the need to control the forces of global capital.

THE POWER OF SOCIAL MOVEMENTS
(AND ITS SECRET)

The supporters of globalization from above control most of the world's governments. They control the global corporations and most of the world's wealth. They have a grip on the minds of people all over the world. It seems inconceivable that they can be effectively challenged.

Yet social movements have overcome equal or even greater concentrations of wealth and power in the past. Colonized peoples from North America to India, and Africa to Vietnam, have thrown out imperial powers with many times their wealth and firepower. The abolitionist movement eliminated slavery in most of the world and the civil rights movement eliminated legal segregation in the United States. In recent decades, mass movements have brought down powerful dictatorships from Poland to the Philippines. A coordinated domestic and global movement abolished South African apartheid. To understand how social movements are able to overcome what seem to be overwhelming forces, we need to take a deeper look at the processes underlying such successes.

How Social Movements Arise

Normally, most people follow life strategies based on adapting to the power relations of their world, not on trying to change them. They do so for a varying mix of reasons, including:

- Belief that existing relations are good and right.
- Belief that changing them is impossible.
- Fear that changing them would lead to something worse.
- An ability to meet their own needs and aspirations within existing power relations.
- Belief that existing power relations can and will change for the better.
- Identification with the dominant groups or with a larger whole—for example, a religion or nation.
- Fear of sanctions for violation of social rules or the will of the powerful.[11]

Most institutions and societies have elaborate systems for assuring sufficient consent or acquiescence to allow their key institutions to function. These means of maintaining a preponderance of power—often referred to as "hegemony"—range from education to media, and from elections to violent repression.[12]

Over time, problems with existing social relationships may accumulate, initiating a process of change. These problems usually affect particular social groups—for example, particular communities, nations, classes, racial, ethnic and gender groups, religious and political groupings, and the like. The process may start with some people internally questioning or rejecting some aspects of the status quo. It becomes a social process as people discover that others are having similar experiences, identifying the same problems, asking the same questions, and being tempted to make the same rejections. Then people begin to identify with those others and to interact with them. This turns what might have been an individual and isolating process into a social one.[13]

Seeing that other people share similar experiences, perceptions, and feelings opens a new set of possibilities. Perhaps collectively we can act in ways that have impacts isolated individuals could never dream of having alone. And if we feel this way, perhaps others do, too.

This group formation process constructs new solidarities.[14] Once a consciousness of the need for solidarity develops, it becomes impossible to say whether participants' motives are altruistic or selfish, because the interest of the individual and the collective interest are no longer in conflict; they are perceived as one.

This process occurs not only in individuals, but also in groups, organizations, and constituencies. Thus form social movements.[15]

Why Social Movements Can Be Powerful

The fact that people develop common aspirations doesn't mean that they can realize them. Why are social movements able to change society? The power of existing social relations is based on the active cooperation of some people and the consent and/or acquiescence of others. It is the activity of people—going to work, paying taxes, buying products, obeying government officials, staying off private property—that continually re-creates the power of the powerful.

Bertolt Brecht dramatized this truth in his poem "German War Primer":

General, your tank is a strong vehicle.

It breaks down a forest and crushes a hundred people.

But it has one fault: it needs a driver.[16]

This dependence gives people a potential power over society—but one that can be realized only if they are prepared to reverse their acquiescence.[17] The old American labor song "Solidarity Forever" captures the tie between the rejection of acquiescence and the development of collective power:

They have taken untold millions

That they never toiled to earn

But without our brain and muscle

Not a single wheel can turn.

We can break their haughty power,

Gain our freedom when we learn

That the union makes us strong.[18]

Social movements can be understood as the collective withdrawal of consent to established institutions.[19] The movement against globalization from above can be understood as the withdrawal of consent from such globalization.

Ideally, democracy provides institutionalized means for all to participate equally in shaping social outcomes. But in the rather common situation in which most people have little effective power over established institutions, even those that claim to be democratic, people can still exercise power through the withdrawal of consent. Indeed, it is a central means through which democratization can be imposed.

Withdrawal of consent can take many forms, such as strikes, boycotts, and civil disobedience. Gene Sharp's *The Methods of Nonviolent Action* lists no fewer than 198 such methods, and no doubt a few have been invented since it was written.[20] Specific social relations create particular forms of consent and its withdrawal. For example, WTO trade rules prohibit city and state selective purchasing laws like the Massachusetts ban on purchases from companies that invest in Burma—making such laws a form of withdrawal of consent from the WTO, in effect an act of governmental civil disobedience.[21] (Several foreign governments threatened to bring charges against the Massachusetts Burma law in the WTO before it was declared unconstitutional by the U.S. Supreme Court in June 2000.)

The World Bank depends on raising funds in the bond market, so critics of the World Bank have organized a campaign against purchase of World Bank bonds, modeled on the successful campaign against investment in apartheid South Africa. Concerted refusal of impoverished debtor countries to continue paying on their debts—for example, through so-called debtors' cartels—would constitute a powerful form of withdrawal of consent from today's global debt bondage.

Just the threat of withdrawal of consent can be an exercise of power. Ruling groups can be forced to make concessions if the alternative is the undermining of their ultimate power sources.[22] The movement for globalization from below has demonstrated that power repeatedly. For example, the World Bank ended funding for India's Narmada Dam when 900 organizations in 37 countries pledged a campaign to defund the Bank unless it canceled its support. And Monsanto found that global concern about genetically engineered organisms so threatened its interests that it agreed to accept the Cartagena Protocol to the Convention on Biological Diversity, allowing GEOs to be regulated.[23]

At any given time, there is a balance of power among social actors.[24] Except in extreme situations like slavery or military occupation, unequal power is reflected not in an unlimited power of one actor over the other. Rather, it is embedded in the set of rules and practices that are mutually accepted, even though they benefit one far more than the other. When the balance of power is changed, subordinate groups can force change in these rules and practices.

The power of the people is a secret that is repeatedly forgotten, to be rediscovered every time a new social movement arises. The ultimate source of power is not the command of those at the top, but the acquiescence of those at the bottom. This reality is hidden behind the machinations of politicians, business leaders, and politics as usual. The latent power of the people is forgotten both because those in power have every reason to suppress its knowledge and because it seems to conflict with everyday experience in normal times. But when the people rediscover it, power structures tremble.

Linking the Nooks and Crannies

New movements often first appear in small, scattered pockets among those who are unprotected, discriminated against, or less subject to the mechanisms of hegemony. They reflect the specific experiences and traditions of the social groups among which they arise. In periods of rapid social change, such movements are likely to develop in many such milieus and to appear very different from each other as a result. In the case of globalization from below, for example, we have seen significant mobilizations by French chefs concerned about preservation of local food traditions, Indian farmers concerned about corporate control of seeds, and American university students concerned about school clothing made in foreign sweatshops. Even if in theory people ultimately have power through withdrawal of consent, how can such disparate groups ever form a force that can exercise that power?

One common model for social change is the formation of a political party that aims to take over the state, whether by reform or by revolution. This model has always been problematic, since it implied the perpetuation of centralized social control, albeit control exercised in the interest of a different group.[25] However, it faces further difficulties in the era of globalization.

Reform and revolution depend on solving problems by means of state power, however acquired. But globalization has outflanked governments at local and national levels, leaving them largely at the mercy of global markets, corporations, and institutions. Dozens of parties in every part of the world have come to power with pledges to overcome the negative effects of globalization, only to submit in a matter of months to the doctrines of neoliberalism and the "discipline of the market." Nor is there a global state to be taken over.[26]

Fortunately, taking state power is far from the only or even the most important means of large-scale social change. An alternative pathway is examined by historical sociologist Michael Mann in *The Sources of Social Power*.[27] The characteristic way that new solutions to social problems emerge, Mann maintains, is neither through revolution nor reform. Rather, new solutions develop in what he calls "interstitial locations"—nooks and crannies in and around the dominant institutions. Those who were initially marginal then link together in ways that allow them to outflank those institutions and force a reorganization of the status quo.

At certain points, people see existing power institutions as blocking goals that could be attained by cooperation that transcends existing institutions. So people develop new networks that outrun them. Such movements create subversive "invisible connections" across state boundaries and the established channels between them.[28] These interstitial networks translate human goals into organizational means.

If such networks link groups with disparate traditions and experiences, they require the construction of what are variously referred to as shared worldviews, paradigms, visions, frames, or ideologies. Such belief systems unite seemingly disparate human beings by claiming that they have meaningful common properties:

> An ideology will emerge as a powerful, autonomous movement when it can put together in a single explanation and organization a number of aspects of existence that have hitherto been marginal, interstitial to the dominant institutions of power.[29]

The emerging belief system becomes a guide for efforts to transform the world. It defines common values and norms, providing the basis for a common program.[30] When a network draws together people and practices from many formerly marginal social spaces and makes it possible for them to act together, it establishes an independent source of power. Ultimately, new power networks may become strong enough to reorganize the dominant institutional configuration.

The rise of labor and socialist movements in the 19th century and of feminist and environmental movements in the 20th century in many ways fit this model of emergence at the margins, linking, and outflanking.[31] So, ironically, does the emergence of globalization from above....

Self-organization in marginal locations and changing the rules of dominant institutions are intimately linked. The rising European bourgeoisie both created their own market institutions and fought to restructure the political system in ways that would allow markets to develop more freely. Labor movements both organized unions and forced governments to protect labor rights, which in turn made it easier to organize unions.

Over time, movements are likely to receive at least partial support from two other sources. Some institutions, often ones that represent similar constituencies and that themselves originated in earlier social movements but have become rigidified, develop a role of at least ambiguous support. And sectors of the dominant elites support reforms and encourage social movements for a variety of reasons, including the need to gain support for system-reforming initiatives and a desire to win popular backing in ultra-elite conflicts.

Social movements may lack the obvious paraphernalia of power: armies, wealth, palaces, temples, and bureaucracies. But by linking from the nooks and crannies, developing a common vision and program, and withdrawing their consent from existing institutions, they can impose norms on states, classes, armies, and other power actors.

The Lilliput Strategy

How do these broad principles of social movement–based change apply to globalization from below? In fact, they describe the very means by which it is being constructed. We call this the Lilliput Strategy, after the tiny Lilliputians in Jonathan Swift's fable *Gulliver's Travels* who captured Gulliver, many times their size, by tying him up with hundreds of threads.

In response to globalization from above, movements are emerging all over the world in social locations that are marginal to the dominant power centers.

These are linking up by means of networks that cut across national borders. They are beginning to develop a sense of solidarity, a common belief system, and a common program. They are utilizing these networks to impose new norms on corporations, governments, and international institutions.

The movement for globalization from below is, in fact, becoming an independent power. It was able, for example, to halt negotiations for the Multilateral Agreement on Investment (MAI), to block the proposed "Millennium Round" of the WTO, and to force the adoption of a treaty on genetically engineered products. Its basic strategy is to say to power holders, "Unless you accede to operating within these norms, you will face threats (from us and from others) that will block your objectives and undermine your power."

The threat to established institutions may be specific and targeted withdrawals of support. For example, student anti-sweatshop protestors have made clear that their campuses will be subject to sit-ins and other forms of disruption until their universities agree to ban the use of school logos on products made in sweatshops. Or, to take a very different example, in the midst of the Battle of Seattle, President Bill Clinton, fearing loss of electoral support from the labor movement, endorsed the use of sanctions to enforce international labor rights.[32] The threat may, alternatively, be a more general social breakdown, often expressed as fear of "social unrest."[33]

The slogan "fix it or nix it," which the movement has often applied to the WTO, IMF, and World Bank, embodies such a threat. It implies that the movement (and the people of the world) will block the globalization process unless power holders conform to appropriate global norms. This process constitutes neither revolution nor conventional "within the system" and "by the rules" reform. Rather, it constitutes a shift in the balance of power.

As the movement grows in power, it can force the modification of institutions or the creation of new ones that embody and/or impose these norms as enforceable rules.[34] For example, the treaties on climate change and on genetic engineering force new practices on corporations, governments, and international institutions that implement norms propounded by the environmental and consumer movements. Student anti-sweatshop activists force their universities to join an organization that bans university logos on products made under conditions that violate specified rules regarding labor conditions. The world criminal court, endorsed by many countries under pressure of the global human rights movement, but resisted by the United States, would enforce norms articulated at the Nuremberg war crimes tribunal.

These new rules in turn create growing space for people to address problems that the previous power configuration made insoluble. Global protection of human rights makes it easier for people to organize locally to address social and environmental problems. Global restrictions on fossil fuels that cause global warming, such as a carbon tax, would make it easier for people to develop renewable energy sources locally.

While the media have focused on global extravaganzas like the Battle of Seattle, these are only the tip of the globalization-from-below iceberg. The

Lilliput Strategy primarily involves the building of solidarity among people at the grassroots. For example:

- Under heavy pressure from the World Bank, the Bolivian government sold off the public water system of its third largest city, Cochabamba, to a subsidiary of the San Francisco–based Bechtel Corporation, which promptly doubled the price of water for people's homes. Early in 2000, the people of Cochabamba rebelled, shutting down the city with general strikes and blockades. The government declared a state of siege and a young protester was shot and killed. Word spread all over the world from the remote Bolivian highlands via the Internet. Hundreds of e-mail messages poured into Bechtel from all over the world demanding that it leave Cochabamba. In the midst of local and global protests, the Bolivian government, which had said that Bechtel must not leave, suddenly reversed itself and signed an accord accepting every demand of the protestors. Meanwhile, a local protest leader was smuggled out of hiding to Washington, D.C., where he addressed the April 16 rally against the IMF and World Bank.[35]

- When the Japanese-owned Bridgestone/Firestone (B/F) demanded 12-hour shifts and a 30 percent wage cut for new workers in its American factories, workers struck. B/F fired them all and replaced them with 2,300 strike-breakers. American workers appealed to Bridgestone/Firestone workers around the world for help. Unions around the world organized "Days of Outrage" protests against B/F. In Argentina, a two-hour "general assembly" of all workers at the gates of the B/F plant halted production while 2,000 workers heard American B/F workers describe the company's conduct. In Brazil, Bridgestone workers staged one-hour work stoppages, then "worked like turtles"—the Brazilian phrase for a slowdown. Unions in Belgium, France, Italy, and Spain met with local Bridgestone managements to demand a settlement. U.S. B/F workers went to Japan and met with Japanese unions, many of whom called for the immediate reinstatement of U.S. workers. Five hundred Japanese unionists marched through the streets of Tokyo, supporting B/F workers from the U.S. In the wake of the worldwide campaign, Bridgestone/Firestone unexpectedly agreed to rehire its locked out American workers.[36]

- In April 2000, AIDS activists, unions, and religious groups were poised to begin a lawsuit and picketing campaign denouncing the Pfizer Corporation as an AIDS profiteer for the high price it charges for AIDS drugs in Africa. Pfizer suddenly announced that it would supply the drug fluconazole, used to control AIDS side effects, for free to any South African with AIDS who could not afford it. A few weeks later, U.S., British, Swiss, and German drug companies announced that they would cut prices on the principal AIDS drugs, anti-retrovirals, by 85 to 90 percent. Meanwhile, when South Africa tried to pass a law allowing it to ignore drug patents in health emergencies, the Clinton administration lobbied hard against it and put South Africa on a watch list that is the first step toward trade sanctions. But then, according to the *New York Times*, the Philadelphia branch of Act Up, the gay advocacy group, decided

to take up South Africa's cause and start heckling Vice President Al Gore, who was in the midst of his primary campaign for the presidency. The banners saying that Mr. Gore was letting Africans die to please American pharmaceutical companies left his campaign chagrined. After media and campaign staff looked into the matter, the administration did an about face and accepted African governments' circumvention of AIDS drug patents.[37]

- Two independent unions, the United Electrical Workers Union (UE) in the United States and the Frente Auténtico del Trabajo (FAT) in Mexico, formed an ongoing Strategic Organizing Alliance in the mid-1990s. At General Electric (GE) in Juarez, FAT obtained the first secret ballot election in Mexican labor history, aided by pressure on GE in the United States. The trinational Echlin Workers Alliance was formed to jointly organize Echlin, a large multinational auto parts corporation, in Canada, Mexico, and the U.S. In cooperation with Mexican unions, U.S. unions brought charges under the NAFTA side agreements for the repression of Echlin workers. A rank-and-file activist from FAT traveled to Milwaukee, Wisconsin, to help UE organize foundry workers of Mexican origin. Workers from each country have repeatedly conducted speaking tours organized by those across the border. U.S. workers helped fund and build a Workers' Center in Juarez. And the Cross-Border Mural Project has developed binational teams that have painted murals celebrating international labor solidarity on both sides of the border.[38]

How Movements Go Wrong

It is nowhere guaranteed that any particular social movement will succeed in using its potential power to realize the hopes and aspirations of its participants or to solve the problems that moved them to action in the first place. There are plenty of pitfalls along the way.

Schism: From Catholic and Protestant Christians to Sunni and Shiite Muslims, from Communists and socialists to separatists and integrationists, social movements are notorious for their tendency to split. They can often turn into warring factions whose antagonisms are focused primarily on each other. Splits often occur over concrete issues but then perpetuate themselves even when the original issues are no longer salient.

Repression: Movements can be eliminated, or at least driven underground, by legal and extralegal repression.

Fading out: The concerns that originally drew people into a movement may recede due to changed conditions. An economic upswing or the opening of new lands has often quieted farmer movements. Or constant frustration may simply lead to discouragement and withdrawal.

Leadership domination: In a mild form, the movement evolves into an institution in which initiative and control pass to a bureaucratized leadership and staff, while the members dutifully pay their dues and act only when told to do so by their leaders. In a more virulent form, leaders establish a tyrannical control over members.

Isolation: Movements may become so focused on their own internal life that they are increasingly irrelevant to the experience and concerns of those who are not already members. Such a movement may last a long time as a sect but be largely irrelevant to anyone except its own members.[39]

Cooptation: A movement may gain substantial benefits for its constituency, its members, or its leaders, but do so in such a way that it ceases to be an independent force and instead comes under the control of sections of the elite.

Leadership sell-out: Less subtly, leaders can simply be bought with money, perks, flattery, opportunities for career advancement, or other enticements.

Sectarian disruption: Movements often fall prey to sects that attempt either to capture or to destroy them. Such sects may emerge from within the movement itself or may invade it from without.

To succeed, globalization from below must avoid these pitfalls, promote movement formation in diverse social locations, establish effective linkages, develop a sense of solidarity, a common worldview, and a shared program, and utilize the power that lies hidden in the withdrawal of consent.

ENDNOTES

1. "Das Kapital Revisited," *Financial Times*, August 31, 1998, p. 14.

2. As far as we have been able to determine, the terms "globalization from above" and "globalization from below" were coined by Richard Falk and first appeared in print in Jeremy Brecher, John Brown Childs, and Jill Cutler eds., *Global Visions Beyond the New World Order* (Boston: South End Press, 1993). This book provides a view of the development of transnational social movements and common vision prior to 1993; it reflects an awareness of the increasing global interconnectedness in the post–Cold War era, but it puts limited emphasis on the global economic integration that was then gathering steam. For movements in the early 1990s responding specifically to economic globalization, see Chapter 5 of *Global Village or Global Pillage*. See also Richard Falk, *Predatory Globalization: A Critique* (Massachusetts Blackwell: Maiden, 1999), especially Chapter 8.

3. Jeremy Brecher, "The Opening Shot of the Second Ecological Revolution," *Chicago Tribune*, August 16, 1988.

4. "Poor Countries Draft Proposal On Poverty," *New York Times,* April 12, 2000. The G–77 currently has 133 member nations.

5. Mark Ritchie, quoted in *Global Village or Global Pillage,* p. 97.

6. According to labor journalist Kim Moody,

> In the last couple of years there have been at least two dozen political general strikes in Europe, Latin America, Asia, and North America. This phenomenon began in 1994. There have been more political mass strikes in the last two or three years than at any time in the 20th century.

   Kim Moody, "Workers in a Lean World," a speech to the Brecht Forum in New York, New York, November 14, 1997. Broadcast on Alternative Radio (tape and transcript available from http://www.Alternativeradio.org).

7. Vandana Shiva, "The Historic Significance of Seattle," December 10, 1999, MAI–NOT Listserv, Public Citizen Global Trade Watch.

8. Howard Zinn, "A Flash of the Possible," *The Progressive* 61:1 (January 2000). Available online at https://secure.Progressive.org/zinn00l.htm.

9. For a portrayal of current struggles as a continuation of historical working class struggles, see Boris Kargarlitsky's recent trilogy *Recasting Marxism, including New Realism, New Barbarism: Socialist Theory in the Era of Globalization* (London: Pluto Press, 1999), *The Twilight of Globalization: Property, State and Capitalism* (London: Pluto Press, 1999), and *The Return of Radicalism: Reshaping the Left Institutions* (London: Pluto Press, 2000).

10. It is often pointed out that globalization is creating a capitalism that in significant ways resembles the capitalism that preceded World War I. It could also be observed that globalization from below in some ways resembles the international socialist movement before World War I. Globalization provides an opportunity to reevaluate some of the key features of the post-1914 left, such as its relationship to nationalism and the nation state; the schisms between social democracy, Communism, and anarchism; and the development of organizational forms adapted to the effort to secure state power via reform or revolution.

11. For a fuller discussion of this subject, with extensive references, see Gene Sharp, *The Politics of Nonviolent Action: Part One: Power and Struggle* (Boston: Porter Sargent, 1973), "Why Do Men Obey?" pp. 16–24.

12. The analysis of "hegemony" is generally associated with the work of Antonio Gramsci. See for example Antonio Gramsci, *The Modern Prince and Other Writings* (New York: International Publishers, 1959).

13. E.P. Thompson describes this process of group formation for the specific case of class. "Class happens when some men, as a result of common experiences (inherited or shared), feel and articulate the identity of their interests as between themselves, and as against other men whose interests are different from (and usually opposed to) theirs." E.P. Thompson, *The Making of the English Working Class* (New York: Vintage, 1996), p. 9.

14. Solidarity can take a number of forms. Peter Waterman defines six meanings of international solidarity:

   ▪ Identity = solidarity of common interest and identity

   ▪ Substitution = standing in for those incapable of standing up for themselves

   ▪ Complementarity = exchange of different needed/desired goods/qualities

- Reciprocity＝exchange over time of identical goods/qualities
- Affinity＝shared cross-border values, feelings, ideas, identities
- Restitution＝acceptance of responsibility for historical wrong

He points out that each of these has certain problems and limitations. For example, identity-based solidarity tends to exclude those who don't share the common identity as defined; substitution can lead to an unequal, patronizing relationship of charity. See Peter Waterman, *Globalization, Social Movements, and the New Internationalisms* (London: Mansell, 1999). Preliminary text available online at http://www.antenna.nl/~waterman/dialogue.html. The process of constructing solidarity is illustrated with numerous labor history examples in Jeremy Brecher, *Strike! Revised and Updated Edition* (Cambridge: South End Press Classics, 1999), and analyzed on p. 284.

15. This highly schematic formulation is based primarily on the study and observation of social movements, combined with theories drawn from many sources, for example, Jean-Paul Sartre, *Critique de la Raison Dialectique [Critique of Dialectical Reason]* (Paris: Gallimard, 1960), and Francesco Alberoni, *Movement and Institution* (New York: Columbia UP, 1984).

16. Bertolt Brecht, *Deutsche Kriegsfibel ["German War Primer"], in Gesammilte Werke* (Berlin: Suhrkamp, 1967), vol. 4, p. 638. The translation by Martin Esslin originally appeared in Jeremy Brecher and Tim Costello, *Common Sense for Hard Times* (New York: Two Continents/Institute for Policy Studies, 1976), p. 240.

17. In the "acquiescent state," people's relation to each other is mediated via the market or common relations to authority. The process of movement creation and group formation to some degree replaces these with direct relations. Sartre analyzes this as the transition from the "series" to the "group" (*Critique de la Raison Dialectique*).

18. Written by Ralph Chaplin.

19. Gene Sharp, who analyzes hundreds of historical examples of nonviolent action in the three volumes of his *Politics of Nonviolent Action* (Boston: Porter Sargent, 1973), concludes that the base of nonviolent action is "the belief that the exercise of power depends on the consent of the ruled who, by withdrawing that consent, can control and even destroy the power of their opponent" (*Part One: Power and Struggle, p. 4*). Sharp emphasizes that nonviolent struggle requires indirect strategies that undermine the opponent's strength rather than annihilate the opponent. (Of course, the picture is made less simple by the fact that "the ruled" are not a homogeneous group, and those who withdraw consent may be defeated by those who do not). This analysis does not apply exclusively to nonviolence. Even in war, victory usually results not from physical annihilation of the enemy but from the withdrawal of support of the population from the war effort ("loss of morale"), defection of political supporters of the war, withdrawal of allies, and change in policy by ruling groups in response to the presence or threat of these factors.

20. Sharp, *Part Two: The Methods of Nonviolent Action* (Boston: Porter Sargent, 1973). See also *Part Three: The Dynamics of Nonviolent Action* (Boston: Porter Sargent, 1973).

21. In constitutional terms, this would be described as a form of nullification.

22. Of course, an irrational ruler may not be deterred from acting to repress a nonviolent movement by the fact that doing so may undermine his or her own power. But given an irrational ruler, violence is no more guaranteed to be an effective deterrent than nonviolence.

23. "United States negotiators gave in to a demand from Europe and most of the rest of the world for what is known as the 'precautionary principle'.... Even Greenpeace, an avowed critic of the technology, issued a statement calling the protocol a 'historic step towards protecting the environment and consumers from the dangers of genetic engineering.'" *St. Louis Post-Dispatch*, January 30, 2000.

    The British Environment Minister, Michael Meacher, said: "For the first time countries will have the right to decide whether they want to import GM products or not when there is less than full scientific evidence. It is official that the environment rules aren't subordinate to the trade rules. It's been one hell of a battle."

    "This protocol is a campaign victory in that it acknowledges that GMOs [genetically modified organisms] are not the same as other crops and products and they require that special measures be taken," said Miriam Mayer of the Malaysia-based Third World Network. The U.S. State Department declined to specify whether the biotechnology company Monsanto had been consulted over the past few days. A State Department source said: "We understand there is no major problem so far as the company is concerned." *The Observer*, January 30, 2000.

24. As Gramsci put it, "The fact of hegemony undoubtedly presupposes that the interests and strivings of the groups over which the hegemony will be exercised are taken account of, that a certain balance of compromises be formed, that, in other words, the leading group makes some sacrifices" (*Modern Prince, p. 154*).

25. This critique has long been elaborated in the anarchist and libertarian socialist traditions, and has more recently been developed by the New Left of the 1960s, the Green movement, and the Mexican Zapatistas.

26. ... This is not to argue that states are no longer of significance, or that political parties and contests for government power have not played an important role in the past and might not today or in the future. Rather, it is to deny that social movements can or should be reduced to such a strategy.

27. Michael Mann, *The Sources of Social Power: Volume 1* (Cambridge: Cambridge UP, 1986), Chapter 1, "Societies as Organized Power Networks."

28. Michael Mann, *Sources of Social Power*, p. 522.

29. Michael Mann, *Sources of Social Power*, p. 21.

30. We use the term *values* to refer to criteria for classifications of good and bad or of better and worse. We use the term *norms* for the application of such values to the behavior of particular classes of actors, thereby specifying how they should act.

31. Over time, the labor and socialist movements of course became increasingly focused on national governments and increasingly contained within national frameworks.

32. This threat, strongly resented by many third world governments, contributed to the deadlocking of the WTO negotiations.

33. For example, provoking such general social unrest was an articulated objective of many U.S. opponents of the Vietnam War after other means of halting it had failed and public opinion had swung against it without visible effect on policy.

34. For a similar perspective on how social movements make change through imposing norms, with recent examples and proposals for the future, see Richard Falk, "Humane Governance for the World: Reviving the Quest," in Jan Nederveen Pieterse, ed., *Global Futures*, pp. 23ff. See also *On Humane Governance: Toward a New Global Politics* (University Park, Penn: Pennsylvania State UP, 1995).

35. "Bolivian Water Plan Dropped After Protests Turn Into Melees," *New York Times*, April 11, 2000. For further information on the Cochabamba water struggle, prepared by Jim Schultz, a Cochabamba resident who played a major role in mobilizing global support for the struggle, visit http://www.Americas.org.

36. *ICEM Info 3* (1996) and *ICEM Info 4* (1996); see also *Labor Notes*, October 1994, July 1996, and December 1996.

37. Donald G. McNeil, Jr., "As Devastating Epidemics Increase, Nations Take On Drug Companies," *New York Times*, July 9, 2000, and *Toronto Star*, May 12, 2000.

38. For information on the FAT–UE alliance, visit the UE website at http://www.ranknfile-ue.org/international.html.

39. Two classic explorations of this dynamic are Robert Michels, *Political Parties: A Sociological Study of the Oligarchical Tendencies of Modern Democracy* (Glencoe: Free Press, 1949), and Sidney Webb and Beatrice Webb, *The History of Trade Unionism*, 2nd ed. (London: Longmans Green, 1902). This process is analyzed by Sartre in terms of the dissolution of a group back into a series. As Alberoni points out, some degree of such re-serialization is probably inevitable, but it can be limited by practices that provide for periodical reconstitution of the group. Those hostile to social movements sometimes maintain that tyranny is their normal or only possible outcome. A classic example is Norman Cohn, *The Pursuit of the Millennium* (London: Seeker and Warburg, 1957). For a discussion of these issues in the context of various left traditions, see Staughton Lynd, "The Webbs, Lenin, Rosa Luxemburg," in *Living Inside Our Hope: A Steadfast Radical's Thoughts on Rebuilding the Movement* (Ithaca: Cornell UP, 1997), pp. 206ff.

34

Globalization and Feminism: Opportunities and Obstacles for Activism in the Global Arena

MYRA MARX FERREE

This article explains how globalization fuels an expansive range of feminist political activities in both women's organizations and other transnational movements for change.

Globalization is the word of the decade. In newspapers as well as scientific journals, globalization is invoked in relation to everything from moviemaking to unemployment. Much of this discussion implies that globalization is a wholly new phenomenon, that this is only a top-down phenomenon that is happening to people rather than also a grassroots process in which individuals and groups are actively engaged, and that there is nothing particularly gendered about it. This book arises from our conviction that none of these three assumptions are true.

A variety of sociological statistics at the macro level suggests the extent of global integration of the early twenty-first century is more like that of the 1910s than of the 1950s. For example, in 1910 levels of global trade measured by imports and exports and of human interconnection in the form of immigration and transnational organizations were at levels very similar to those we experience today. Two violent world wars and a long cold war reduced these international ties to their low point in the 1950s and 1960s. It may be more accurate to see the end of the cold war as allowing the tide to turn back toward greater global interaction in 2010–2010.

To be sure, many linkages between states and across national boundaries have been created only relatively recently. The European Union is one of the most spectacular of these current experiments in reshaping the meaning of sovereignty, but the African Union is also an important regional form of integration. These pacts follow in the footsteps of other, older links such as the World Council of Churches and the United Nations that continue to be important. Such

SOURCE: From Myra Marx Ferree, "Globalization and Feminism: Opportunities and Obstacles for Activism in the Global Arena," *Global Feminism: Transnational Women's Activism, Organizing, and Human Rights*, Myra Marx Ferree and Aili Mari Tripp, eds. (2006) New York: New York University Press, 2006. Used with permission.

continuing global associations should be understood in the content of other, now-obsolete efforts to integrate political and economic life across national borders, be it the Warsaw Pact or the British colonial system. The commitments, perspectives, and processes that connect the globe today are different in interesting ways from what has gone before, but they are not unprecedented in their scope or consequences, including their facilitation of feminist organization. Comparing 2005 to 1955 and 1905 suggests that feminist mobilization has always been increased by greater globalization.

One way in which global integration today does differ from that of the past is the extent to which it involves ordinary citizens and social movements, not merely governments and elites. Despite the typical assumption that globalization is a massive force bearing down on helpless populations, to look at the actual process is to see a great variety of social actors—including many who are not educational or political elites—engaging in diverse types of integrative work. Social movements of many kinds are finding a voice alongside more privileged actors such as states and corporations. Certainly there are structures and processes at work here that are far larger than any one individual, group, or even state can control, but this has always been characteristic of the world since the age of global navigation and the emergence of industrialization. What is more striking in the present moment is the intersection of the global with the local, and the expansion of popular, decentralized, and democratic forms of interpreting and responding to the top-down challenges posed by a world economy.

Moreover, rather than hierarchical colonial world system or the dueling blocs of the cold war, the reconfiguration of the world order is arising today from multiple locations and pulling in diverse directions. Because "the West" is no longer held together by its anticommunist mobilization, Europe and the United States are discovering new tensions and differences in their relationship. The "third world" is no longer merely defined by its history of colonization but by its own diversity, regionally, economically, and politically. Democratic India and authoritarian Pakistan, prosperous Singapore and economically ravaged Zimbabwe all came into the twentieth century as part of the British Empire, but they enter the twenty-first century with very different concerns. World bodies such as the UN are faced with new conflicts that include citizens challenging their national governments for democratic participation, ethnic conflicts within states, and gender conflicts fed by religious fundamentalisms, as well as the more familiar tensions among national and class interests. Globalization is today as much about the multiplicity of centers of power as it is about increases in their interrelationship.

From these diverse local centers, a variety of nongovernmental groups are engaging in the complex process of political renegotiation that hides under the label of globalization. Social movements like Attac, an international mobilization for democratic control of financial markets and their institutions that was founded in France in 1998, raise questions about the justice of international debt management and call for a "Tobin tax" (a fee placed on economic transactions to help defray the costs of development). Such groups are listened to by governments from Iceland to South Africa, although they are less well known in the United States. The World Social Forum connects social justice activists globally, allowing for a sharing of tactics and resource. Democracy movements in Ukraine, China, and Syria have used both mass media and Internet connectivity to draw popular support from abroad in struggles with their own governments. Globalization is also a form of political mobilization, and this grassroots involvement is also growing in scope and significance in many parts of the world.

Among the social actors most mobilized in the context of global opportunity structures are women's movements worldwide. We emphasize that women's global mobilization is neither something wholly new and unprecedented nor unconnected to the variety of local and regional conflicts that are part of the process of reshaping the world system. Gender is very much a part of the structure of the social order globally. Gender is therefore also part of what is being remade in the current reconfiguration of power relations. As with other aspects of this global reorganization, this restructuring involves women and men in a variety of local and transnational settings. Some of these women's movements are feminist, but others are not.

This chapter has three specific goals and sections. First, I offer a conceptual definition of both feminism and women's movements, and an argument about why it is important to distinguish between them. Second, I discuss the transnational opportunity structure that affects how even local feminists act, and I raise some questions about what its most promising and most dangerous features may be....

FEMINISM AND WOMEN'S MOVEMENTS: A DIFFERENCE THAT MAKES A DIFFERENCE

Although some scholars use the terms "feminism" and "women's movement" interchangeably, this usage creates certain problems. In some contexts, it makes it seem doubtful that men can be feminists, since how can they be members of a "woman's movement"? In other contexts it can seem problematic to apply the label "feminist" to activist women, whether because they refuse to use this term for themselves or because the women's movement in which they are engaged has other goals, even ones in opposition to any change toward greater gender equality. When women mobilize, as they do, to pursue a wide variety of interests, are all such "women's movements" automatically to be considered feminist?

To make clearer just what kinds of activism are feminist, it is helpful to separate this concept from that of a women's movement. Organizing women explicitly as women to make social change is what makes a "women's movement." It is defined as such because of the *constituency* being organized, not the specific targets of the activists' change efforts at any particular time. The movement, as an organizational strategy, addresses its constituents as women, mothers, sisters, daughters. By using the language of gender, it constructs women as a distinctive interest group, even when it may define the interests that this group shares as diverse and not necessarily centered on gender. Naming "women" as a constituency to be mobilized and building a strategy, organization, and politics around issues defined as being particularly "women's" concerns are the two factors that make a women's movement not a statistical head count of the gender of the membership, though typically women are the activists in such movements. This definition of "women's movement" explicitly recognizes that many mobilizations of women as women start out with a non–gender-directed goal, such as peace, antiracism, or social justice, and only later develop an interest in changing gender relations.

Activism for the purpose of challenging and changing women's subordination to men is what defines "feminism." Feminism is a *goal*, a target for social change, a purpose informing activism, not a constituency or a strategy. Feminist mobilizations are informed by feminist theory, beliefs, and practices, but they may take place in a variety of organizational contexts, from women's movements to positions within governments. Feminism as a goal often informs all or part of the agenda of mixed-gender organizations such as socialist, pacifist, and democratization movements.

Because feminism challenges all of gender relations, it also addresses those norms and processes of gender construction and oppression that differentially advantage some women and men relative to others, such as devaluing "sissy" men or the women who do care work for others. There is no claim being made that one or another particular aspect of gender relations, be it paid work or sexuality, motherhood or militarization, is the best, most "radical," or most authentic feminism. Feminism as a goal can be adopted by individuals of any gender, as well as by groups with any degree of institutionalization, from informal, face-to-face, temporary associations to a legally constituted national or transnational governing body.

Feminist activists and activism typically are embedded in organizations and institutions with multiple goals. To have a feminist goal is in no way inconsistent with having other political and social goals as well. The question of where feminism stands on the list of priorities of any individual or group is an empirical one. It is not true by definition that a person or group that calls itself feminist necessarily puts this particular goal in first place, since in practice it could be discovered to be displaced by other values (such as achieving or redistributing power or wealth, defending racial privilege, or fighting racial discrimination). Nor is it true by definition that a person or group that does not call itself feminist does not have feminist goals, since the identity can carry other connotations in a local setting (whether of radicalism or exclusivity or cultural difference) that an activist may seek to avoid by choosing another label.

These two definitions together generate a dynamic picture of both feminism and women's movements. On the one hand, women's movements are mobilizations understood to be in a process of flux in which feminism may be becoming more or less of a priority issue for them. Regardless of their goals, mobilizations that use gender to mobilize women are likely to bring their constituents into more explicitly political activities, empower women to challenge limitations on their roles and lives, and create networks among women that enhance their ability to recognize existing gender relations as oppressive and in need of change. Thus the question of when and how women's movements contribute to increases in feminism is a meaningful one.

On the other hand, feminism circulates within and among movements, takes more or less priority among their goals, and may generate new social movements, including women's movements. Successful feminist mobilization creates more places and spaces for feminism to accomplish its aims, within movements and within institutional power structures. Thus, for example, feminism can percolate into organized medicine, where activists may then construct women's movement associations of doctors, nurses, or patients, develop new tools to recognize and treat illnesses that affect women and men differently, and make institutions deliver services more appropriately to women in their communities. Feminist mobilizations often intersect with other forms of transformative struggles. Activists originally inspired by feminism may expand their goals to challenge racism, colonialism, and other oppressions, and activists with other primary agendas may be persuaded to adopt feminism as one of their objectives, especially as feminist activists show them how mutually supportive all these goals may be. Thus it is also a meaningful question to ask how feminism contributes to creating and expanding social movements, including women's movements.

As a consequence of both these processes, feminism and women's movements dynamically affect each other. In this set of changing relations, to restrict analysis to only those temporary phases in which women's movements have chosen to focus exclusively on challenging gender subordination or seeking equality with men of their own group marginalizes the ongoing intersectional elements of both. Distinguishing between feminism and women's movements, and then relating them empirically, moves the multiplicity of constituencies and dynamic changes in goals among activists "from margin to center" among the questions for analysis. When and why do women's movements embrace feminist goals—and when not? When and why do feminists choose to work in women's movements rather than in mixed-gendered ones or policymaking institutions—and when not? When and why do democratization, peace or economic justice movements make feminism part of their agenda—and when not? These are important questions that can only be asked, let alone answered, if there is a clear definitional distinction between feminism and women's movements. The scope of feminist theory and its overall social critique is also obscured if the difference between feminism and women's movements is not made explicit. For some feminists, feminism means simultaneously combating other forms of political and social subordination, since for many women, embracing the goal of equality with the men of their class, race, or nation would mean accepting a still-oppressed status. For some feminists, feminism means recognizing ways in which male-dominated institutions have promoted values fundamentally destructive for all people, such as militarism, environmental exploitation, or competitive global capitalism, and associating the alternative values and social relations with women and women-led groups. To define feminism in a way that limits its applicability only to those mobilizations that *exclusively* focus on challenging women's subordination to men would exclude both these types of feminism.

When analysts do this, they discover that the groups that are left to study are typically mobilizations of relatively privileged women who are seeking access to the opportunities provided by social, political, and economic institutions to men of their nationality, class, race, ethnicity, and religion.[1] The middle-class, white, Western bias observed in studies of "feminism" is at least in part a result of such an inappropriately narrow and static definition of the object of study ("feminism"). Defining feminism should not be confounded with other criteria such as the preferred constituency addressed (women or both genders), the organizational form preferred (social movement, community group, state or transnational authority), the strategy pursued (working inside or outside institutions, more or less collectively, with transgressive or demonstrative protest activities or not), or the priority feminism takes in relation to other goals (antiracism, environmentalism, pacifism, neoliberalism, etc.). Feminists do many different things in real political contexts in order to accomplish their goals, and working in and through women's movements can be very important strategically. But especially when trying to see just how feminism as a goal is being advanced in and through a variety of transnational strategies, it becomes self-defeating to presuppose that only women's movements can be the carriers of feminism.

Moreover, by stressing how feminism as a goal is characteristically combined with other goals and making its relative priority a question open to empirical examination, this approach more readily looks at the influence of the transnational opportunity structure upon both feminism and women's movements. "Political opportunity structure" (POS) is the preferred term among scholars interested in the positive opportunities and the obstacles provided by a specific political and social structure. Globalization is made concretely meaningful by seeing it as a process that increases the importance and level of integration of transnational political structures. At this transnational level, the POS can vary substantially from that provided at the local level alone. Thus Zapatista rebels reach out through the Internet for support from people and groups spread around the world to counter the repressive power exercised locally by the Mexican government.

The transnational opportunity structure is a political context that seems open to feminism, particularly as it takes up the discourses of human rights and development. What other goals are combined with feminism in which local contexts, and how does that help or hinder these ideas to travel transnationally? For example, if feminism is connected to the defense of class privilege, and upper-middle-class women's ability to enter the paid labor market is given priority over migrant women's ability to earn a living wage by their domestic work, then feminism is not going to be an appealing identity for those who do not already enjoy economic advantages.

The *intersectionality* of social movements characterizes them and shapes how they position themselves in the transnational arena in which they operate. Intersectionality means that privilege and oppression, and movements to defend and combat these relations, are not in fact singular. No one has a gender but not a race, a nationality but not a gender, an education but not an age. The location of people and groups within relations of production, reproduction, and representation (relations that are organized worldwide in terms of gender inequality) is inherently multiple. These multiple social locations are often—not, as is often assumed, atypically—contradictory. Organizations as well as individuals hold multiple positions in regard to social relations of power and injustice, and typically enjoy privilege on some dimensions even while they struggle with oppression on another. This multiplicity and the contradictions to which it gives rise are rarely acknowledged theoretically. As Ferree and Roth (1998) argue, scholars of social movements have instead tended to construct ideal-typical movements, envisioning these as composed of ideal-typical constituents: thus "worker's" movements as of and for indigenous men, "feminist" movements as of and for white, middle-class women.[2] The reality is of course, much more complex, but it only emerges clearly when the goals and constituents of movements are acknowledged as distinct.

In sum, this book approaches feminism as one important goal of social change. It asks the question of how feminism is being related to women's movements and other organization strategies that are being pursued locally and in transnational spaces, as well as to the various other goals that specific women and men have when they engage in social and political activities. And it looks especially at globalization as a process that is potentially empowering as well as disempowering women as they look for effective strategies to make feminist social change, including sometimes building women's movements.

TRANSNATIONAL OPPORTUNITY STRUCTURES: LOOKING FOR LEVERS TO MOVE THE WORLD

Women's movements are far from the only tools that feminists have taken up to try to challenge and change male domination. Globalization in the sense of integration means speedy flows of ideas across great distances. This has contributed to the sharing of strategies that also reach beyond classic women's movements, protest demonstrations, and projects. Three groups of strategies for making feminist change have spread like wildfire through the world system: developing a "women's policy machinery" within state institutions, building an issue advocacy network outside of formal institutions, and developing women's movement practices that are knowledge-creating, many of which link policy machineries with advocacy networks to multiply political effectiveness. None of these is without its problematic aspects in the transnational system.

First, women's policy machinery has now been put in place in most countries of the world, nearly all of which has come into existence since the first UN Conference on Women in Mexico City in 1975. Such policy machinery includes specific national, local, or regional administrative structures that are targeted to women as a politically relevant group. Women's policy machinery includes ministries of women's affairs, agencies charged with "mainstreaming" gender perspectives into policy and/or bringing women into administrative positions, and programs designed to ensure that women receive a certain share of seats in elected and/or appointed bodies, from parliaments to corporate boards of directors. Women's policy machinery, unlike a women's movement, is formally embedded in state or transnational structures that have institutionalized authority. Policy machineries differ widely in their form and effectiveness, from the old but weak and bureaucratically low-level Women's Bureau in the U.S. Department of Labor to the Ministry for Women's Affairs in France.[3]

Women's policy machinery is a mechanism by which gender inequality can be addressed, but it offers no guarantee that this is how it will be used. The competing goals of those who occupy the positions that this machinery creates as well as the different interests of those to whom they are accountable— typically authorities above them as well as constituents below—make for a mixed picture of what the machinery could produce. The term "policy machinery" itself is one that arose within administrative elites and diffused among activists, but it is not a bad image to use in considering the consequences of these structural innovations. Rather than achieving feminist goals by the very fact of their existence, they are tools, like levers, that require active use—there needs to be pressure put on the lever for it to budge anything within the system of male power. Paradoxically, sometimes the creation of women's policy machinery seems to be mistaken for an end in itself or a substitution for active mobilization to exert pressure for change, and thus in practice can lead to demobilization by the women's movements that helped to create them....

Second, globalization has facilitated the emergence of feminism as a goal in a wide variety of issue advocacy networks active at the transnational level. Overtly feminist discourse is heard in a variety of nongovernmental organizations that operate across national borders, working on a huge variety of issues from HIV/AIDS to literacy to economic restructuring, and in contexts as different as the World Bank and the World Social Forum. Gender equity as a principle has been taken up in networks concerned with health, peace, and social justice, as well as in networks organized directly to deal with issues seen as especially affecting women, such as trafficking in human beings, prostitution and other forms of sex work, and the use of genetic and reproductive technologies.

Many of these issues cut across national boundaries, and the networks constructed to deal with them are not organized as much on the basis of nationality as was true of their predecessors in the early twentieth century. A typical organization of a hundred years ago was "inter-national" in the sense of multiple national organizations belonging to a coordinating umbrella organization to which each national member group sent representatives. By contrast, in a world today characterized by Internet linkages, cheap airfares, and widespread telephone service, more fluid networks made up of individuals and organizations from many parts of the world actually interacting with each other more routinely can supplement or even supplant the conventional, hierarchical style of international nongovernmental organizing. NGOs are ever more diverse in their form and can be transnational in membership (individuals and groups not representing nations but belonging regardless of nationality).

NGOs are also linked in wider transnational networks around certain issues and values, as Keck and Sikkink (1998) pointed out, and coordinate the pressure the groups bring on national governments, as Swider shows here in the case of migrant and labor groups in Hong Kong and Bagi⊠ shows with regard to NGOs operating in the former Yugoslavia. Such networks are thus becoming potentially powerful transnational actors in their own right. Rather than one unitary principle of feminism being the basis for networking, as the International Council of Women adopted at the beginning of the last century, the actual political work of such NGOs and networks is differentiated and issue-specific. The flexibility and issue focus of networks on specific problems, from the access of women to scientific professions to the work conditions of migrant domestic workers or female genital cutting, makes them politically able to span a wider range of activist groups. Paradoxically, while feminism has entered a great many of these networks, the very variety of their goals fragments feminist attention and makes women's movements as such seem exclusionary, overly broad, and less attractive forms of mobilization. Networks instead tend to combine paid professionals and unpaid local activists, men and women, inside and outside of government, and in many countries.

Global terrorism and "national security" are also increasingly recognized as being intertwined and gendered issues. This feminist concern can take the form of considering how religious fundamentalism, control over women's bodies, national identity, and male pride and privilege are being negotiated and renegotiated in diverse transnational as well as national settings. Both fundamentalists (Christian, Islamic, Jewish, and Hindu) and those who challenge them are linked in networks that may include state as well as nonstate actors. Among the interesting questions that this increasingly global conflict raises is how and when feminist principles become co-opted in the national interests of either liberal modernist states or religious fundamentalism.

In the cold war era, the communist states co-opted the idea of women's liberation as an accomplishment of state socialism, which allowed the communist countries to divert attention from the ways in which women in fact were far from liberated under their regimes, on the one hand, and on the other hand placed Western countries in the position of resisting feminism as godless, antifamily, and a threat to (Western) civilization. Interestingly enough, in the current global "war on terror" rhetoric it is the Western democracies that attempt co-opt feminism as one of their greatest accomplishments. The oppression of women is framed as religious, family-based, and a threat to (Western) civilization, which is now defined as the champion of secular modernity and the value of equal rights for all. Diverting attention from the way that women continued to be far from liberated in Western capitalist democracies is one discursive accomplishment of this strategy, and if it succeeds, it could be a demobilizing factor for feminist women's movements.

Thus the strategic use of transnational networks has both a material side in the flow of resources and support for issues they spread globally across national borders and discursive side in the way that issues are framed and conflicts organized on a global level....

The third level with which feminists have tried to change the world is with knowledge-creation strategies. Women's movements have been prolific producers of "new words" to name old problems from sexual harassment, acquaintance rape, wife beating, the double day/shift, and the nanny chain. Women's movements have been important places for the development of transnational feminist theory and identity, creating the free spaces that foster ideological innovation and strategic inventions, like the women's policy machinery of the 1990s and the shelters for battered women of the 1980s. Creating the space to produce new feminist analyses of gender and of gender systems' effects on both women and men, the many national women's movements and the journals, magazines, and women's studies programs to which they gave rise have developed feminist theory....

Conferences share this knowledge, none more spectacularly than the 1995 Beijing Fourth Women's World Conference and NGO Forum. Ideas such as "gender mainstreaming" and "gender budgeting" become developed through the active participation of feminists engaged in knowledge production work in their own countries and transnationally. These ideas then become part of the shared language and competences that women's movements and women's policy machineries in many countries adopt and use. Conferences organized on a transnational basis in and across many disciplines offer social support to women to keep them actively pursuing feminist goals in their scholarship and carrying their theory out into practice with activist groups and transnational networks. Knowledge and its creation become a sustaining aspect of the work of making feminist change, and this work especially blurs the distinction among those in policy machinery, in movements, in social service, and in academia. Evaluation research accompanies social change projects, and feminist theory informs statistical data collection.

All over the world, women's associations fund and conduct studies, disseminate reports, encourage discussion, and train researchers and policymakers to develop greater awareness of gender inequities and greater commitment to redressing them. Lobbying, monitoring, funding demonstration projects, assessing best practices, and encouraging new networks are all activities in which feminist women's movements are increasingly engaged as they become more institutionalized as policy relevant actors in their own right.

Knowledge work links policymakers and social movements, serving as a powerful strategy for spreading feminism. But feminist ideas can spread without any accompanying feminist identity. Feminist women's movements struggle to create and sustain feminist identities that women will find meaningful for themselves, and through such identities, movements give meaning to even the losing battles that they fight. As crucibles of identity, community, and commitment for feminists, women's movements can play a critical role in sustaining activism across time (generations) and space (geography). However, feminist women's movements do not just provide the sometimes-comfortable homes where valiant feminists can return and refresh themselves but are themselves at times sites of tremendous diversity and conflict.

Thus the decline in popular mobilization in the form of autonomous feminist women's movement organizations and mass demonstrations can be partly attributed to the crucible for conflict that they can offer. Feminist identity is a highly charged and much-debated concept, and for some networks and organizations it may be more convenient and effective to simply avoid the issue. But the heat of conflict in feminist women's movements has also been accompanied by the light of developing feminist theory and the warmth of a feminist community in times of struggle. If feminism becomes so diffused in networks and policy institutions that women's movements themselves fade out as an active part of the picture, there paradoxically might emerge a time in which we have feminism without feminists.

GLOBAL FEMINISM, SITUATED ACTIVISM: PERSPECTIVES ON POWER AND SOCIAL CHANGE

Although the present wave of globalization is different than the one that crested in the early twentieth century, some questions about the relationships among feminism, women's movements, and globalization persist. How can women's movements manage the challenges of diversity, of generational succession, and of organizational institutionalization that are posed by becoming a more fragmented field of special interest groups that share a concern with women's equal rights but differ in so many other regards? How can the inequality of resources around the world be used to create constructive flows of support? What conditions foster democratic participation transnationally and build solidarity for addressing problems not one's own? ....

Globalization can work to women's advantage—as especially seen in the UN—but also unleash forces of inequality that will further disadvantage women. Just what feminism means and what women's movements do for women are therefore questions not merely for theory but for the practice of the next decades to determine.

ENDNOTES

1. See, for example, studies using such a mixed and static definition as Margolis 1993; Chafetz and Dworkin 1986; and the critiques in Gluck 1998 and Buechler 2000.

2. This approach also offers an alternative to Molyneux's model of women's pragmatic and strategic gender interests, and does not presume that movements, constituencies to which they strategically appeal, and the interests of these constituents can be theoretically known by analysts with access to the "correct" understanding of the social structure, but instead works from the idea that interests, constituents, and movements all need to be socially constructed. See Ferree and Mueller 2004 for a more developed discussion of this point.

3. See the discussion of this process especially in Europe in Stetson and Mazur's *Comparative State Feminism* (1995).

BIBLIOGRAPHY

Buechler, Steven. 2000. *Social Movements in Advanced Capitalism*. New York: Oxford University Press.

Chafetz, Janet, S., and Anthony G. Dworkin. 1986. *Female Revolt: Women's Movements in World and Historical Perspective*. Totowa, N.J.: Rowman and Allanheld.

Ferree, Myra Marx, and Silke Roth. 1998. "Kollektive Identität und Organizations-kulturen: Theorien neuer sozialer Bewegungen aus amerikanischer Perspektive (Collective identity and organizational culture: An American perspective on the new social movements)." *Forschungsjournal Neue Soziale Bewegungen* 11: 80–91.

Ferree, Myra Marx, and Carl McClurg Mueller. 2004. "Feminism and the Women's Movement: A Global Perspective." In D.A. Snow, S.A. Soule, and H. Kriesi, eds., *The Blackwell Companion to Social Movements*. Malden, Mass.: Blackwell.

Gluck, Sherna Berger. 1998. "Whose Feminism, Whose History." In N.A. Naples, ed., *Community Activism and Feminist Politics: Organizing Across Race, Class, and Gender*, 31–56. New York: Routledge.

Keck, Margaret E., and Kathryn Sikkink. 1998. *Activists Beyond Borders: Advocacy Networks in International Politics*. Ithaca, N.Y.: Cornell University Press.

Margolis, Diane R. 1993. "Women's Movements Around the World: Cross-Cultural Comparisons." *Gender and Society* 7:379–399.

Stetson, Dorothy McBride, and Amy Mazur. 1995. *Comparative State Feminism*. Thousand Oaks, Calif.: Sage.

35

Students Against Sweatshops

BHUMIKA MUCHHALA

The following interview captures Bhumika Muchhala's experience in the Students Against Sweatshops movement. Muchhala also offers his views about where the movement stands in relation to the larger struggle for social justice.

Could you tell us about your background, and how you came to be an anti-sweatshop activist?

I was born in India but grew up in Jakarta from the age of five. My father, an Indian accountant, worked for multinational companies there. I attended an American international school that was completely Eurocentric—they never taught us anything about Indonesian language, culture, or politics. I learned Bahasa from hanging out with the street vendors. I used to sneak out to the street corner and eat *boso*, noodle soup. I came to the U.S. to study at Carnegie Mellon University, in Pittsburgh, and also took a lot of classes at the University of Pittsburgh itself. I had been apolitical as a teenager. But that changed during the ousting of Suharto in 1998—I turned on the TV and watched as my city went up in flames. It shook every fibre in my being to see all these buildings that I knew so well on fire, to see the riots in the streets. I kept hearing references to the International Monetary Fund, and student activists crying out against Suharto's corruption and cronyism, and Chinese-Indonesian dominance of the economy. It was then that I started reading the paper and searching the internet to learn more about the IMF. I learned a lot from the website for Global Exchange, the human-rights and environmental NGO set up in 1988. I decided to double my major, to learn about U.S. foreign policy and U.S. imperialism.

I became involved with Students in Solidarity, a large group of student activists at the University of Pittsburgh, and also got in touch with Robin Alexander at United Electric, an independent union. She had started an organization called PLANTA (Pittsburgh Labor Action Network for the Americas), which introduced me to labour organizing, and all the issues surrounding it. At the same time, Students in Solidarity were doing actions with the janitors at the AT&T buildings in downtown Pittsburgh, as well as with campus workers at the UP. Through an organization called Pugwash, I also got involved in debates on

SOURCE: From Bhumika Michhala, "Students Against Sweatshops." In *A Movement of Movements: Is Another World Really Possible?* Tom Mertes (ed.). New York: Verso 2004, pp. 192–201. Used with permission.

ethical issues in science with students in public policy, science, engineering and environmental departments.

But my main entry into activism came in my senior year, after a summer at Global Exchange. I applied by a fluke, because I really liked the website. They invited me to do an internship, which mostly involved logistical work for the speaking tours of international activists. I translated some information from Bahasa to English for their Gap campaign. I also petitioned outside Gap stores. Back at college, I launched a Gap campaign in Pittsburgh with Robin Alexander from UE. We did a number of Gap actions, as well as a fair-trade coffee campaign. I began to feel, though, that there were a lot of problems with the whole approach to Gap. Many people wanted to turn it into an all-out boycott, but that means people losing their jobs. Far more important is to find a way to make these corporations accountable, and responsible. We don't want them out of Indonesia, we want them to treat their workers with respect and adhere to codes of conduct. It's more about understanding the power structures—the racism, imperialism and neo-colonialism—than seeing the world in black and white terms. I also remember that my initial reaction to the Gap protests was that my friends in Indonesia would never do anything like this—not simply because they were more interested in hanging out at the mall, but because they don't have the do-gooding impulses of many activists here. I wasn't driven by the same sentiment—I wanted more debate, more facts.

How did you end up in Seattle?

I was going to a lot of Global Exchange workshops, including those on the WTO, that were run in conjunction with JustAct. I got involved with national recruiting and organizing, and decided to go out to Seattle because I knew a lot of my colleagues were going to be there. I didn't do direct action, though, because I'm not a citizen. After that, I compiled an oral history of students who were active around the WTO protests—*Student Voices: One Year After Seattle*—while working at the Institute for Policy Studies. The original plan was to conduct a few interviews as preparation for a workshop we put on with STARC—Students Transforming and Reforming Corporations. That was cancelled, but I was asked to do a report instead. At first, my goal was to interview 15 students, but that turned into 30, then 50—I ended up with 60-plus students. Looking at the report now, what strikes me is how blind it was to the demographics: almost all the students were white. At the same time, that makes it a pretty representative cross-section of the students who were at Seattle, or the April 16 demonstrations in Washington D.C., or at the Republican National Convention.

How did you become involved with Students Against Sweatshops?

My involvement began with an eight-month trip to Indonesia. I was approached to do some fact-finding by the Workers' Rights Consortium, an organization set up in 2000 by student activists, labour experts and university administrations to pressure the manufacturers of clothing bearing college logos into adhering to a code of conduct for their workers. It was conceived as an alternative to the Clinton administration's Fair Labor Association, set up in 1996, in which corporate interests predominate. Currently, 116 colleges are affiliated, and pay dues to fund the

WRC's activities. Its board of directors consists of five independent labour-rights experts, including Linda Chavez-Thompson of the AFL–CIO and Mark Barenberg, a Columbia law professor; five representatives from universities; and five from USAS. It also has an advisory council, which includes U.S.-based NGOs, academics, George Miller—a Democratic congressman—and international representatives.

I was technically an independent researcher—the WRC provided me with a place to stay in Indonesia, an office and transportation, but no salary. The AFL–CIO's Solidarity Centre was a point of contact for people who would introduce me to the various Indonesian trade unions. I interviewed more than 200 workers, over a period of eight months, in assembly plants for Reebok, Nike, Champion, Gear, Gap, Banana Republic—mostly U.S. multinationals, but also a lot of European labels and knock-offs. Initially, it was a startling experience, but I found it easy to strike up a rapport with the workers, because they were mainly women between seventeen and twenty-six—close to my age—who were curious, fun and energetic. They came from all over Indonesia—from Sumatra, Sulawesi, Kalimantan, Java. Local chiefs often pick young women to be sent to work in industrial centres, and the women consider themselves lucky and honoured. They make friends and learn about Indonesia's patchwork of different cultures. They tend to get 600,000 rupiah a month—around $60—and send up to a third of that home to their villages, which means they're living on little more than a dollar a day. They usually run out of money before the end of the month, but they're very resourceful—friends lend money to each other, women's cooperatives pool their incomes, and everyone makes incredible economies, cutting down an already meagre diet.

What happens to the women after the age of 26?

They just get worn down: they develop arthritis or respiratory problems. In the hat factories, for example, they burn a lot of coal and use thinners and other solvents. Once they can no longer work, they are laid off and return to the village, or else become street vendors, selling vegetables, cigarettes or shampoo from stalls or kiosks; others become homeless. Some dream of returning to school, others get married and settle down to raise a family.

What did the workers think was most exploitative about the process?

They were very aware that they were being exploited—they would work for 15 hours at a stretch and come home with stomach pains and headaches. Some of them told me about having to work till 2 or 6 in the morning, having to clean toilets, about suffering verbal and sexual abuse. They were in no sense passive victims, but they responded to some of the questions with astonishment—they had never really considered air temperature in the factory as something that they were deprived of or denied. They were amused and surprised that somebody cared so much about tiny, mundane details of their lives, such as what the bathrooms were like. But when I started explaining to them how the monitoring process works, and how we could negotiate with the retailers who had power over their supervisors, they became more interested.

What did you, and the Workers' Rights Consortium, learn from the experience?

I found we had a common interest, along with Indonesian and U.S. unions and activists, in working out how to channel this power through the current corporate structure—how to effect real changes among the Korean management, U.S. retailers, contractors and institutional purchasers such as universities, and secure real improvements for the workers. Being able to meet the workers and Indonesian union organizers face-to-face was an important step. The WRC hierarchy are busy flying from one meeting to another, and don't have time to spend weeks or months in one place; activists and union organizers in producing regions, meanwhile, can't easily be reached by phone, and tend not to have email, so it's hard to make people aware of what we're trying to do without making contact in person. I made a few connexions, and after I left, four USAS activists flew in and started operating in some of the same factories, which has helped to keep the momentum going.

You became active in USAS on returning from your stint with the WRC. Can you tell us about the origins and structure of USAS?

There had been various campaigns against sweatshops, starting with groups of immigrant workers in the U.S. garment industry in New York, California and Texas in the late 1980s; there were also campaigns in the early 1990s by the National Labor Committee, and the Coalition to Eliminate Sweatshop Conditions in California. Students began to focus on the issue of sweatshop labour during summer internships at the AFL–CIO in 1996, and at UNITE, the United Needle and Textile Workers Union. Some Duke students, including Tico Almeida, who spent the summer of 1997 researching the question for UNITE, returned to campus and began lobbying the university to make manufacturers of university apparel sign up to a code of conduct. The campaign was successful, and encouraged students who had been thinking along similar lines to start campaigns on their own campuses. USAS was founded in the spring of 1998, as an informal network of campus anti-sweatshop groups.

The first national conference was held in New York in July 1998. Over 200 campus delegates attended the second conference in 1999, when it was decided to set up a permanent office in Washington, D.C. Our national conferences—now held in August—last three days, and feature keynote speakers, workshops on labour action and anti-white supremacy training, as well as panel discussions and meetings of the various caucuses.

As far as structure is concerned, the leadership consists of a coordinating committee, democratically elected at the annual conference, plus the student representatives to the WRC. There are seven regional representatives, four from the identity caucuses—women, people of colour, working-class people, LGBTQ—and three members-at-large. In addition to the committee, there are seven regional organizers who report to the committee, and coordinate with organizers on individual campuses. Then, there is the national office with three permanent staff—a field organizer, a programme coordinator, and a person responsible for fundraising and communications. There are also standing committees on individual issues, such as international solidarity, alliance-building and solidarity with farmworkers. As programme coordinator, I primarily liaise with the international solidarity committee—organizing letter-writing and solidarity campaigns with workers in the Gap factory in El Salvador, for instance.

Does the national office determine what your campaigns are going to be, or do groups at particular universities become aware of an issue, which then percolates upwards?

It's primarily campus-based—different groups decide to do different things. Some are involved in mobilizing research or teaching assistants for rallies, some have taken part in local living-wage campaigns for campus workers, notably those employed by Sodexho–Marriott. Other campaigns have come about through contacts initially made by USAS activists. For example, in late 2000 USAS members were part of a delegation that went to the Kukdong factory in Puebla, and which also included people from AFL–CIO, United Electric and the Mexican union Frente Auténtico de Trabajo. The following January around 850 workers at the factory went on strike to protest the sacking of five of their co-workers, who were trying to organize an independent union. That spring, USAS sent out more activists to talk to the workers and find out how best to support their action. Nike and Reebok contract to the Korean-run *maquiladora*, so USAS organized pickets of their stores in the U.S., put pressure on university administrations and commissioned the WRC to investigate. Within two months, the workers had been reinstated, and by September the governor of Puebla made a public promise to give recognition to the Kukdong workers' new union, SITEKIM.

A similar process took place with the campaign at the New Era cap factory in Derby, New York, where, in early 2001, two-thirds of the workforce was laid off after affiliating with the Communication Workers of America. USAS sent a delegation there in March 2001, and then began working in tandem with CWA, getting colleges to cut contracts with New Era, putting pressure on major-league baseball teams, for whom New Era are the exclusive supplier. After a long strike, the workers came to a bargaining agreement with the company, and were reinstated—it was a big victory. USAS had a similar success with a cap factory in the Dominican Republic called BJ&B. Apart from that, there have been campaigns against Taco Bell—working with the Coalition of Immokalee Workers in Florida—and Mount Olive pickles, as well as a general campaign against cap producers such as Nike, Adidas, Reebok and Gap over the past few years.

What about the composition of USAS?

The anti-sweatshop coalition is pretty specific to a certain class and culture. There is a considerable level of working-class to upper-class diversity in USAS, but the majority are middle class, suburban—rabble-rousing in actions, but they get good grades. In a 1999 survey of USAS by a researcher called Peter Siu, more than a third of activists stated their family income was over $100,000, and only 8 percent said it was under $40,000. As with the mobilizations at Seattle and elsewhere, it's predominantly a white movement. Though the conditions in sweatshops resonate with Latinos and the Asian diaspora, these people aren't yet as politically active on campuses—perhaps because they don't feel comfortable in organizing culture. Black students are more focused on civil rights, and they often have other priorities that occupy them in their own neighbourhoods—for working-class students of colour, the prison-industrial complex hits home more than the IMF or World Bank. USAS does have good relations with the Prison Moratorium Project, and maintains a presence at their annual conference; but beyond that, it's left up to individual campuses to decide which struggles to adopt. At the moment, USAS is trying to start up dialogue on the subject of race and culture, but so far it's proceeded along the lines of "how can we recruit more people of colour?" Personally, I find that culturally insensitive and tokenstic.

You mentioned working with the CWA. What are USAS's relations with the unions like?

On the New Era campaign we definitely worked hand-in-hand with the CWA—a progressive union compared with some others—but for the most part we work pretty autonomously. The unions have a strong presence at our national conferences, and AFL–CIO and UNITE make important financial contributions—the former gave $50,000 in the academic year 2000–2001. It is quite a contentious issue. The AFL does take up a lot of the centre-ground of our campaign work, and people often ask us if we're being used as their youth wing, pointing to the fact that many USAS students go on to become organizers in the Service Employees International Union or Hotel Employees and Restaurant Employees Union. There's also the question of whether we're being steered in a particular direction—some people in USAS feel the AFL is protectionist, which is not something we would want to be associated with. It's true that part of the struggle at New Era was to protect the workers' jobs, since the company threatened to shift production overseas if they made trouble. But similar conditions applied at the Kukdong factory in Puebla, where we made Nike and Reebok promise not to "cut and run." As I mentioned earlier with regard to the Gap boycott, it's not about shutting down manufacturing in the developing world, it's about making companies treat their workers with respect.

Are there tensions between the unions and USAS with regard to international versus national campaigns?

Yes, because USAS chapters try to address both. A lot of activists look at international campaigns such as Kukdong, and ask, "what about the workers on our own campuses, serving us food, cutting our lawns or cleaning our homes?" Living-wage campaigns have been very prominent on scores of campuses for this reason. Our rank and file are, after all, predominantly white, upper-middle class, liberal college students attending elite institutions; their engagement with labour issues is in many ways the product of privilege, and they make use of their status to focus media attention on those issues.

What other campaigns would you see as models?

One that has brought enormous inspiration is the campaign against the Narmada Dam—Medha Patkar and Arundhati Roy have been very influential. Of U.S.–based campaigns, I would name those focusing on the prison-industrial complex, especially in California. USAS is also a member of the National Student Youth Peace Coalition, which was formed in the wake of September 11, and opposed the assaults on Afghanistan and Iraq.

36

The Rise of Food Democracy

BRIAN HALWELL

This article gives us a glimpse of local resistance movements around the world that are emerging to challenge the global food system.

The National Touring Association, one of the largest lobbying groups in Norway, representing walkers, hikers and campers, recently joined forces with the nation's one and only celebrity chef to develop a line of foods made from indigenous ingredients to stock the country's network of camping huts. For instance, someone staying in a mountain cottage in Jotunheimen National Park would dine on cured reindeer heart, sour cream porridge and small potatoes grown only in those mountain valleys. Sekem, Egypt's largest organic food producer, has developed a line of breads, dried fruits, herbs, sauces and other items made entirely from ingredients grown in the country. The brand is recognized by 70 percent of Egyptians, and sales have doubled each of the last five years.

In Zimbabwe, six women realized that their husbands, who are peanut farmers, were literally getting paid peanuts for their crop while they bought pricey imported peanut butter. These women decided to invest in a grinder and are now producing a popular line of peanut butter from local nuts that sells for 15 percent less than mainstream brands. In Nebraska, in the United States, a group of local farmers got together and opened a farmers' grocery that stocks only foods raised in that state. They found suppliers of bacon and baked beans, sour cream and sauerkraut, and virtually all major grocery items, all from Nebraska.

What ties together these disparate enterprises from around the world? At a time when our food often travels farther than ever before, they are all evidence of "food democracy" erupting from an imperialistic food landscape. At first blush, food democracy may seem a little grandiose—a strange combination of words. But if you doubt the existence of power relations in the realm of food, consider a point made by Frances and Anna Lappé in their book *Hope's Edge* (see *UN Chronicle*, Issue 3, 2001). The typical supermarket contains no fewer than 30,000 items, about half of them produced by ten multinational food and beverage companies, with 117 men and 21 women forming the boards of directors of those companies. In other words, although the plethora of products you see at a

SOURCE: From "The Rise of Food Democracy" by Brian Halwell, *UN Chronicle Online Edition*, Vol. XLII, #1, January 2005. Used with permission.

typical supermarket gives the appearance of abundant choice, much of the variety is more a matter of branding than of true agricultural variety and, rather than coming from thousands of farmers producing different local varieties, they have been globally standardized and selected for maximum profit by just a few powerful executives. Food from far-flung places has become the norm in much of the United States and the rest of the world. The value of international trade in food has tripled since 1961, while the tonnage of food shipped between countries has grown fourfold during a time when populations only doubled. For example, apples in Des Moines supermarkets come from China, even though there are apple orchards in Iowa; potatoes in Lima's supermarkets come from the United States, even though Peru boasts more varieties of potato than any other country.

The long-distance food system offers unprecedented and unparalleled choice to paying consumers—any food, any time, anywhere. At the same time, this astounding choice is laden with contradictions. Ecologist and writer Gary Nabhan wonders "what culinary melodies are being drowned out by the noise of that transnational vending machine," which often runs roughshod over local cuisines, varieties and agriculture. The choice offered by the global vending machine is often illusory, defined by infinite flavouring, packaging and marketing reformulations of largely the same raw ingredients (consider the hundreds of available breakfast cereals). The taste of products that are always available but usually out of season often leaves something to be desired.

Long-distance travel requires more packaging, refrigeration and fuel, and generates huge amounts of waste and pollution. Instead of dealing directly with their neighbours, farmers sell into a remote and complex food chain of which they are a tiny part and are paid accordingly. A whole constellation of relationships within the food shed—between neighbours, between farmers and local processors, between farmers and consumers—is lost in the process. Farmers producing for export often find themselves hungry as they sacrifice the output of their land to feed foreign mouths, while poor urbanites in both the First and Third Worlds find themselves living in neighbourhoods unable to attract most supermarkets and other food shops, and thus without healthy food choices. Products enduring long-distance transport and long-term storage depend on preservatives and additives and encounter all sorts of opportunities for contamination on their journey from farm to plate. The supposed efficiencies of the long-distance chain leave many people malnourished and underserved at both ends of the chain.

The changing nature of our food in many ways signals what the changing global economic structure means for the environment, our health and the tenor of our lives. The quality, taste and vitality of foods are profoundly affected by how and where they are produced and how they arrive at our tables. Food touches us so deeply that threats to local food traditions have sometimes provoked strong, even violent, responses. José Bové, the French shepherd who smacked his tractor into a McDonald's restaurant to fight what he called "culinary imperialism," is one of the better-known symbols in a nascent global movement to protect and invigorate local food sheds.

It is a movement to restore rural areas, enrich poor nations, return whole-some foods to cities and reconnect suburbanites with their land by reclaiming lawns, abandoned lots and golf courses to use as local farms, orchards and gardens.

Local food is pushing through the cracks in the long-distance food system: rising fuel and transportation costs; the near extinction of family farms; loss of farmland to spreading suburbs; concerns about the quality and safety of food; and the craving for some closer connection to it. Eating local allows people to reclaim the pleasures of face-to-face interactions around food and the security that comes from knowing what one is eating. It might be the best defense against hazards intentionally or unintentionally introduced in the food supply, including E-coli bacteria, genetically modified foods, pesticide residues and bio-warfare agents. In an era of climate change and water shortages, having farmers nearby might be the best hedge against other unexpected shocks. On a more sensual level, locally grown and in-season food served fresh has a definite taste advan-tage—one of the reasons this movement has attracted the attention of chefs, food critics and discriminating consumers around the world.

The local alternative also offers huge economic opportunities. A study by the New Economics Foundation in London found that every £10 spent at a local food business is worth £25 for that area, compared with just £14 when the same amount is spent in a supermarket. That is, a pound (or dollar, peso or rupee) spent locally generates nearly twice as much income for the local econ-omy. The farmer buys a drink at the local pub; the pub owner gets a car tune-up at the local mechanic; the mechanic brings a shirt to the local tailor; the tailor buys some bread at the local bakery; the baker buys wheat for bread and fruit for muffins from the local farmer. When these businesses are not locally owned, money leaves the community at every transaction.

This sort of multiplier is perhaps most important in the developing world where the vast majority of people are still employed in agriculture. In West Africa, for example, each $1 of new income for a farmer yields an average increase to other workers in the local economy, ranging from $1.96 in Niger to $2.88 in Burkina Faso. No equivalent local increases occur when people spend money on imported foods. While the idea of complete food self-sufficiency may be impractical for rich and poor nations alike, greater self-sufficiency can buffer them against the whims of international markets. To the extent that food pro-duction and distribution are relocated in the community under local ownership, more money will circulate in the local community to generate more jobs and income.

But here's what makes these declarations of food independence, despite their small size, so threatening to the agricultural status quo. They are built around certain distinctions—geographic characteristics—that global trade agreements are trying so hard to eliminate. These agreements, whether the European Union Trade Zone or the North American Free Trade Agreement, depend on erasing borders and geographic distinctions.... Multinational food companies that source the cheapest ingredients they can find also depend on erasing these distinctions....

Look around and you can glimpse the change worldwide. Farmers in Hawaii are uprooting their pineapple plantations to sow vegetables in hopes of replacing the imported salads at resorts and hotels. School districts throughout Italy have launched an impressive effort to make sure cafeterias are serving a Mediterranean diet by contracting with nearby farmers. At the rarefied levels of the World Trade Organization, officials are beginning to make room for nations to feed themselves, realizing that this might be the best hope for poor nations that cannot afford to import their sustenance. Even some of the world's biggest food companies are starting to embrace these values, a reality that raises some unsettling questions and awesome opportunities for local food advocates. Recently, officials at both Sysco—the world's largest food-service provider—and Kaiser Permanente—the largest health care provider in the United States—declared their dependence on small local farmers for certain products they cannot get anywhere else. These changes will unfold in a million different ways, but the general path will look familiar. Farmers will plant a greater diversity of crops. Less will be shipped as bulk commodity and more will be packaged, canned and prepared to be sold nearby. Small food businesses will emerge to do this work, governments will encourage new businesses, and shoppers seeking pleasure and reassurance will eat deliberately and inquire about the origins of their food. Communities worldwide all possess the capacity to regain this control and this makes the simple idea of eating local so powerful. These communities have a choice, and they are choosing instead to eat here.

37

Big Oil Wreaks Havoc in the Amazon, But Communities Are Fighting Back

KELLY HEARN

Giant oil corporations have polluted Peruvian communities. Here, Kelly Hearn describes some of the strategies local residents use in seeking environmental justice.

On Wednesday the Ninth Circuit Court of Appeals in San Francisco heard oral arguments in *Tomas Maynas Carijano v. Occidental Petroleum*, a case in which the defendant resides just miles from the courthouse in a plush office building and the plaintiff in a wooden hut in the Peruvian Amazon.

At issue before the court is whether a U.S. district court was right to send the pollution and public health lawsuit against the California-based oil company to Peru rather than keep it in the U.S. where it was filed. But, considering the unprecedented oil boom in the Peruvian Amazon, there is more at stake right now than just this lawsuit.

I've never crashed at Oxy's headquarters but I have slept in the hut of the plaintiff, known as Apu Tomas, and many like them in the Peruvian Amazon. I drank ayahuasca with him. He fed me a tiny fire-roasted bird I watched him kill with a blowgun. By boatplane, pick-up, cargo boat and canoe I traveled through what Peru's oil maps know as Block 1A, where Occidental in the 1970s built a network of pipes and pumping stations. In 2000, Oxy sold the dwindling oil field to the Argentine company Pluspetrol. But its massive infrastructure remains, especially giant wastewater pipes that engineers for years used to dump billions of gallons of toxic wastewater directly into rivers and streams without warning natives who play, drink and swim in the water.

Today many of those people have poisonous levels of lead in their blood (the Oxy suit is actually being brought on behalf of Carijano and over 20 other natives), and some have died. I often wonder how the accountants and marketers and engineers would talk if they had to live drinking directly out of rivers and fishing and hunting for protein sources. This isn't to say that I totally buy the myth of the ecologically pure native, but I don't have to in order to know that

SOURCE: From Kelly Hearn, "Big Oil Wreaks Havoc in the Amazon, But Communities are Fighting Back," AlterNet (March 11, 2010). Used with permission.

the systems and tenets of our global system (the U.S.–backed version) simply don't cut it, simply can't deal with the problems we face.

Oxy wants the trial in Peru. Plaintiffs say that while they'll fight anywhere, their chances of getting relief seem slimmer in one venue than another.

"I want the case to stay in the United States," Apu Tomas (Apu means leader or chief) told me when I visited his village. "Natives can never get justice in Peru."

People who watch this sort of thing see parallels to a similar case in the Ecuadorian Amazon where 30,000 ethnic natives are suing Chevron for polluting their rainforest homes. The plaintiffs in that case years ago filed suit in the United States, but Chevron convinced a judge that it belongs in Ecuador. After years of bitter litigation, the Ecuadorian courts appear poised to slap a $26 billion judgment on the company later this year. But Chevron is now indicating that it won't comply with the judgment of the Ecuadorian court because it was denied due process. Send us down there. No, things didn't work out. Take us back.

Oxy apparently thinks it has a better chance outside the U.S. It might be right. Many Peruvians are happy with the impressive economic growth of recent years, growth that's been dependent in no small degree on an unprecedented rainforest oil boom steered by Peru's pro-business president, Alan Garcia. And the Apu is right in indicating that the Peruvian government (as well as Peru's elite class) isn't known for falling over itself to help natives. If the appellant judges in California don't reverse or remand the lower court's decision, the plaintiffs hope at least they will somehow ensure that Oxy is made to pay up should it lose in Peru.

But going to Peru might not serve the behemoth company well because of the strong showing populism has made in South America in recent years. This is tied to an increasing willingness of natives to stand up and be counted. In Peru last summer, Garcia's white elite government got its bell rung by a nationwide indigenous strike in which natives (especially in the Amazon regions) took to the streets to protest land reforms linked to the U.S. Free Trade Agreement. People died in clashes with the police. There was talk of revolution if demands weren't met. I met with the man the government blamed it all on, an educated and politically powerful Indian named Alberto Pizango. He told me that Indians are tired of 500 years of getting kicked around. "It's our moment," he said.

The oil boom in the Peruvian Amazon also makes this lawsuit particularly important. Thanks in part to outsiders, natives are learning to use GPS units, Google Earth, blogs and lawsuits to fight back. But most would agree that the lawsuits are really what change Big Oil's behavior.

Carter Beasley is an engineer who has worked on South American oil projects for 25 years. He once told me that these kinds of suits are making oil companies pay far more attention to how things get done. In the past, they hid their sins in the Amazon's remoteness.

The Occidental case was brought after a Peruvian rights group, Racimos de Ungurahui, joined with EarthRights International and Amazon Watch to issue a 2007 report that documented health problems resulting from Oxy's toxic legacy. I interviewed Oxy's spokesperson Richard Kline in 2008.

"We are aware of no credible data of negative community health impacts resulting from Occidental's operations in Peru," Kline told me. He said the report was full of "inflammatory misstatements, unfounded allegations and unsupported conclusions" and that it failed to provide basic information that would help determine whether oil operations contributed to the alleged environmental and health problems.

I've spoken to plenty of public health experts who tell me that it's hard proving illnesses are causally linked to oil production because natives live in grinding poverty with poor sanitation habits, nutrition and health education. Oil companies eagerly latch on. "Natives are already sick because they are poor," the reasoning goes. "So our pollution can't be identified as the cause." But what about presumption? In Maine, where I am a volunteer firefighter, the state legislature has passed a law decreeing certain kinds of cancers are presumed to be part of our job if someone has spent a certain number of years in the service. Until now, we had to prove cancer resulted from the job (good luck with that). If the idea of presumption works here in Maine, why not in cases where oil companies have spewed out toxins by the billions of gallons?

Some people have the wrong idea that natives are constantly ducking in the bush getting ready to spear an oil worker. Many natives want oil development and they only ask a few simple things in return. Do it right. Clean up any spills. Live by your promises. Give us a fair stake instead of trying to buy us off with the modern equivalent of beads and mirrors. Natives are poor. They want and deserve a better, healthier life, something more than a constant struggle to survive. Life is hard in the jungle, despite the image many Americans have seen on ecotourism calendars of happy natives picking low-hanging fruit.

I'd be willing to bet that Oxy's chief executive has more money than all of Peru's natives put together. It is hard for me to understand how one of the world's richest and most powerful companies can dump billions of gallons of toxins into the environment and raise their hands and say "We hate this, we really do, but we sold all this stuff to another company and we aren't responsible."

Several years ago, at the height of Ecuador's oil boom, oil companies were taking over the traditional lands of Amazonian Cofan Indians until Randy Borman, the son of white U.S. missionaries (who had lived his life as a Cofan) organized them into gun-toting groups that kidnapped oil workers. The Cofan got their land back. A few years ago, I spoke on the phone with Borman and he told me that "companies only understand force. That's just how it is."

I hope the oil outsiders and developers learn to do what they do in an environmentally sound way, in ways that respect the people who have lived on the land for countless generations. I hope the people of Peru can benefit from the jobs and oil royalties generated by environmentally sound extraction methods (like horizontal drilling) and that companies will consider natives as real stakeholders. And I hope that Oxy has to pay big. I hope its gazillionaire executives have smaller bonus checks and are forced to one day explain to their children why their massive company wasn't able to correct such a giant injustice.

REFLECTION QUESTIONS FOR CHAPTER 9

1. Can "bottom-up" social movements actually be successful? Have there been instances in U.S. history where the powerful have been thwarted by the powerless? Any examples from world history? What appear to be the conditions present when these movements have been successful? Are any of those conditions present now in the United States? In other parts of the world?

2. David Brooks, writing in *The New York Times* (November 11, 2004), stated that globalization, for the most part, is working to reduce poverty. He quotes a World Bank report that economic growth is producing a "spectacular" decline in poverty in East and South Asia. Other areas are improving (except for sub-Saharan Africa) but not as rapidly. His explanation: globalization, with lower trade barriers, ensuring property rights, and free economic activity, is causing international trade to surge. In his words, "free trade reduces world suffering." If this is true, then why all the protest by the critics of globalization? Social movements for worldwide economic justice and environmental safeguards should be drying up. Is Brooks right, or is he missing something? How would Brecher, Costello, and Smith respond to Brooks?

3. Ferree's discussion of gender in transnational social movements identifies three "levers to move the world." What does she mean by *levers* and why are they important?

4. How does Muchhala's account of his own experiences in the Students Against Sweatshops movement reflect some of the problems of transnational organizing and resistance?

5. How does Halwell account for the success of local food markets? Do you agree that food democracy can promote justice in the new global age?

6. What do Peruvian Rights groups want and how are they fighting back?

Chapter 10

Rethinking Globalization

M ay 2010 witnessed a global meltdown. Greece faced default on its massive debts, held by European banks. The euro and the shares of European banks fell sharply. The more affluent European nations had to bail out the weaker ones in an effort to stabilize the euro. All of this fueled the sudden 1,000-point drop in the Dow on Wall Street. These events led Thomas L. Friedman to observe, "[We are living] in an increasingly integrated world where [we are guided] by the simple credo… 'Lost there, felt here.'"[1]

This credo of globalization—"lost there, felt here"—fits not only for the world of finance, but also applies to the environment, jobs, and migration patterns. This chapter builds on this basic concept as it considers whether globalization unifies or divides the world; the positive and negative consequences of transformative technology; the objections of groups that resist the reach of globalization; and how we might meet the challenges of the twenty-first century.

Martin Jacques begins by asking whether the effect of globalization has been to promote less respectful and more intolerant attitudes, especially in the United States, toward other cultures, religions, and societies. By obliterating distance with technology, Jacques argues, we have obtained an "illusion of intimacy" that actually denigrates non-Western cultures and encourages a lack of respect for differences. Morozov picks up this theme, arguing that contrary to its promise, the Internet has not ushered in a new era of freedom, civic engagement, political activism to overthrow dictatorial regimes, and perpetual peace. Indeed, some worry that U.S. companies such as Google, Twitter, and Facebook are creating "information sovereignty," another form of U.S. domination.

Naomi Klein's article considers the many groups that oppose various aspects of globalization, for example free trade but not the free movement of people; the Americanization of the world; the long reach of multinational corporations; and the promotion of free market ideology by the International Monetary Fund and the World Trade Organization (cutting taxes, privatizing services, deregulated markets, and opposition to unions). Opponents of globalization seek to promote local democracy and encourage global labor unions, the regulation of transnational corporations, and the promotion of human diversity in culture, values, and economic models.

As you read the articles in this section and reflect on the articles throughout this book, consider this by economist Jeffrey Sacks, who defines the challenge of the twenty-first century as facing up to "the reality that humanity shares a common fate on a crowded planet." We must end our view of the world as a struggle of "us" vs. "them," he argues, and instead seek global solutions—together. He makes a powerful argument: "Our generation's greatest challenges—in environment, demography, poverty and global politics—are also our most exciting opportunity ... The challenge is to turn fragile and unfulfilled global commitments into real solutions."[2]

ENDNOTES

1. Friedman, Thomas L. 2010. "A Question for Lydia." *New York Times* (May 14). Online: http://www.nytimes.com/2010/05/16/opinion/16friedman.html.

2. Sachs, Jeffrey D. 2008. "#1 Common Wealth: 10 Ideas That Are Changing the World," *Time* (March 24):40.

38

We Are Globalised, but Have No Real Intimacy with the Rest of the World

MARTIN JACQUES

Jacques argues that globalization has brought with it a new kind of hubris—that our way is the only way, and everyone should be like us.

I have just read Ruth Benedict's *The Chrysanthemum and the Sword*. It is a classic. Published in 1947, it analyses the nature of Japanese culture. Almost 60 years and many books later, it remains a seminal work. Like all great works of scholarship, the book manages to transcend the time and era in which it was written, ageing in certain obvious respects, but retaining much of its insight and relevance. If you want to make sense of Japan, Benedict's book is as good a place to start as any. Here, though, I am interested in the origins and purpose of the book.

In June 1944, as the American offensive against Japan began to bear fruit, Benedict, a cultural anthropologist, was assigned by the U.S. office of war administration to work on a project to try and understand Japan as the U.S. began to contemplate the challenge that would be posed by its defeat, occupation and subsequent administration. Her book is written with a complete absence of judgmental attitude or sense of superiority, which one might expect; she treats Japan's culture as of equal merit, virtue and logic to that of the U.S. In other words, its tone and approach could not be more different from the present U.S. attitude towards Iraq or that country's arrogant and condescending manner towards the rest of the world.

This prompts a deeper question: has the world, since then, gone backwards? Has the effect of globalisation been to promote a less respectful and more intolerant attitude in the west, and certainly on the part of the U.S., towards other cultures, religions and societies? This contradicts the widely held view that globalisation has made the world smaller and everyone more knowing. The answer, at least in some respects, is in the affirmative—with untold consequences lying in wait for us. But more of that later. First, why and how has globalisation had this effect?

Of course, it can rightly be argued that European colonialism embodied a fundamental intolerance, a belief that the role of European nations was to bring "civilized values" to the natives, wherever they might be. It made no pretence, however, at seeking to make their countries like ours: their enlightenment, as the colonial attitude would have it, depended on our physical presence. In no instance, for example, were they regarded as suitable for democracy, expect where there was racial affinity, with white settler majorities, as in Australia and Canada. In contrast, the underlying assumption with globalisation is that the whole world is moving in the same direction, towards the same destination: it is becoming, and should become, more and more like the west. Where once democracy was not suitable for anyone else, now everyone is required to adopt it, with all its western-style accoutrements.

In short, globalisation has brought with it a new kind of western hubris— present in Europe in a relatively benign form, manifest in the U.S. in the belligerent manner befitting a superpower: that western values and arrangements should be those of the world; that they are of universal application and merit. At the heart of globalisation is a new kind of intolerance in the west towards other cultures, traditions and values, less brutal than in the era of colonialism, but more comprehensive and totalitarian. The idea that each culture is possessed of its own specific wisdom and characteristics, its own novelty and uniqueness, born of its own individual struggle over thousands of years to cope with nature and circumstance, has been drowned out by the hue and cry that the world is now one, that the western model—neoliberal markets, democracy and the rest— is the template for all.

The new attitude is driven by many factors. The emergence of an increasingly globalised market has engendered a belief that we are all consumers now, all of a basically similar identity, with our Big Macs, mobile phones and jeans. In this kind of reductionist thinking, the distance between buying habits and cultural/political mores is close to zero: the latter simply follows from the former. Nor is this kind of thinking confined to the business world, even if it remains the heartland. This is also now an integral part of popular common sense, and more resonant and potent as an international language because consumption has become the mass ideology of western societies. The fact that television and tourism have made the whole world accessible has created the illusion that we enjoy intimate knowledge of other places, when we barely scratch their surface. For the vast majority, the knowledge of Thailand or Sri Lanka acquired through tourism consists of little more than the whereabouts of the beach.

Then there is the phenomenon of Davos Man, the creation of an overwhelmingly western-weighted global elite, which thinks it knows all about these things because it describes itself as global and rubs shoulders on such occasions with a small number of hand-picked outsiders. Nor should we neglect its media concomitant, the commentariat—columnists who wax lyrical on these things even if their knowledge of the world is firmly bounded by the borders of the west. A couple of days at a conference in Egypt, India or Malaysia makes instant experts of them. So is much of modern western opinion made.

The net effect of all this is a lack of knowledge of and respect for difference. Globalisation has obliterated distance, not just physically but also, most dangerously, mentally. It creates the illusion of intimacy when, in fact, the mental distances have changed little. It has concertinaed the world without engendering the necessary respect, recognition and tolerance that must accompany it. Globalisation is itself an exemplar of the problem. Goods and capital may move far more freely than ever before, but the movement of labour has barely changed. Jeans may be inanimate, but migrants are the personification of difference. Everywhere, migration is a changed political issue. In the modern era of globalization, everything is allowed to move except people.

After three decades of headlong globalisation, the world finds itself in dangerous and uncharted waters. Globalisation has fostered the illusion of intimacy while intolerance remains as powerful and unyielding as ever—or rather, has intensified, because the western expectation is now that everyone should be like us. And when they palpably are not, as in the case of the Islamic world, then a militant intolerance rapidly rises to the surface. The wave of Islamophobia in the west—among the people and the intelligentsia alike—is a classic example of this new intolerance. When I wrote a recent article for these pages on the Danish cartoons, arguing the Europe had to learn a new way of relating to the world, I got nearly 400 emails in response. Over half of these were negative and many were frightening in their intolerance especially those from the U.S., which were often reminiscent in their tone of the worst days of the 1930s.

We live in a world that we are much more intimate with and yet, at the same time, also much more intolerant of—unless, that is, it conforms to our way of thinking. It is the western condition of globalisation, and its paradox of intimacy and intolerance suggests that the western reaction to the remorseless rise of the non-west will be far from benign.

39

Think Again: The Internet

EVGENY MOROZOV

Morozov wonders: Do the marvels of the Internet translate into social justice? He concludes that web technology alone does not make for a better world.

"THE INTERNET HAS BEEN A FORCE FOR GOOD."

No. In the days when the Internet was young, our hopes were high. As with any budding love affair, we wanted to believe our newfound object of fascination could change the world. The Internet was lauded as the ultimate tool to foster tolerance, destroy nationalism, and transform the planet into one great wired global village. Writing in 1994, a group of digital aficionados led by Esther Dyson and Alvin Toffler published a manifesto modestly subtitled "A Magna Carta for the Knowledge Age" that promised the rise of "electronic neighborhoods bound together not by geography but by shared interest." Nicholas Negroponte, then the famed head of the MIT Media Lab, dramatically predicted in 1997 that the Internet would shatter borders between nations and usher in a new era of world peace.

Well, the Internet as we know it has now been around for two decades, and it has certainly been transformative. The amount of goods and services available online is staggering. Communicating across borders is simpler than ever: hefty international phone bills have been replaced by inexpensive subscriptions to Skype, while Google Translate helps readers navigate Web pages in Spanish, Mandarin, Maltese, and more than 40 other languages. But just as earlier generations were disappointed to see that neither the telegraph nor the radio delivered on the world-changing promises made by their most ardent cheerleaders, we haven't seen an Internet-powered rise in global peace, love, and liberty.

And we're not likely to. Many of the translational networks fostered by the Internet arguably worsen—rather than improve—the world as we know it. At a recent gathering devoted to stamping out the illicit trade in endangered animals, for instance, the Internet was singled out as the main driver behind the increased global commerce in protected species. Today's Internet is a world where

SOURCE: From Evgeny Morozov, "Think Again: The Internet." *Foreign Policy* (May/June 2010). Used with permission.

homophobic activists in Serbia are turning to Facebook to organize against gay rights, and where social conservatives in Saudi Arabia are setting up online equivalents of the Committee for the promotion of Virtue and the Prevention of Vice. So much for the "freedom to connect" lauded by U.S. Secretary of State Hillary Clinton in her much-ballyhooed speech on the Internet and human rights. Sadly enough, a networked world is not inherently a more just world.

"TWITTER WILL UNDERMINE DICTATORS."

Wrong. Tweets don't overthrow governments; people do. And what we've learned so far is that social networking sites can be both helpful and harmful to activists operating from inside authoritarian regimes. Cheerleaders of today's rapidly proliferating virtual protests point out that online services such as Twitter, Flickr, and YouTube have made it much easier to circulate information that in the past had been strictly controlled by the state—especially gruesome photos and videos and evidence of abuses by police and the courts. Think of the Burmese dissidents who distributed cell-phone photos documenting how police suppressed protests, or opposition bloggers in Russia who launched Shpik.info as a Wikipedia-like site that allows anyone to upload photos, names, and contact details of purported "enemies of democracy"—judges, police officers, even some politicians—who are complicit in muzzling free speech. British Prime Minister Gordon Brown famously declared last year that the Rwandan genocide would have been impossible in the age of Twitter.

But does more information really translate into more power to right wrongs? Not necessarily. Neither the Iranian nor the Burmese regime has crumbled under the pressure of pixilated photos of human rights abuses circulated on social networking sites. Indeed, the Iranian authorities have been as eager to take advantage of the Internet as their green-clad opponents. After last year's protests in Tehran, Iranian authorities launched a website that publishes photos from the protests, urging the public to identify the unruly protesters by name. Relying on photos and videos uploaded to Flickr and YouTube by protesters and their Western sympathizers, the secret police now have a large pool of incriminating evidence. Neither Twitter nor Facebook provides the security required for a successful revolution, and they might even serve as an early warning system for authoritarian rulers. Had East Germans been tweeting about their feelings in 1989, who knows what the Stasi would have done to shut down dissent?

Even when Twitter and Facebook do help score partial victories, a betting man wouldn't put odds on the same tick working twice. Take the favorite poster child of digital utopians: in early 2008 a Facebook group started by a 33-year-old Colombian engineer culminated in massive protests, with up to 2 million people marching in Bogotá's streets to demonstrate against the brutality of Marxist FARC rebels. (A *New York Times* article about the protests gushed: "Facebook has helped bring public protest to Colombia, a country with no real history of mass demonstrations.") However, when the very same "digital revolutionaries" last September tried to organize a similar march against Venezuelan leader and FARC sponsor Hugo Chávez, they floundered.

The reasons why follow-up campaigns fail often have nothing to do with Facebook or Twitter, and everything to do with the more general problems of organizing and sustaining a political movement. Internet enthusiasts argue that the Web has made organizing easier. But this is only partially true; taking full advantage of online organizing requires a well-disciplined movement with clearly defined goals, hierarchies, and operational procedures (think of Barack Obama's presidential campaign). But if a political movement is disorganized and unfocused, the Internet might only expose and publicize its vulnerabilities and ratchet up the rancor of internecine conflicts. This alas, sounds much like Iran's disorganized green movement.

"GOOGLE DEFENDS INTERNET FREEDOM."

Only when convenient. If the world's human rights community had to choose its favorite Fortune 500 company, Google—the world's overwhelming leader in Internet search and a trendsetter in everything from global mapping to social networking—would be a top contender. Decrying the Chinese government's censorship demands, Google recently decided to move its Chinese search engine to Hong Kong and promised to spare no effort to protect identities of Chinese dissidents who use Gmail. Much of the Western world applauded, as Google seemed to live up to its "don't be evil" corporate motto.

Let's remember that Google, like any company, is motivated by profit rather than some higher purpose. The company entered China not to spread the gospel of Internet freedom, but to sell ads in what is now the world's largest online market. Only four years after agreeing to censor its search results did it refuse to do so any longer. Yet had it managed to make greater inroads among Chinese consumers, does anyone doubt that its decision to defy Beijing would have been much more difficult?

Sometimes Google really does operate on principle. In early March, Google executives held a joint event with Freedom House, bringing bloggers from the Middle East to Washington to participate in a series of talks on such topics as "digital media's power in social movements" and "political parties and elections 2.0." Last summer, Google stood up to protect Cyxymu, a Georgian blogger who found himself the target of intense cyberattacks—supposedly from Russian nationalists unhappy with his take on the 2008 Russia–Georgia war—by keeping his Google-hosted blog online. Following the incident, the company's public-policy blog even boasted of Google's commitment to "giving a voice to 'digital refugees.'"

But the company's reputation as a defender of Internet freedom is decidedly mixed. For example, its Internet filtering process in Thailand—driven by the country's strict laws against insulting the monarchy—is not particularly transparent and draws much criticism for the country's netizens. In India, Google faces understandable government pressure to remove extremist and nationalist content from Orkut, its social networking site; yet some Indian critics charge that Google is overzealous in its self-censorship because it fears losing access to the vast Indian market. Google's defense of Internet freedom is, ultimately, a pragmatically principled stance, with the rules often applied on a case-by-case basis. It would be somewhat naïve—and, perhaps, even dangerous—to expect Google to become the new Radio Free Europe.

"THE INTERNET MAKES GOVERNMENTS MORE ACCOUNTABLE."

Not necessarily. Many Internet enthusiasts on both sides of the Atlantic who were previously uninterested in policy debates have eagerly taken on the challenge of playing government watchdog, spending days and nights digitizing public data and uploading it into online databases. From Britain's They Work for You to Kenya's Mzalendo to various projects affiliated with the U.S.–based Sunlight Foundation such as MAPLight.org, a host of new independent websites has begun monitoring parliamentary activity, with some even offering comparisons between parliamentarians' voting records and campaign promises.

But have such efforts resulted in better or more honest politics? The results, so far, are quite mixed. Even the most idealistic geeks are beginning to understand that entrenched political and institutional pathologies—not technological shortfalls—are the greatest barriers to more open and participatory politics. Technology doesn't necessarily pry more information from closed regimes; rather, it allows more people access to information that is available. Governments still maintain great sway in determining what kinds of data get released. So far, even the Obama administration, the self-proclaimed champion of "open government," draws criticism from transparency groups for releasing information about population counts for horses and burros while hoarding more sensitive data on oil and gas leases.

And even when the most detailed data get released, it does not always lead to reformed policies, as Lawrence Lessig pointed out in his trenchant *New Republic* cover story last year. Establishing meaningful connections between information, transparency, and accountability will require more than just tinkering with spreadsheets; it will require building healthy democratic institutions and effective systems of checks and balances. The Internet can help, but only to an extent: it's political will, not more info, that is still too often missing.

"THE INTERNET BOOSTS POLITICAL PARTICIPATION."

Define it. The Internet has certainly created new avenues for exchanging opinions and ideas, but we don't yet know whether this will boost the global appeal and practice of democracy. Where some see a renewal of civic engagement, others see "slacktivism," the new favorite pejorative for the shallow, peripheral, and fluid political campaigning that seems to thrive on the Internet—sometimes at the expense of more effective real-world campaigning. And where some applaud new online campaigns purportedly aimed at increasing civic participation, such as Estonia's planned 2011 launch of voting via text-messaging, others, myself included, doubt whether the hassle of showing up at a polling place once every two or four years is really what makes disengaged citizens avoid the political process.

The debate over the Internet's impact on participation echoes a much earlier controversy about the ambiguous social and political effects of cable television. Long before blogs were invented, scholars and pundits were arguing over whether the boob tube was turning voters into passive, apolitical entertainment maniacs who, when given greater choice, favored James Bond flicks and *Happy Days* reruns over nightly news broadcasts—or whether it was turning them into hyperactive, obsessive citizens who watch C-SPAN nonstop. The argument then, and now, was that American-style democracy was turning into niche markets for politics, with the entertainment-obsessed masses opting out, on TV and at the polling booth, and news junkies looking for ever-quicker fixes in the sped-up news cycle. The Internet is cable television on steroids; both tuning in and tuning out of political discourse have never been easier.

Another danger is that even the news we read will come increasingly from selective sources, such as our Facebook friends, which might decrease the range of views to which we're exposed. Three-quarters of Americans who consume their news online say they receive at least some of it through forwarded emails or posts on social networking sites, according to a 2010 study by the Pew Research Center's Internet & American Life Project. Currently, less than 10 percent of Americans report relying on just one media platform. But that could easily change as traditional news sources lose market share to the Web.

"THE INTERNET IS KILLING FOREIGN NEWS."

Only if we let it. You won't hear this from most Western news organizations, which today are fighting for their financial survival and closing foreign bureaus, but we've never had faster access to more world news than we do today. Aggregators like Google News might be disrupting the business models of CNN and the *New York Times*, forcing substantial cutbacks in one particularly costly form of news-gathering—foreign correspondents—but they have also equalized the playing field for thousands of niche and country-specific news sources, helping them to reach global audiences. How many people would be reading AllAfrica.com or the Asia Times Online were it not for Google News? While we decry the Internet's role in destroying the business model that supported old-school foreign reporting, we should also celebrate the Web's unequivocally positive effects on the quality of research about global affairs done today on the periphery of the news business. The instantaneous fact-checking, ability to continuously follow a story from multiple sources, and extensive newspaper archives that are now freely available were unimaginable even 15 years ago.

The real danger in the changing face of foreign news is the absence of intelligent and respected moderators. The Internet may be a paradise for well-informed news junkies, but is a confusing news junkyard for the rest of us. Even fairly sophisticated readers might not know the difference between the *Global Times*, a nationalist Chinese daily produced under the auspices of the Communist Party, and the *Epoch Times*, another China-related daily published by the Falun Gong dissident group.

"THE INTERNET BRINGS US CLOSER TOGETHER."

No. Geography still matters. In her best-selling 1997 book *The Death of Distance*, the *Economist's* then senior editor Frances Cairncross predicted that the Internet-powered communications revolution would "increase understanding, foster tolerance, and ultimately promote worldwide peace." But pronouncing the death of distance was premature.

Even in a networked world, the hunger for consumer goods and information is still taste-dependent, and location remains a fairly reliable proxy for taste. A 2006 study published in the *Journal of International Economics*, for instance, found that for certain digital products—such as music, games, and pornography—each 1 percent increase in physical distance from the United States reduced by 3.25 percent the number of visits an American would make to a particular website.

Not only user preferences, but also government and corporate actions—motivated as often by cost and copyright as by political agendas—might mean the end of the era of the single Internet. That is to say, the days in which everyone can visit the same websites regardless of geographic location might be waning, even in the "free" world. We are seeing more attempts, mostly by corporations and their lawyers, to keep foreign nationals off certain Web properties. For instance, digital content that is available to British via BBC's innovative iPlayer is increasingly unavailable to Germans. Norwegians can already access 50,000 copyrighted books online free through the country's Bookshelf initiative, but one has to be in Norway to do so—the government is footing the annual $900,000 bill for licensing fees and doesn't plan to subsidize the rest of the world.

Moreover, many celebrated Internet pioneers—Google, Twitter, Facebook—are U.S. companies that other governments increasingly fear as political agents. Chinese, Cuban, Iranian, and even Turkish politicians are already talking up "information sovereignty"—a euphemism for replacing services provided by Western Internet companies with their own more limited but somewhat easier to control products, further splintering the World Wide Web into numerous national Internets. The age of the Splinternet beckons.

Two decades in, the Internet has neither brought down dictators nor eliminated borders. It has certainly not ushered in a post-political age of rational and data-driven policymaking. It has sped up and amplified many existing forces at work in the world, often making politics more combustible and unpredictable. Increasingly, the Internet looks like a hypercharged version of the real world, with all of its promise and perils, while the cyber utopia that the early Web enthusiasts predicted seems ever more illusory.

40

Reclaiming the Commons

NAOMI KLEIN

Naomi Klein describes the anti-globalization movement as a broadening series of struggles against various aspects of privatization. The different strands of these struggles are linked by the concept of "a single world with many worlds in it."

What is the "anti-globalization movement"?[1] I put the phrase in quote marks because I immediately have two doubts about it. Is it really a movement? If it is a movement, is it anti-globalization?

Let me start with the first issue. We can easily convince ourselves it is a movement by talking it into existence at a forum like this—I spend far too much time at them—acting as if we can see it, hold it in our hands. Of course, we *have* seen it—and we know it's come back in Quebec, and on the U.S.–Mexican border during the Summit of the Americas and the discussion for a hemispheric Free Trade Area. But then we leave rooms like this, go home, watch some TV, do a little shopping, and any sense that it exists disappears, and we feel like maybe we're going nuts. Seattle—was that a movement or a collective hallucination? To most of us here, Seattle meant a kind of coming-out party for a global resistance movement, or the "globalization of hope," as someone described it during the World Social Forum at Porto Alegre. But to everyone else Seattle still means limitless frothy coffee, Asian-fusion cuisine, e-commerce billionaires and sappy Meg Ryan movies. Or perhaps it is both, and Seattle bred the other Seattle—and now they awkwardly coexist.

This movement we sometimes conjure into being goes by many names: anti-corporate, anti-capitalist, anti-free trade, anti-imperialist. Many say that it started in Seattle. Others maintain it began five hundred years ago—when colonialists first told indigenous peoples that they were going to have to do things differently if they were to "develop" or be eligible for "trade." Others again say it began on 1 January 1994 when the Zapatistas launched their uprising with the words Ya Basta! on the night NAFTA became law in Mexico. It all depends on whom you ask. But I think it is more accurate to picture a movement of many movements—coalitions of coalitions. Thousands of groups today are all working against forces whose common thread is what might broadly be

SOURCE: From Naomi Klein, "Reclaiming the Commons." *New Left Review* 9 (May/June 2001).

described as the privatization of every aspect of life, and the transformation of every activity and value into a commodity. We often speak of the privatization of education, of healthcare, of natural resources. But the process is much vaster. It includes the way powerful ideas are turned into advertising slogans and public streets into shopping malls; new generations being target-marketed at birth; schools being invaded by ads; basic human necessities like water being sold as commodities; basic labor rights being rolled back; genes are patented and designer babies loom; seeds are genetically altered and bought; politicians are bought and altered.

At the same time there are oppositional threads, taking form in many different campaigns and movements. The spirit they share is a radical reclaiming of the commons. As our communal spaces—town squares, streets, school, farms, plants—are displaced by the ballooning marketplace, a spirit of resistance is taking hold around the world. People are reclaiming bits of nature and of culture, and saying "this is going to be public space." American students are kicking ads out the classrooms. European environmentalists and ravers are throwing parties at busy intersections. Landless Thai peasants are planting organic vegetables on over-irrigated golf courses. Bolivian workers are reversing the privatization of their water supply. Outfits like Napster have been creating a kind of commons on the internet where kids can swap music with each other, rather than buying it from multinational record companies. Billboards have been liberated and independent media networks set up. Protests are multiplying. In Porto Alegre, during the World Social Forum, José Bové, often caricatured as only a hammer of McDonald's, travelled with local activists from the Movimento Sem Terra to a nearby Monsanto test site, where they destroyed three hectares of genetically modified soya beans. But the protest did not stop there. The MST has occupied the land and members are now planting their own organic crops on it, vowing to turn the farm into a model of sustainable agriculture. In short, activists aren't waiting for the revolution, they are acting right now, where they live, where they study, where they work, where they farm.

But some formal proposals are also emerging whose aim is to turn such radical reclamations of the commons into law. When NAFTA and the like were cooked up, there was much talk of adding on "side agreements" to the free trade agenda, which were supposed to encompass the environment, labor and human rights. Now the fight-back is about taking them out. José Bové—along with the Via Campesina, a global association of small farmers—has launched a campaign to remove food safety and agricultural products from all trade agreements, under the slogan "The World Is Not for Sale." They want to draw a line around the commons. Maude Barlow, director of the Council of Canadians, which has more members than most political parties in Canada, has argued that water isn't a private good and shouldn't be in any trade agreement. There is a lot of support for this idea, especially in Europe since the recent food scares. Typically these anti-privatization campaigns get under way on their own. But they also periodically converge—that's what happened in Seattle, Prague, Washington, Davos, Porto Alegre and Quebec.

BEYOND THE BORDERS

What this means is that the discourse has shifted. During the battles against NAFTA, there emerged the first signs of a coalition between organized labor, environmentalists, farmers and consumer groups within the countries concerned. In Canada most of us felt we are fighting to keep something distinctive about our nation from "Americanization." In the United States, the talk was very protectionist: workers were worried that Mexicans would "steal" away "our" jobs and drive down "our" environmental standards. All the while, the voices of Mexicans opposed to the deal were virtually off the public radar—yet these were the strongest voices of all. But only a few years later, the debate over trade has been transformed. The fight against globalization has morphed into a struggle against corporatization and, for some, against capitalism itself. It has also become a fight for democracy. Maude Barlow spearheaded the campaign against NAFTA in Canada twelve years ago. Since NAFTA became law, she's been working with organizers and activists from other countries, and anarchists suspicious of the state in her own country. She was once seen as very much the face of a Canadian nationalism. Today she has moved away from the discourse. "I've changed," she says. "I used to see this fight as saving a nation. Now I see it as saving democracy." This is a cause that transcends nationality and state borders. The real news out of Seattle is that organizers around the world are beginning to see their local and national struggles—for better-funded public schools, against union-busting and casualization, for family farms, and against the widening gap between rich and poor—through a global lens. That is the most significant shift we have seen in years.

How did this happen? Who or what convened this new international people's movement? Who sent out the memos? Who built these complex coalitions? It is tempting to pretend that someone did dream up a master plan for mobilization at Seattle. But I think it was much more a matter of large-scale coincidence. A lot of smaller groups organized to get themselves there and then found to their surprise just how broad and diverse a coalition they had become part of. Still, if there is one force we can thank for bringing this front into being, it is the multinational corporations. As one of the organizers of Reclaim the Streets has remarked, we should be grateful to the CEOs for helping us see the problems more quickly. Thanks to sheer imperialist ambition of the corporate project at this moment in history—the boundless drive for profit, liberated by trade deregulation, and the wave of mergers and buy-outs, liberated by weakened anti-trust laws—multinationals have grown so blindingly rich, so vast in their holdings, so global in their reach, that they have created our coalitions for us.

Around the world, activists are piggy-backing on the ready-made infrastructures supplied by global corporations. This can mean cross-border unionization, but also cross-sector organizing—among workers, environmentalists, consumers, even prisoners, who may all have different relationships to one multinational. So you can build a single campaign or coalition around a single brand like General Electric. Thanks to Monsanto, farmers in India are working with environmentalists and consumers around the world to develop direct-action strategies that cut off genetically modified foods in the fields and in the supermarkets. Thanks to Shell Oil and Chevron, human rights activists in Nigeria, democrats in Europe, and environmentalists in North America have united in a fight against the unsustainability of the oil industry. Thanks to the catering giant Sodexho–Marriott's decision to invest in a Corrections Corporation of America, university students are able to protest against the exploding U.S. for-profit prison industry simply by boycotting the food in their campus cafeterias. Other targets include pharmaceutical companies who are trying to inhibit the production and distribution of low-cost AIDS drugs, and fast-food chains. Recently, students and farm workers in Florida have joined forces around Taco Bell. In the St. Petersburg area, field hands—many of them immigrants from Mexico—are paid an average of $7,500 a year to pick tomatoes and onions. Due to a loophole in the law, they have no bargaining power: the farm bosses refuse even to talk with them about wages. When they started to look into who bought what they pick, they found that Taco Bell was the largest purchaser of the local tomatoes. So they launched the campaign *Yo No Quiero Taco Bell* together with students, to boycott Taco Bell on university campuses.

It is Nike, of course, that has most helped to pioneer this new brand of activist synergy. Students facing a corporate take-over of their campuses by the Nike swoosh have linked up with workers making its branded campus apparel, as well as with parents concerned at the commercialization of youth and church groups campaigning against child labor—all united by their different relationships to a common global enemy. Exposing the underbelly of high-gloss consumer brands has provided the early narratives of this movement, a sort of call-and-response to the very different narratives these companies tell every day about themselves through advertising and public relations. Citigroup offers another prime target, as North America's largest financial institution with innumerable holdings, which deals with some of the worst corporate malefactors around. The campaign against it handily knits together dozens of issues—from clear-cut logging in California to oil-and-pipeline schemes in Chad and Cameroon. These projects are only a start. But they are creating a new sort of activist: "Nike is a gateway drug," in the words of Oregon student activist Sarah Jacobson. By focusing on corporations, organizers can demonstrate graphically how so many issues of social, ecological and economic justice are interconnected. No activist I've met believes that the world economy can be changed one corporation at a time, but the campaigns have opened a door into the arcane world of international trade and finance. Where they are leading is to the central institutions that write the rules of global commerce: the WTO, the IMF, the FTAA, and for some the market itself. Here too the unifying threat is privatization—the loss of the commons. The next round of WTO negotiations is designed to extend the reach of commodification still further. Through side agreements like GATS (General Agreement on Trade and Services) and TRIPS (Trade-Related Aspects of Intellectual Property Rights), the aim is to get still tougher protection of property rights on seeds and drug patents, and to marketize services like health care, education and water-supply.

The biggest challenge facing us is to distill all of this into a message that is widely accessible. Many campaigners understand the connections binding together the various issues almost intuitively—much as Subcomandante Marcos says, "Zapatismo isn't an ideology, it's an intuition." But to outsiders, the mere scope of modern protests can be a bit mystifying. If you eavesdrop on the movement from the outside, which is what most people do, you are liable to hear what seems to be a cacophony of disjointed slogans, a jumbled laundry list of disparate grievances without clear goals. At the Democratic National Convention in Los Angeles last year, I remember being outside the Staples Center during the Rage Against the Machine concert, just before I almost got shot, and thinking there were slogans for everything everywhere, to the point of absurdity.

MAINSTREAM FAILURES

This kind of impression is reinforced by the decentralized, non-hierarchical structure of the movement, which always disconcerts the traditional media. Well-organized press conferences are rare, there is no charismatic leadership, protests tend to pile on top of each other. Rather than forming a pyramid, as most movements do, with leaders up on top and followers down below, it looks more like an elaborate web. In part, this web-like structure is the result of internet-based organizing. But it is also a response to the very political realities that sparked the protests in the first place: the utter failure of traditional party politics. All over the world, citizens have worked to elect social democratic and workers' parties, only to watch them plead impotence in the face of market forces and IMF dictates. In these conditions, modern activists are not so naïve as to believe change will come from electoral politics. That's why they are more interested in challenging the structures that make democracy toothless, like the IMF's structural adjustment policies, the WTO's ability to override national sovereignty, corrupt campaign financing, and so on. This is not just making a virtue of necessity. It responds at the ideological level to an understanding that globalization is in essence a crisis in representative democracy. What has caused this crisis? One of the basic reasons for it is the way power and decision-making have been handed along to points ever further away from citizens: from local to provincial, from provincial to national, from national to international institutions, that lack all transparency or accountability. What is the solution? To articulate an alternative, participatory democracy.

If you think about the nature of the complaints raised against the World Trade Organization, it is that governments around the world have embraced an economic model that involves much more than opening borders to goods and services. This is why it is not useful to use the language of anti-globalization. Most people do not really know what globalization is, and the term makes the movement extremely vulnerable to stock dismissals like: "If you are against trade and globalization why do you drink coffee?" Whereas in reality the movement is a rejection of what is being bundled along with trade and so-called globalization—against the set of transformative political policies that every country in the world has been told they must accept in order to make themselves hospitable to investment. I call this package

"McGovernment." This happy meal of cutting taxes, privatizing services, liberalizing regulations, busting unions—what is this diet in aid of? To remove anything standing in the way of the market. Let the free market roll, and every other problem will apparently be solved in the trickle down. This isn't about trade. It's about using trade to enforce the McGovernment recipe.

So the question we are asking today, in the run-up to the FTAA, is not: Are you for or against trade? The question is: Do we have the right to negotiate the terms of our relationship to foreign capital and investment? Can we decide how we want to protect ourselves from the dangers inherent in deregulated markets—or do we have to contract out those decisions? These problems will become much more acute once we are in a recession, because during the economic boom so much has been destroyed of what was left of our social safety net. During a period of low unemployment, people did not worry much about that. They are likely to be much more concerned in the very near future. The most controversial issues facing the WTO are these questions about self-determination. For example, does Canada have the right to ban a harmful gasoline additive without being sued by a foreign chemical company? Not according to the WTO's ruling in favor of the Ethyl Corporation. Does Mexico have the right to deny a permit for a hazardous toxic-waste disposal site? Not according to Metalclad, the U.S. company now suing the Mexican government for $16.7 million in damages under NAFTA. Does France have the right to ban hormone-treated beef from entering the country? Not according to the United States, which retaliated by banning French imports like Roquefort cheese—prompting a cheese-maker called Bové to dismantle a McDonald's. Americans thought he just didn't like hamburgers. Does Argentina have to cut its public sector to qualify for foreign loans? Yes, according to the IMF—sparking general strikes against the social consequences. It's the same issue everywhere: trading away democracy in exchange for foreign capital.

On smaller scales, the same struggles for self-determination and sustainability are being waged against World Bank dams, clear-cut logging, cash-crop factory farming, and resource extraction on contested indigenous lands. Most people in these movements are not against trade or industrial development. What they are fighting for is the right of local communities to have a say in how their resources are used, to make sure that the people who live on the land benefit directly from its development. These campaigns are a response not to trade but to a trade-off that is now five hundred years old: the sacrifice of democratic control and self-determination to foreign investment and the panacea of economic growth. The challenge they now face is to shift a discourse around the vague notion of globalization into a specific debate about democracy. In a period of "unprecedented prosperity," people were told they had no choice but to slash public spending, revoke labour laws, rescind environmental protections—deemed illegal trade barriers—defund schools, not build affordable housing. All this was necessary to make us trade-ready, investment-friendly, world-competitive. Imagine what joys await us during a recession.

We need to be able to show that globalization—this version of globalization—has been built on the back of local human welfare. Too often, these connections between global and local are not made. Instead we sometimes seem to have two activist solitudes. On the one hand, there are the international anti-globalization activists who may be enjoying a triumphant mood, but seem to be fighting far-away issues, unconnected to people's day-to-day struggles. They are often seen as elistists: white middle-class kids with dreadlocks. On the other hand, there are community activists fighting daily struggles for survival, or for the preservation of the most elementary public services, who are often feeling burnt-out and demoralized. They are saying: What in the hell are you guys so excited about?

The only clear way forward is for these two forces to merge. What is now the anti-globalization movement must turn into thousands of local movements, fighting the way neoliberal politics are playing out on the ground: homelessness, wage stagnation, rent escalation, police violence, prison explosion, criminalization of migrant workers, and on and on. These are also struggles about all kinds of prosaic issues: the right to decide where the local garbage goes, to have good public schools, to be supplied with clean water. At the same time, the local movements fighting privatization and deregulation on the ground need to link their campaigns into one large global movement, which can show where their particular issues fit into an international economic agenda being enforced around the world. If that connection isn't made, people will continue to be demoralized. What we need is to formulate a political framework that can both take on corporate power and control, and empower local organizing and self-determination. That has to be a framework that encourages, celebrates and fiercely protects the right to diversity—cultural diversity, ecological diversity, agricultural diversity—and yes, political diversity as well: different ways of doing politics. Communities must have the right to plan and manage their schools, their services, their natural settings, according to their own lights. Of course, this is only possible within a framework of national and international standards—of public education, fossil-fuel emissions, and so on. But the goal should not be better far-away rules and rulers, it should be close-up democracy on the ground.

The Zapatistas have a phrase for this. They call it "one world with many worlds in it." Some have criticized this as a New Age non-answer. They want a plan. "We know what the market wants to do with those spaces, what do you want to do? Where's your scheme?" I think we shouldn't be afraid to say: "That's not up to us." We need to have some trust in people's ability to rule themselves, to make the decisions that are best for them. We need to show some humility wherenow there is so much arrogance and paternalism. To believe in human diversity and local democracy is anything but wishy-washy. Everything in McGovernment conspires against them. Neoliberal economics is biased at every level towards centralization, consolidation, homogenization. It is a war waged on diversity. Against it, we need a movement of radical change, committed to a single world with many worlds in it, that stands for "the one no and the many yesses."

ENDNOTE

1.   This is a transcript of a talk given at the Center for Social Theory and Comparative History at UCLA in April 2001.

REFLECTION QUESTIONS FOR CHAPTER 10

The articles in this chapter challenge us to rethink globalization. Here are some questions to ponder at the conclusion of this chapter, and the book as a whole:

1. Are the arguments more persuasive for "globalization is good" or for "globalization is bad"? Why?

2. Should the values of the West—democracy, freedom, and capitalism—be adopted universally, or is there room and tolerance for differences?

3. Are there creative ways to use the power of globalization to promote the common good?

4. Should we move away from nationalism to consider the consequences of globalization for the planet?

Websites

www.g8alternatives.org.uk
G8 Alternatives is a Scottish broad-based coalition that coordinates protests at the various G8 (leaders of the leading governments in transnational economic activities) summits.

www.developmentgap.org
The Development Gap is an organization promoting a just and sustainable alternative to free trade.

www.canadianlabour.ca
The Canadian Labour Congress represents most Canadian labor unions. It publishes a newsletter on the effects of NAFTA.

www.globalsolutions.org
The goal of Citizens for Global Solutions is to build peace, justice, and freedom in a democratically governed world.

www.cleanclothes.org
The Clean Clothes Campaign is an effort based in the Netherlands to promote fair labor practices in the apparel industry.

www.canadians.org
The Council of Canadians is devoted to advancing alternatives to corporate-style free trade and other issues facing Canada.

www.epi.org
The Economic Policy Institute publishes reports on international and domestic economic issues.

www.equalexchange.coop
Equal Exchange is a worker-owned cooperative dedicated to fair trade with small-scale farmers in the developing world.

http://www.change.org/50years
The 50 Years is Enough Network is a coalition of U.S. citizens' groups linked to groups in more than 50 countries whose goal is to reform the World Bank and International Monetary Fund.

www.foe.org
Friends of the Earth focuses on the environmental impact of globalization.

www.globalexchange.org
Global Exchange is an organization that works across borders to help build democracies, battle racism and inequality, and evolve a sustainable future.

www.foodfirst.org
The Institute for Food and Development Policy (FoodFirst) is an education-for-action

organization working to reduce hunger and poverty throughout the world.

www.ifg.org
The International Forum on Globalization sponsors education and research on the global economy.

www.imf.org
The International Monetary Fund provides information about the processes of globalization, focusing on their positive consequences.

www.laborrights.org
The International Labor Rights Forum is an advocacy group focused on strengthening enforcement of international labor rights.

http://multinationalmonitor.org
The Multinational Resource Center is home to the *Multinational Monitor,* a magazine that focuses on the negative aspects of transnational corporations.

www.tni.org
The Transnational Institute is an international network of activist researchers working on solutions to global problems.

www.forumsocialmundial.org.br
The World Social Forum is an open meeting place where groups and movements opposed to corporate global domination and imperial military domination come together to share their experiences, debate ideas, formulate proposals, and network for effective action.

www.worldwatch.org
The Worldwatch Institute promotes environmentally sustainable development by providing information on global environmental threats.

www.counterpunch.org
Counterpunch is an U.S.–based leftist political organization focused on creating an online progressive community and fostering internet activism.

www.commondreams.org
Common Dreams is a nonprofit organization focused on creating an online progressive community and fostering internet activism.

www.alternet.org
AlterNet is an award-winning news magazine and online community that generates original journalism and filters the best of hundreds of other independent media sources.

www.huffingtonpost.com
The *Huffington Post* is a progressive internet newspaper that covers, for example, politics, media, and religion.

www.motherjones.com
Mother Jones is a nonprofit news organization that specializes in investigative, political, and social justice reporting.

www.inthesetimes.com
In These Times is a nonprofit and independent newsmagazine committed to political and economic democracy and opposed to the dominance of transnational corporations.

www.currenthistory.com
Current History was founded by the *New York Times* in 1914 as a journal of contemporary world affairs.

www.guardian.co.uk
The *Guardian* is a British newspaper that was founded in 1821. It focuses on international news and politics.